Table of Atomic W

	Symbol	Atomic Number	Atomic Weight
Mercury	Hg	80	200.59
Molybdenum	Mo	42	95.94
Neodymium	Nd	60	144.24
Neon	Ne	10	20.179
Neptunium	Np	93	237.0482
Nickel	Ni	28	58.70
Niobium	Nb	41	92.9064
Nitrogen	N	7	14.0067
Nobelium	No	102	(255)
Osmium	Os	76	190.2
Oxygen	O	8	15.9994
Palladium	Pd	46	106.4
Phosphorus	P	15	30.9738
Platinum	Pt	78	195.09
Plutonium	Pu	94	(244)
Polonium	Po	84	(210)
Potassium	K	19	39.098
Praseodymium	Pr	59	140.9077
Promethium	Pm	61	(147)
Protactinium	Pa	91	231.0359
Radium	Ra	88	226.0254
Radon	Rn	86	(222)
Rhenium	Re	75	186.207
Rhodium	Rh	45	102.9055
Rubidium	Rb	37	85.4678
Ruthenium	Ru	44	101.07
Samarium	Sm	62	150.4
Scandium	Sc	21	44.9559
Selenium	Se	34	78.96
Silicon	Si	14	28.086
Silver	Ag	47	107.868
Sodium	Na	11	22.9898
Strontium	Sr	38	87.62
Sulfur	S	16	32.06
Tantalum	Ta	73	180.9479
Technetium	Tc	43	98.9062
Tellurium	Te	52	127.60
Terbium	Tb	65	158.9254
Thallium	Tl	81	204.37
Thorium	Th	90	232.0381
Thulium	Tm	69	168.9342
Tin	Sn	50	118.69
Titanium	Ti	22	47.90
Tungsten	W	74	183.85
Uranium	U	92	238.029
Vanadium	V	23	50.9414
Xenon	Xe	54	131.30
Ytterbium	Yb	70	173.04
Yttrium	Y	39	88.9059
Zinc	Zn	30	65.38
Zirconium	Zr	40	91.22

Qualities and Quantities

PREPARATION FOR COLLEGE CHEMISTRY

Qualities and

Quantities

PREPARATION FOR COLLEGE CHEMISTRY

RICHARD M. McCURDY
Los Angeles Pierce College

HARCOURT BRACE JOVANOVICH, INC.
NEW YORK CHICAGO SAN FRANCISCO ATLANTA

To my teachers
and
To my students,
especially
Danny H.
who lives in this book

Photo credits appear on page 316.

ISBN: 0-15-574100-4

Library of Congress Catalog Card Number: 75-2955

Printed in the United States of America

to the instructor

This book is intended to serve as a preparative text for students who require one or more years of college chemistry, but who have either never had high school chemistry or have forgotten it.

I believe that the average student must bring the following knowledge and skills to the modern general chemistry course:

1. computational skills involving mass, volume, density, and percent composition;
2. understanding of applications of the ideal gas law;
3. sufficient knowledge of atomic structure to recognize the origins and meaning of bonding and molecular structure;
4. familiarity with the nomenclature of common inorganic substances;
5. knowledge of equation writing and balancing;
6. ability to perform weight/weight and weight/volume computations from equations and solution stoichiometry;
7. knowledge of the various categories of chemical reactions; and
8. some knowledge of the physics of energy transformations.

Students with a sound grasp of these concepts and skills can then benefit from a meaningful treatment of atomic and molecular structure, thermodynamics, chemical equilibrium, complex-ion bonding, and nuclear chemistry in a subsequent course.

The organization of this book enables students with no prior mathematical preparation except arithemtic to solve chemical problems successfully. The problem-solving method is used consistently throughout the text. The method emphasizes those properties that can be measured and the linear relationships that exist between them. Half the problems in each set are answered in factored form, so students can check the method as well as the answers.

The gas laws and composition are presented early because they provide immediate applications on which students can test their problem-solving skills. Students usually have no difficulty accepting the idea of atoms as particles in motion that may occur in certain number combinations. The derivation of kinetic molecular theory is valuable for the quantitative picture it gives of the effects of temperature, pressure, and numbers of molecules, but it is not necessary for continuity and may be omitted. The early introduction of the gas laws and composition has an added benefit in that it provides the subject matter for relevant laboratory experience.

In deference to those who prefer a different order of presentation, there is sufficient overlap in Chapters Four through Eight for atomic structure or composition to be introduced first.

The last six chapters, in addition to providing considerable drill in computational skills, present some intuitive ideas of thermodynamics and reaction mechanisms so that the student will have some basis for understanding the quantitative treatment in subsequent courses.

Although the contents of this text is solely my responsibility, I gratefully acknowledge that the final form is the product of thoughtful consideration by many fine educators. I would like to note especially the contributions of Frances Collins, University of Massachusetts at Amherst; Edward C. Fohn, Green River Community College; Melvin W. Hanna, University of Colorado; John J. Healey, Chabot College; James MacDonald, DeAnza College; and Michael J. Millam, Phoenix College. I also feel exceptionally fortunate to have had the experience of working with the highly professional editorial, design, and production staff of Harcourt Brace Jovanovich, Inc.

RICHARD M. McCURDY

to the student

This book was written primarily to help you learn chemistry. You may find that the skills and attitudes you develop will be useful in other fields as well. The basic concepts of chemistry are easy to understand, and their application to the solution of problems in chemical arithmetic is even easier. The amount of study and practice required depends on how efficiently you work.

Your efficiency will improve if you learn to use words precisely. Always be able to reply with an exact answer to the questions, "What am I doing?" and "What do I want to know?" For example, saying you are flying a kite is not being exact; you are really holding a string.

Develop your awareness of the relationships that exist between one observable fact and another. You know that if you hold the string so that the kite is at a certain angle with the wind it will fly. So you are expressing the *result* of your action, rather than the action itself. Not knowing the relationship could result in your holding the string unsuccessfully all day.

In a similar way the scientist who measures the temperature of the contents of a reaction vessel knows that what is really being measured is the volume of the column of mercury in a thermometer compared to the volume at some reference temperature. Knowing this, the scientist can be aware of the possibilities of error. The manufacturer may not have put the right amount of mercury in the thermometer, or the scale could be wrong. The scientist who knows the relationship between what is being measured and what is wanted can check to see if it is satisfied.

You can succeed in chemistry by (1) labeling things for what they are; (2) learning the relationships that exist between physical quantities; and (3) using the relationships to find the desired result from the given data. The frosting on the cake, though, is knowing why these things are true and why they work. The explanations, insofar as we know them, are presented in this book in a nonmathematical form. Additional references are provided at the end of some chapters for those of you who wish to explore any of these topics in greater depth.

Above all, be thoughtful. Be critical. Test every new idea to see if it is true within your experience.

RICHARD M. McCURDY

contents

45 **3** *form and substance*

63 **4** *gases*

93 **5** *composition*

atomic structure **6** 123

models of compounds **7** 143

chemical nomenclature **8** 165

181 **9** *chemical reactions and equations*

205 **10** *calculations from equations*

221 **11** *solutions*

acids and bases 12 253

oxidation and reduction 13 271

energy 14 287

Qualities and Quantities

PREPARATION FOR COLLEGE CHEMISTRY

"Science is the attempt to make the chaotic diversity of our sense-experience correspond to a logically uniform system of thought." (Albert Einstein, *Science, 91,* 1940)

After reading this chapter you should be aware of

1. The ways in which science differs from other branches of human knowledge;
2. The meaning of a scientific law;
3. The difference between empirical laws and fundamental laws;
4. How theories are formed from hypotheses and laws;
5. The meaning and uses of experiments;
6. The meaning of technology;
7. The role of chemistry in science and technology;
8. The reasons more scientific research is necessary in order to prevent ecological disaster.

1 introduction

What is science? 1.0

Science has been called many names, some of them downright uncomplimentary. What is there about this abstract noun that people fear and worship?

According to Webster's, *science* is

1. *knowledge* gained by *study* and *practice.*
2. any department of *systematized* knowledge.
3. a branch of study concerned with the *observation* and *classification* of facts, especially with the establishment of *verifiable general laws. . . .*
4. specific accumulated knowledge, systematized and formulated with reference to the *discovery* of general *truths* or the operation of general laws.

That looks pretty formidable. No wonder people are afraid of it. Before you decide that science has nothing for you, let us look at the key words in that definition to see what deeper meaning they might have.

1.1 *Science is knowledge obtained by study and practice*

A concert pianist, an airline pilot, a karate expert, and a pro football quarterback could rightly claim that they are scientists as much as a chemist or physician, according to the above definition. To be a proper science, there must be an accumulation of knowledge about a given subject, and the knowledge must have a predictive value. The scientist must be able to make reliable predictions about the outcome of a given situation within the subject field of the science.

The accuracy of the prediction depends on how much is known about the situation. Astronomers can make extremely reliable predictions about the motions of the bodies of the solar system because there are a limited number of objects in the solar system. Chemists and physicists can make fairly accurate predictions about the behavior of matter in the laboratory because the number of conditions to be controlled is limited.

As the number of complicating factors increases, the predictions become less accurate and are valid for shorter times. The weather scientist (meteorologist) must weigh hundreds of factors to predict the weather, and the accuracy of the predictions suffers, especially over longer periods of time. Physicians and social scientists must struggle with many variables, including the unpredictable whims of individuals. Medical and social scientists may rely upon experience that is true much of the time, but they still have to use intuition and inspired guess-work for many of their decisions.

Finally we come to the quarterback. In spite of all his study and practice all he can do is guess what is going to happen and hope for the best. This is as it should be. It is the unpredictability—the element of surprise—of art, music, and sports that appeals to us. These human activities are not sciences, and they would be spoiled for us if they were.

1.2 *Science is systematized knowledge*

A box of facts is information of a sort, but not in usable form. It is like an unassembled jigsaw puzzle—the facts are there but the picture is not clear. The process of systematizing the knowledge is similar to assembling the jigsaw puzzle. One must find out which facts are related to other facts piece by piece until a pattern becomes clear.

An automobile is a system of mechanical parts. A mechanic must have a systematic knowledge of an automobile in order to avoid trying to put a window crank into the carburetor. The mechanic must know where each part goes, how it relates to the parts it is connected to, and its place in the overall function of the car. **EXAMPLE 1.1**

A yeast cell is a complicated structure of protein, fat, nucleoprotein, and assorted small molecules. The list of yeast cell components does not tell how it functions. To *know* a yeast cell one must be able to state the function of each component and its relationship to the activities of the organism in fermentation and reproducing itself. **EXAMPLE 1.2**

Science is . . . concerned with the observation and classification of facts . . . 1.3

This is like separating the contents of a box containing several jigsaw puzzles. The relevant details must be separated from the irrelevant to find the causes of an event. Observed facts must first be collected. Then all the observations that affect the results are put together. In a laboratory scientists can contrive a situation that more or less copies the natural event, but in such a way that they can control all the factors that might change the outcome. The creation of situations for the purpose of observing them is *experimentation.*

Louis Pasteur, a French chemist, was commissioned by the beer and wine industry to find out why some batches of their products spoiled. The role of yeasts in fermentation was not known at that time. Pasteur found by observation that good and bad wines contained different microorganisms. By excluding the bad organisms and inoculating with the good ones, he could get good wine every time. This was good news to the French and great fun for Pasteur. **EXAMPLE 1.3**

. . . especially with the establishment of verifiable general laws. . . . 1.4

A *physical law* is a description, usually in mathematical form, of the relationships that exist between the parts of a system or between the system and its surroundings. A physical law tells what changes would be expected under a certain set of known conditions. The mathematical form is useful because it tells not only what change is expected but also how much.

A physical law is a statement of experience. Generally an experience becomes a law if the same conditions can be repeated again and again until the probability that the same result will always be obtained is overwhelming.

EXAMPLE 1.4

(a) **The Law of Levers.** Experiment shows that a 5 pound (lb) weight applied to a rigid beam 10 feet (ft) from a balance point (a fulcrum) will just lift a 50 lb weight placed 1 ft from the fulcrum.

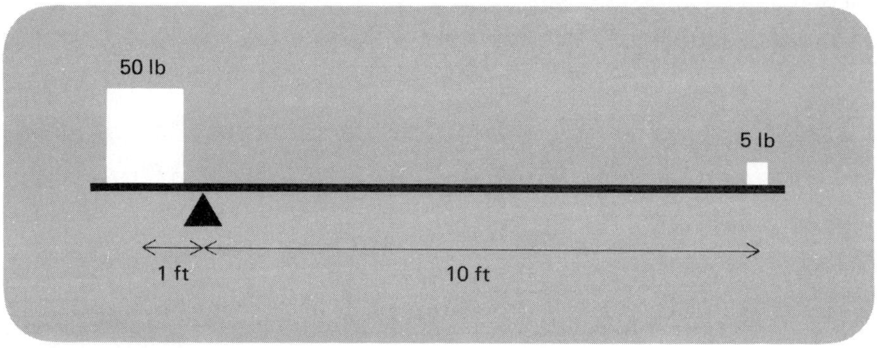

Figure 1.1 The law of levers: Force on left times distance from fulcrum equals force on right times distance from fulcrum.

A few more experiments produce the following results:

Force applied	through a distance of	will just balance	an opposite force of	feet from fulcrum
8 lb	5 ft		20 lb	2 ft
10 lb	8 ft		40 lb	2 ft
20 lb	6 ft		60 lb	2 ft

Still more experiments confirm what we now suspect: The applied force multiplied by its distance from the fulcrum equals the force times the distance in the opposite direction. We can express this statement as the equation

$$W_1 \times d_1 = W_2 \times d_2$$

where W equals force and d is distance from fulcrum.

More careful experiments will show that we must pay attention to the direction of the forces, and we must include the weight of the beam in our calculations. With these added refinements the law holds exactly under all conditions.

(b) **Hooke's Law: The Law of Springs.** Experience teaches us that most things will stretch when pulled, and some things stretch more than others. If the article being pulled returns to its original dimensions when the force is released, the substance is *elastic*. Springs, rubber bands, and nylon hose are elastic.

To discover the relationship between the stretching force (tension) and the distance the article stretches we must perform an experiment. A particular spring might give the following values:

Tension Force	Change in Length
2 lb	4 in.
4 lb	8 in.
10 lb	20 in.
100 lb	50 in. (lost elasticity)

Before we became impatient and overloaded the spring, we could see that the distance the spring stretched was directly proportional to the load on it. In fact, this spring increased 2 inches in length for each pound of tension. It is equally true to say that $\frac{1}{2}$ pound is supported per 1 inch of stretch. We may therefore write

$$\text{force (lb)} = \frac{1\ \text{lb}}{2\ \text{in.}} \times \text{distance (in.)}$$

If we run more experiments, we find that this law does not hold under all conditions. Temperature changes will change the values; and if we exceed a certain load, the spring will lose its elasticity and the law will not hold anymore.

How are laws discovered? 1.5

Most people discover at an early age that if they do not keep their center of gravity almost directly above their feet they will land (a) on their diaper, or (b) on their face. The child who pulls over a lamp to have it fall on him discovers that this law is generally true of lamps, too. The child who does not experiment with walking never discovers the above law. The requirements for discovery are (1) *observation* and (2) correct *generalization* from the observation. The purposeful creation of situations to observe is experimentation.

When the experimenter has made enough observations to see a pattern of similar occurrences, the law may be generalized to include all such happenings with the same circumstances. Of course children do not consciously measure their center of gravity and how far it is removed from the vertical. They learn to feel the difference. In the beginning, people found laws by observing qualitatively what events or things were related to each other. In a similar fashion, but with more thought, a man may have discovered that, although a log would keep him afloat in the water, a hollowed-out log would do a better job and also carry his lunch.

The exact relationship between the volume of the log canoe and the load it would carry depended on the development of measuring devices. Archimedes deduced that the load carried was exactly equal to the weight of the water displaced by the canoe. The weight of the displaced liquid was its volume times its density. Now *there's* a law! It not only states that a dependence or relationship exists between two things (volume and weight), but it also tells how much. And it generally holds true for all objects in all liquids, whether floating or not, so long as the objects do not dissolve in the liquid. Figure 1.2 illustrates how this law operates.

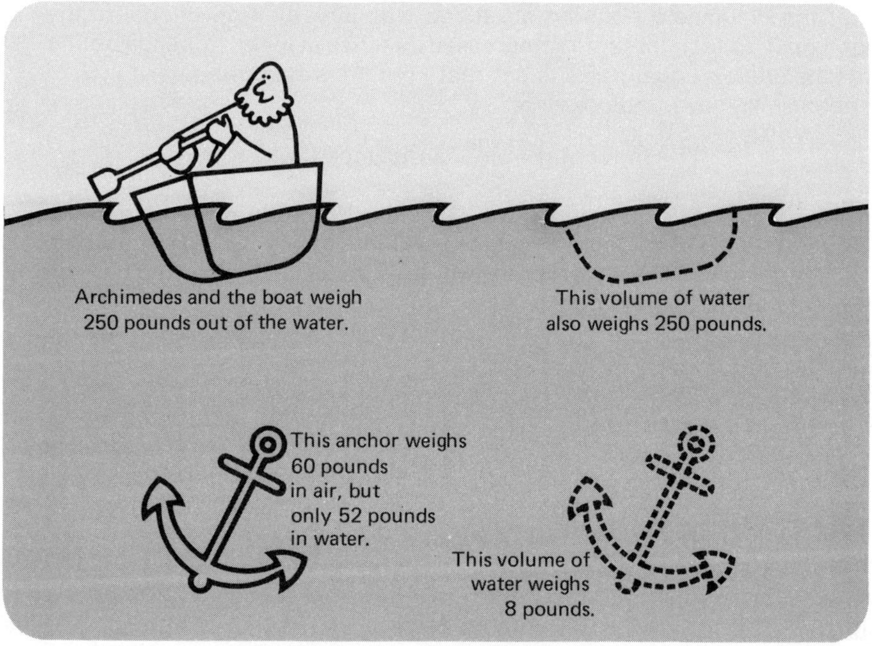

Figure 1.2 Archimedes' law.

1.6 *Fundamental and empirical laws*

Some laws express the relationship between two or more things in a simple way, and no exceptions have ever been found to them. In fact, more and more careful measurements on these laws seem to affirm that they hold exactly under all conceivable conditions. These laws are based upon experience, like all laws, but the experience is so universal and general that the probability that an exception will be found is remote. The confidence we place in the laws of gravity and electromagnetism and the conservation laws of mass/energy, momentum, and charge is so great that they can be considered *fundamental laws.*

Other laws describe relationships that hold approximately within a limited range of conditions. The force necessary to stretch a spring or the air resistance of an automobile versus its speed are approximate laws, which depend on more factors than the engineer cares to contemplate, much less measure. Later in this book we will find that the gas laws predict the behavior of gases to a good approximation, unless the conditions of temperature and pressure exceed certain limits. Beyond these limits the laws are not valid. Laws that are approximate or that have a limited range of validity are called *empirical laws.* They are useful for building bridges, airplanes, and assorted gadgets that work within the limits of specified conditions.

Theories 1.7

In this book we will examine the essential ideas of atomic theory, kinetic molecular theory, quantum theory, and others. What are theories, and why are they useful?

A *theory* is an explanation of a complex event in terms of an imaginary model that obeys all the known fundamental laws. The postulated model is imaginary because the real parts of the system are too small or numerous to observe directly. The hypothetical model is presumed to be a simplified version of the real system. If the model is simplified enough, its behavior can be deduced from the logical application of the laws of physics and mechanics.

If the deduced behavior of the model is like the observed behavior of the real system the explanation has a good start, but it is not yet a theory. A theory must predict something that has *not* been observed, but that *is* observable. To have a good theory a scientist must be able to say, "According to my theory, there must be a planet beyond Saturn. Look for it." Or, "According to my theory, mass can be converted to energy. You experimenters ought to be able to find a way to do it."

A theory that predicts something contrary to the observed facts or something unobservable is worthless.

Technology 1.8

Technology, or applied science, is the application of the knowledge of physical laws. The result of scientific experiment is knowledge; the result of technology is a material product or a service. Most research is devoted to finding new products and materials to make life more comfortable, and, not so incidentally, to make a profit for the inventor. Technology pays for itself and also supports pure science.

Much of the criticism of science and technology in recent years has come as a reaction to premature or misapplied use of the products of technology. In the 1950s it was discovered that the indiscriminate use of DDT insecticide was decimating certain bird species. In the 1960s hundreds of children were born deformed because their mothers had taken an otherwise harmless sleeping pill during a critical time in their pregnancies. In the 1970s we are still trying to find a solution to the suffocating air pollution caused by automobile engines. The list of horrible examples is uncomfortably long.

Since all productive work that relies on a systematic application of physical laws is technology, it does not make sense to condemn technology as a whole. We would quickly starve or be reduced to gatherers of wild nuts and fruits if all technology were banned.

The difficulty here is not science; it is ignorance, deliberate or not, of the consequences of the actions. The cure for ignorance is knowledge. We must take the time to search out the consequences of a proposed action and then decide whether the cost is worth the benefits.

1.9 *Chemistry*

Chemistry is that branch of physical science concerned with the structure and composition of all substances and with changes in their properties. What is it made of, how is it put together, and what changes will occur if . . . are the questions a chemist asks. Some chemists are occupied in finding or making new substances to do a specific job better. A polymer chemist, for example, might be making and testing new materials for synthetic fabrics or tougher rubber for tires. An organic chemist might be synthesizing variations of a natural antibiotic or a hormone in order to eliminate undesirable effects connected with the natural product or to make the substance more effective. A biochemist might study the many reactions occurring in a living cell in the hope of discovering the causes of malfunctions.

Chemists monitor your water supply, your foods, and your medicines. They are the watchdogs of the environment, too, warning us of the kind and amount of pollutants present.

REVIEW QUESTIONS

1. In your own words, what is a science?

2. In what ways is a physical law different from history?

3. When does a human activity change from an art to a technology? Discuss this in terms of a specific activity, such as fishing.

4. List the objects within your reach that are the direct product of chemical technology.

5. What is the difference between observation and experimentation?

6. What is a physical law?

7. Compare and contrast the law of levers with the law of stretched springs (Examples 1.4a and b).

8. What is a theory?

9. What value or usefulness does a theory have?

SUGGESTED READING

Hans Reichenbach, *The Rise of Scientific Philosophy* (Berkeley: University of California Press, 1951).

EXERCISES

1. What is the next move?

a. b. c. choose from these

2. What is the next number?

(a) 1, 2, 4, 8, __

(b) 3,15; 4.5, 22.5; 6, 30; 7, __

(c) 3, 10; 4, 7.5; 6, 5; 2, 15; 12, __

3. What is the next row?

1
1 1
1 2 1
1 3 3 1
1 4 6 4 1
_ _ _ _ _ _

4. Four objects were weighed in air and weighed while suspended completely submerged in water.

		Wt in air	Wt in water
Iron cube	1 in. × 1 in. × 1 in.	0.272 lb	0.236 lb
Aluminum block	1 in. × 1 in. × 3 in.	0.294 lb	0.185 lb
Glass sphere	2.26 in. diameter	1.00 lb	0.783 lb
Beer can	2.5 in. diameter × 4.5 in. high	0.850 lb	0.055 lb

What law relates the weight in air to the weight in water?

5. Just for fun, can you find the rule that enables you to name the number your friend is thinking of, if you are told only in which groups (A, B, C, D) the number is found?

A		B		C		D	
1	9	2	10	4	12	8	12
3	11	3	11	5	13	9	13
5	13	6	14	6	14	10	14
7	15	7	15	7	15	11	15

6. Make up 5 sets containing 31 numbers obeying such a law.

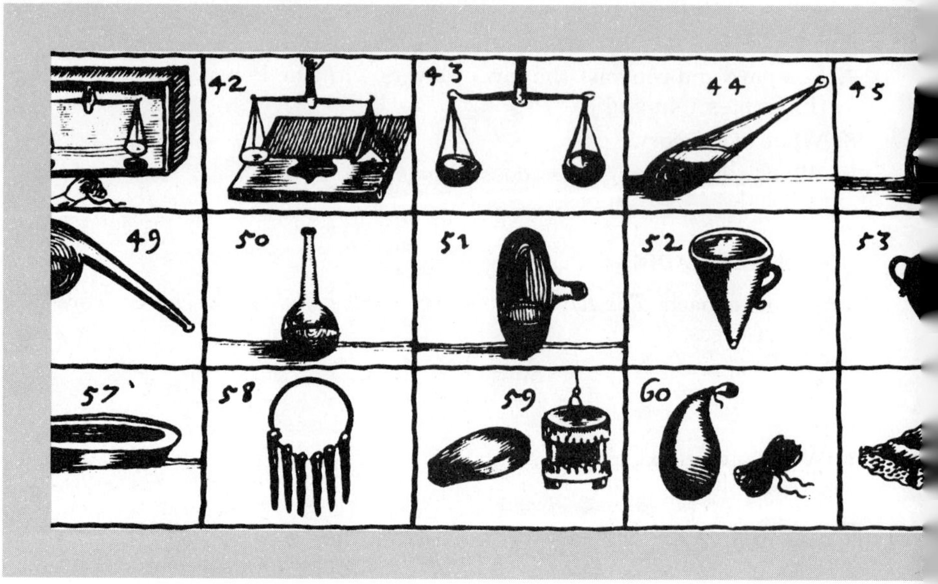

We observe the world through measurements and we infer meaning to those measurements by the relations that hold between them.

After reading this chapter you should be able to

1. Describe how to make a measurement;
2. Determine the number of significant figures in a measurement;
3. Write a number in scientific notation;
4. Describe the standards of mass, space, and time in the metric system;
5. Draw a graph from a set of data;
6. Write a linear equation from a description of the relationship;
7. Distinguish extensive properties from intensive properties;
8. Calculate one quantity from another related quantity;
9. Demonstrate the use of the chain rule;
10. Use percent in calculations of the part or the whole;
11. Use density in calculations of mass and volume.

measurements and calculations 2

Measurements 2.0

A tribe that can get along without numbers is either very primitive or in a state where material things are of no consequence. In other circumstances, people typically engage in commerce as soon as they have a surplus of something. Commerce required the invention of numbers just as agriculture and landownership required the invention of geometry. It was not long before commercial necessities spawned the middleman, an individual who often had to describe goods he did not have to a buyer who had not seen them.

Consider the problems of a sporting goods representative describing a baseball to a Melanesian. The sales person must compare each characteristic of the baseball to something that is familiar to the Melanesian until a composite mental picture is produced. "The baseball is shaped *like* a turtle egg, only bigger. It has a tough skin *like* a wart hog, except it is not hairy. It is hard *like* a coconut, but it is more resilient." By the time they come to what to do with it, communications might be at a hilarious stage.

The beginnings of experimental science in the eighteenth century brought about a need for accurate descriptions of experiments and apparatus so that other natural philosophers could duplicate the procedures and obtain the same results. The use of parts of the body such as hands or feet to describe length, or the weight of common objects like grains of wheat or stones to describe weight could only convey an approximate idea of the quantities involved. It was necessary to provide some standards of comparison which could be reproduced accurately anywhere in the world.

Properties

Those characteristics or attributes that describe an object or a substance are known as its properties. The *physical properties* of an object include its mass, size, temperature, how it acts under electrical and mechanical forces, and any other behavior that does not involve a change in its composition. The *chemical properties* of a substance predict how the composition of the substance changes when it interacts with other matter. Although there are dozens of ways to describe the physical properties of something, virtually every way can be derived from four independent units. We must have standards for length, mass, time, and temperature.

What do we do when we measure something?

The process of comparing a property of a thing to the same characteristic of a standard is called *measurement.* We measure the length of a material thing by counting the number of standard lengths laid end to end in a straight line that traverses the object. We measure the mass of an object by counting the number of standard masses that experience the same weight due to the earth's gravity.

Every measurement starts with a real object, whose behavior can be compared to a standard by some measuring device. If the direct comparison is impossible or inconvenient because of size or distance (like the distance to the moon or the mass of an atom), the desired quantity must be calculated from an indirect measurement by some known or assumed physical law.

EXAMPLE 2.1 The distance to the moon can be inferred from the *time* it takes for a radar impulse to travel from the earth to the moon and back. We measure the period of time between transmission and reception of a radio signal to be 2.58 seconds (sec). We suppose that the signal leaves the earth, bounces off the moon, and returns at a constant speed, 186,000 miles per second. The distance to the moon should be

$$\tfrac{1}{2} \times 2.58 \text{ sec} \times 186{,}000 \, \frac{\text{mi}}{\text{sec}} = 240{,}000 \text{ mi}$$

if all our assumptions are correct.

Three steps are required in any measuring process:

1. A standard unit quantity is chosen for comparison;

2. The same characteristic of the sample and the standard are compared with a suitable measuring device;

3. The number of standard units and subunits equivalent to the sample are counted and recorded.

If a measurement cannot be made directly, it can often be made by indirect means if a law or relationship exists between the desired quantity and the actual measurement. In this book we will be seeking the answers to the questions,

WHAT CAN I MEASURE?
HOW IS MY MEASUREMENT RELATED TO WHAT I WANT TO KNOW?

Measurements and uncertainty 2.1

A physical quantity can never be measured exactly. An exact measurement means that the difference between the measured value and the "true" value can be made as small as desired. A machinist, let us say, might be satisfied that a part is exact if it fits a hole within one-thousandth of an inch. A supervisor might insist that the discrepancy not exceed one ten-thousandth of an inch. The customer may want the pieces to fit within one millionth of an inch. As long as there is a finite difference, no matter how small, the measurement is not exact.

"Well," you may say, "that is all right for a definition but I will be satisfied that a measurement is exact if I can't see any difference." But the judgment of whether the uncertainty in a value makes a difference or not depends on how reliable any value based on that measurement must be.

EXAMPLE 2.2

Suppose it is necessary to make a 300-mile (mi) trip in a car with a 20-gallon (gal) gas tank. The car averages 15 miles per gallon, more or less. That is, the gas mileage may be as low as 14 miles per gallon and as high as 16 miles per gallon. There are no filling stations open between the starting point and the destination.

At the lower mileage the car will travel

$$20 \text{ gal} \times 14 \frac{\text{mi}}{\text{gal}} = 280 \text{ mi}$$

At the higher mileage the car will travel

$$20 \text{ gal} \times 16 \frac{\text{mi}}{\text{gal}} = 320 \text{ mi}$$

How far are you willing to walk?

Unless the possibility of a 20-mile walk makes no difference, it is necessary to know the gas mileage of the car more precisely. If it is *known* that the car will get 15.1 miles per gallon, the trip can be made. If the mileage is only 14.9 miles per gallon, some other alternative must be chosen.

The analysis of how the uncertainty in measurements affects the answers derived from them is too complicated for this book. Some simple rules, however, will enable you to judge the reliability of your measurements and calculations.

RULE 1. A measurement cannot be more precise than the instrument used to make the measurement. *Precision* is the amount of variation in the observed value of a measurement repeated several times. An instrument is usually, but not always, precise to the smallest division readable on the instrument. That is, a measurement made with an instrument inscribed in tenths of a unit can be reported to the nearest tenth of a unit, but not to one-hundredth of a unit.

RULE 2. The value of measurements that are less precise (have more variation) than the smallest division readable can be obtained by averaging the results of several measurements.

EXAMPLE 2.3 Several measurements of the diameter of a cylinder in centimeters (cm) gave the following results:

$$\begin{array}{r} 1.268 \text{ cm} \\ 1.267 \text{ cm} \\ 1.271 \text{ cm} \\ 1.270 \text{ cm} \\ \underline{1.271 \text{ cm}} \\ \text{Sum} = 6.347 \text{ cm} \end{array}$$

The average is the sum of the individual values divided by the number of measurements.

$$\frac{6.347 \text{ cm}}{5} = 1.269 \text{ cm}$$

In this case the most probable true value is 1.269 cm. There is some uncertainty about the 9 in the last place.

RULE 3. An answer derived by calculation from a measurement cannot have more significant figures (see the next section) than the measurement with the least number of significant figures in the calculation.

A precise measurement may be reproducible to several significant figures, yet the value could be wrong. The *accuracy* of a measurement is the degree to which the measurement matches the *true value*. Instruments must be calibrated for accuracy by comparing their observed values against a known standard. If the instrument measures the value of a known standard truly and precisely, it presumably will report other measurements truly as well and is said to be accurate.

Significant figures and scientific notation 2.2

The significant figures of a measurement are those numbers that can be repeated with confidence plus one. That is, the last number (reading from left to right) is assumed to be somewhat uncertain. In example 2.3, it is certain that the answer lies between 1.26 and 1.28, and it is most probably at 1.269 plus or minus a few thousandths. This number has four significant figures.

Sometimes a measurement is given with a specific estimate of uncertainty. For example, the size of ball bearings in a particular lot might be given as 0.500 ± 0.002 inches in diameter. An estimate of the uncertainty can be given when many measurements have been made and the spread of values is known.

If only one measurement can be made, the degree of uncertainty is given by the number of significant figures. It is assumed that the last figure is only an estimate which may vary several units more or less. A problem arises when it is necessary to use zeros to show the size of the number.

The number 2000 has four significant figures if it is certain that the value lies between 1995 and 2005. If the value is only known to be between 1500, say, and 2500, the number only has two significant figures.

This difficulty in interpretation is avoided in scientific work by expressing the number as a decimal number between one and ten, multiplied by a power of ten. All of the numbers written are considered to be significant. Table 2.1 shows how to write some numbers as powers of ten.

Table 2.1
Powers of ten

Number	Expressed as a Power of Ten
100,000	10^5
10,000	10^4
1000	10^3
100	10^2
10	10^1
1	10^0
0.1	10^{-1}
0.01	10^{-2}
0.001	10^{-3}

EXAMPLE 2.4

1. The number 2000 is 2×10^3. If the measurement justifies four significant figures, it should be written

$$2.000 \times 10^3$$

The zeros after the decimal point are significant.

2. The number 0.00132 is 1.32×10^{-3}. It has three significant figures.
3. The number 20 is 2.0×10^{1} (two significant figures); the number 0.123 is 1.23×10^{-1} (three significant figures). If these numbers are multiplied together, the answer must be rounded off to *two* significant figures. (Remember Rule 3.)

$$(2.0 \times 10^{1}) \times (1.23 \times 10^{-1}) = 2.46 \times 10^{0} = 2.5$$

2.3 *The International System (SI) of weights and measures (metric)*

About the time the French were establishing Liberty, Equality, and Fraternity in their nation, they also decided to add some rationality to their measuring system. They picked one standard for length, one for mass, and one for time, and then expressed all subdivisions and multiples of these standards in powers of ten. Each fraction or multiple of the standard is indicated by a prefix to the standard unit (see Table 2.2).

Table 2.2
Common metric prefixes

Number	Multiplier	Prefix
one trillion	10^{12}	tera-
one billion	10^{9}	giga-
one million	10^{6}	mega-
one thousand	10^{3}	kilo-
one	10^{0}	no prefix
one hundredth	10^{-2}	centi-
one thousandth	10^{-3}	milli-
one millionth	10^{-6}	micro-
one billionth	10^{-9}	nano-
one trillionth	10^{-12}	pico-
one thousand trillionth	10^{-15}	femto-

These prefixes may be applied to any unit in the SI system.

2.4 *Length*

The international unit of length, the *meter,* was originally defined to be the distance between two scratches on a platinum bar kept at zero degrees Celsius. Today the requirements of advanced space technology are more stringent than can be satisfied by reading the distance between two scratches. The new definition of the meter uses the wave properties of light as the measuring stick. With the new standard, lengths can be compared with an uncertainty of only one or two parts per hundred million. This corresponds to an uncertainty of one or two inches in the distance between New York and Los Angeles, or ten to twenty feet in the distance between the earth and the moon. The diagram at the right lists some of the things that people have measured. Note that each level in the diagram represents a distance 1000 times bigger than the one below it.

The standard of length in the United States is the *inch,* defined as exactly 2.54 centimeters. A movement is under way to change the system of weights and measures used in the United States to the metric system. Table 2.3 compares metric and U.S. units of length. A more complete table of conversion factors is in the appendix.

Table 2.3

Conversion factors for English and metric length units

1 kilometer (km)	= 1000 meters (m)
1 meter	= 100 centimeters
1 meter	= 1000 millimeters (mm)
1 meter	= 39.37 inches
1 inch	= 2.54 centimeters
1 statute mile	= 1.609 kilometers

Surfaces and volumes 2.5

Our concepts of surface area and volume derive from units of length. We imagine a surface as a region swept out by a line in some sort of sidewise motion, and we calculate the area by multiplying the length of the line by the distance it moves. The units of area are the square of the length units; thus,

$$3 \text{ cm} \times 3 \text{ cm} = 9 \text{ cm}^2$$

Orders of magnitude of length

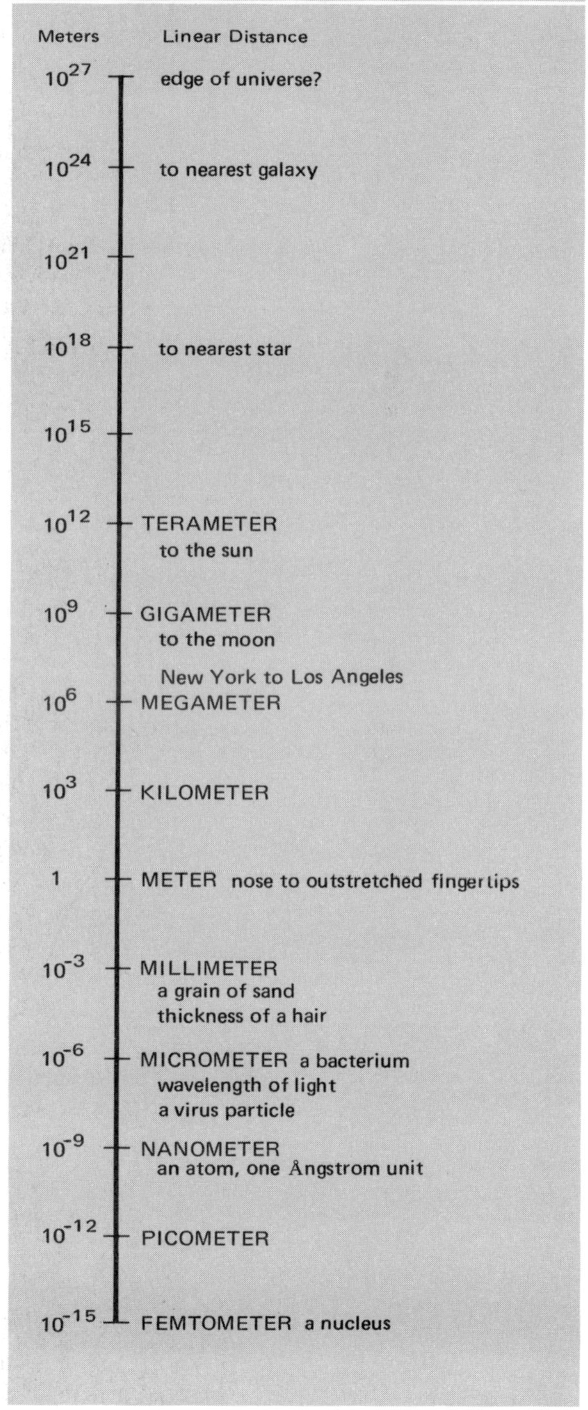

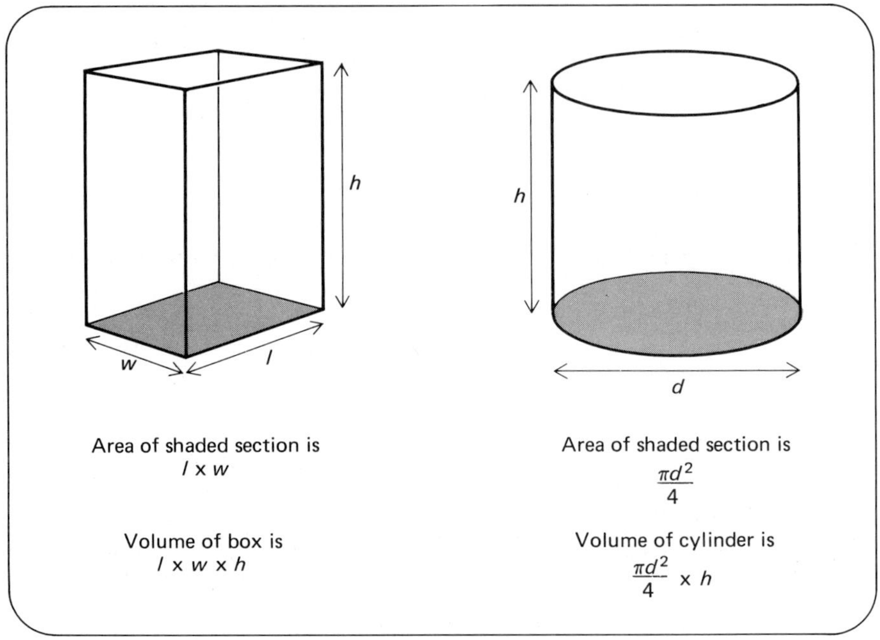

Figure 2.1

In a similar fashion we imagine a volume as a region of space swept out by a surface, and we calculate it by multiplying the area of the surface by the distance through which it moves. Volume units are the cube of length units; thus,

$$5\ \text{cm}^2 \times 3\ \text{cm} = 15\ \text{cm}^3$$

Figure 2.1 gives the relations of the volume of some regular objects to their dimensions.

The volumes of solid objects are usually given in cubic measure (e.g., cubic meters, cubic centimeters, cubic inches). For historical reasons, it is customary to use a different set of units for liquids. The metric unit of volume for fluids is the *liter* (pronounced "leeter"). It is defined as exactly one thousandth of a cubic meter. A cube 10 centimeters or 3.937 inches on each edge occupies a volume of 1 liter. It is about 6% bigger than a quart. Table 2.4 gives some English and metric volume units.

Table 2.4

Conversion factors for English and metric volume units

1 cubic meter (m^3)	= 1000 liters
1 liter	= 1000 milliliters (ml)
1 milliliter	= 1 cubic centimeter (cm^3 or cc)
1 quart (qt)	= 946 milliliters
1 gallon (gal)	= 231 cubic inches ($in.^3$)
1 cubic inch	= 16.4 cubic centimeters

Measuring volumes

Cylinders are the easiest liquid volume measuring devices to manufacture and calibrate. When the diameter is constant the volume depends on the length, and the capacity can be indicated by lines etched on the side of the cylinder. Calibrated cylinders (sometimes called "graduates" or "graduated cylinders") come in a variety of sizes and functions. Thermometers, hypodermic syringes, micropipets, burets, and even the old tin measuring cup are examples of calibrated cylinders of widely differing capacity and precision.

Containers designed to contain a specific volume of fluid have only one calibration mark and are called volumetric flasks.

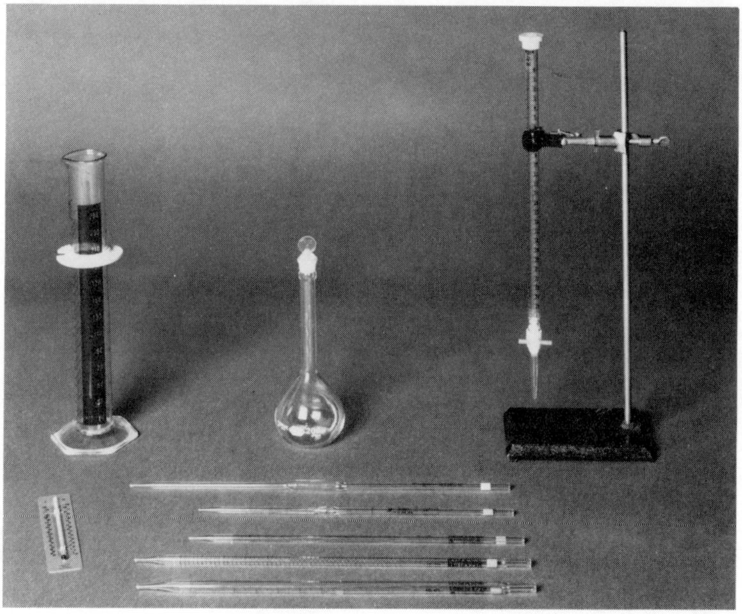

Mass 2.6

The mass of an object is the amount of matter present in it. *Mass* is the property of matter that interacts with gravity and exhibits inertia. *Inertia* is the tendency for things to stay put if they are at rest and to continue moving in a straight line if they are moving. For chemical purposes the quantity of matter is always given by the mass because it does not change with conditions.

The standard unit of mass is the *kilogram.* The standard kilogram is the mass of a cylinder of platinum-iridium kept in Sevres, France. Copies of it, together with sets of subdivisions, have been made so that every country can have standards with which to calibrate their weigh-

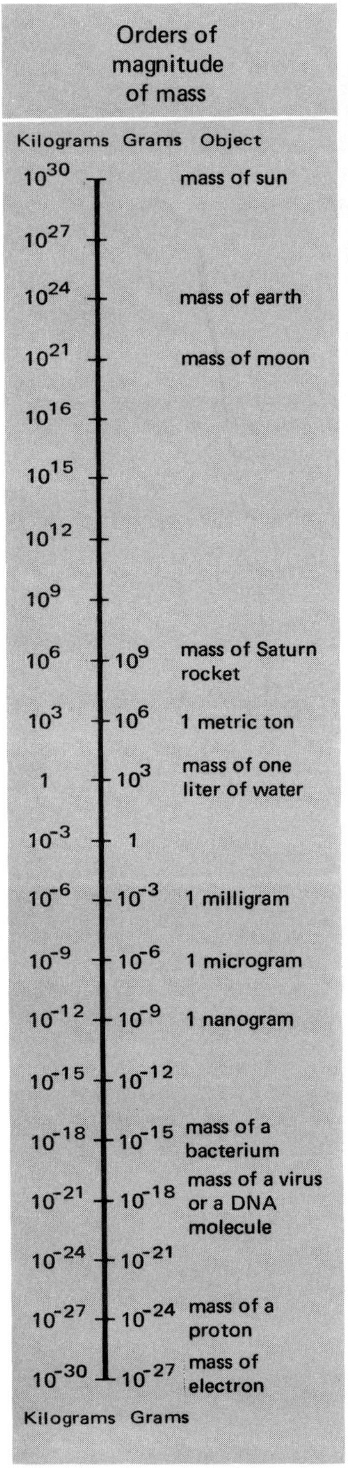

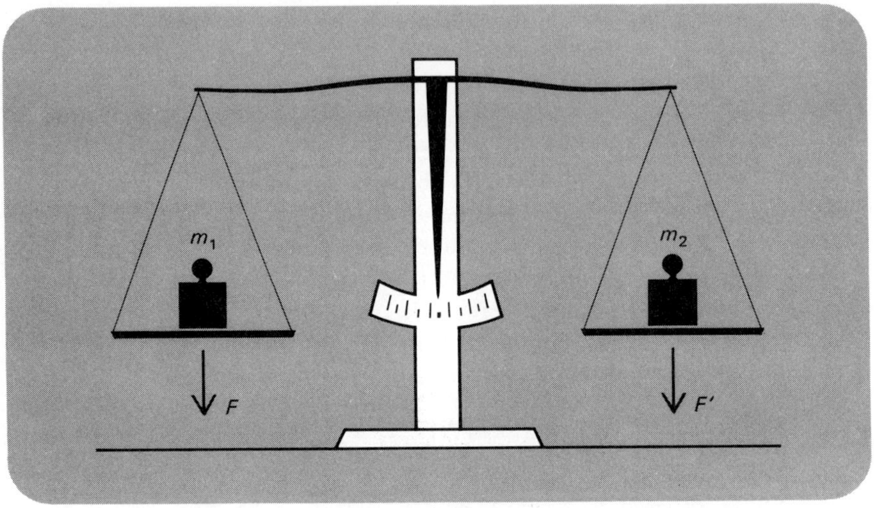

$$F = mg$$

Figure 2.2 Force (F) equals mass (m) times the earth's gravity constant (g). If F equals F' the balance is horizontal and $mg = m'g$. Therefore $m = m'$.

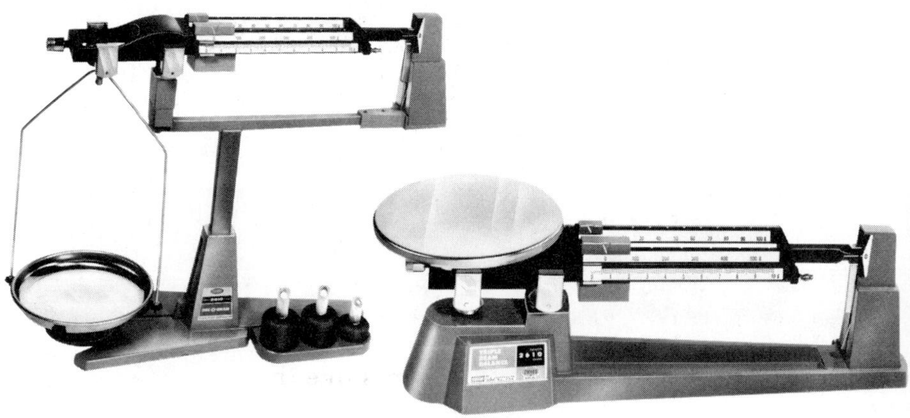

Table 2.5

Conversion factors of English and metric mass units

1 kilogram (kg) = 1000 grams (g)
1 gram = 1000 milligrams (mg)
1 kilogram = 2.205 pounds avoirdupois (lb (avdp))
1 pound (avdp) = 454 grams

ing devices. The masses of some objects of interest are given for comparison in the diagram at the left. Each level is 1000 times more massive than the one below it.

How to compare masses

Two objects on each end of a beam supported exactly in the middle will not move if they are subject to the same force. The force an object "feels" in the earth's gravitational field is directly proportional to its mass. The measure of this force is called the *weight* of the object. The earth's gravitational field varies with location and altitude, but it can be considered constant in any one place. So if the force on the standard is equal to the force on the subject measured in the same location, the two masses must be equal. See Figure 2.2. The instrument designed to measure the equality of gravity forces is called a *balance.*

Table 2.5 shows some mass conversion factors.

Time 2.7

Time is what happens while waiting for something to happen. In other words, it is the period between events. If the events happen regularly enough they can be used to "keep" time. The periods between successive high noons at one place were once used to define time, but it appears that the earth's rotation is not regular enough for modern needs. Modern clocks use vibrations of crystals, flip-flops of molecules, and frequencies of certain colors of light to divide time into extremely small intervals. The frequency of a monochromatic (one color) beam of light is the number of waves of light that pass per unit of time. If that number is defined, the unit of time is defined. The unit of time in the SI system is the *second.*

A comparison of the time periods of certain events is given on page 24. Some events happen regularly, like heartbeats and vibrations of molecules and waves. The time given is the period between events. Other events have a duration. Events like the formation of the earth and the founding of United States of America have happened only once, and the time period given is the period since the event.

Temperature 2.8

A temperature scale has been defined by declaring the hotness of freezing water to be equal to zero degrees and the hotness of boiling water to be 100 degrees. This scale is called the Celsius scale. Temperatures other than zero or one hundred can be measured if we can find some property of a substance that changes in a uniform way with temperature. The volume of mercury is such a property. If we have precisely

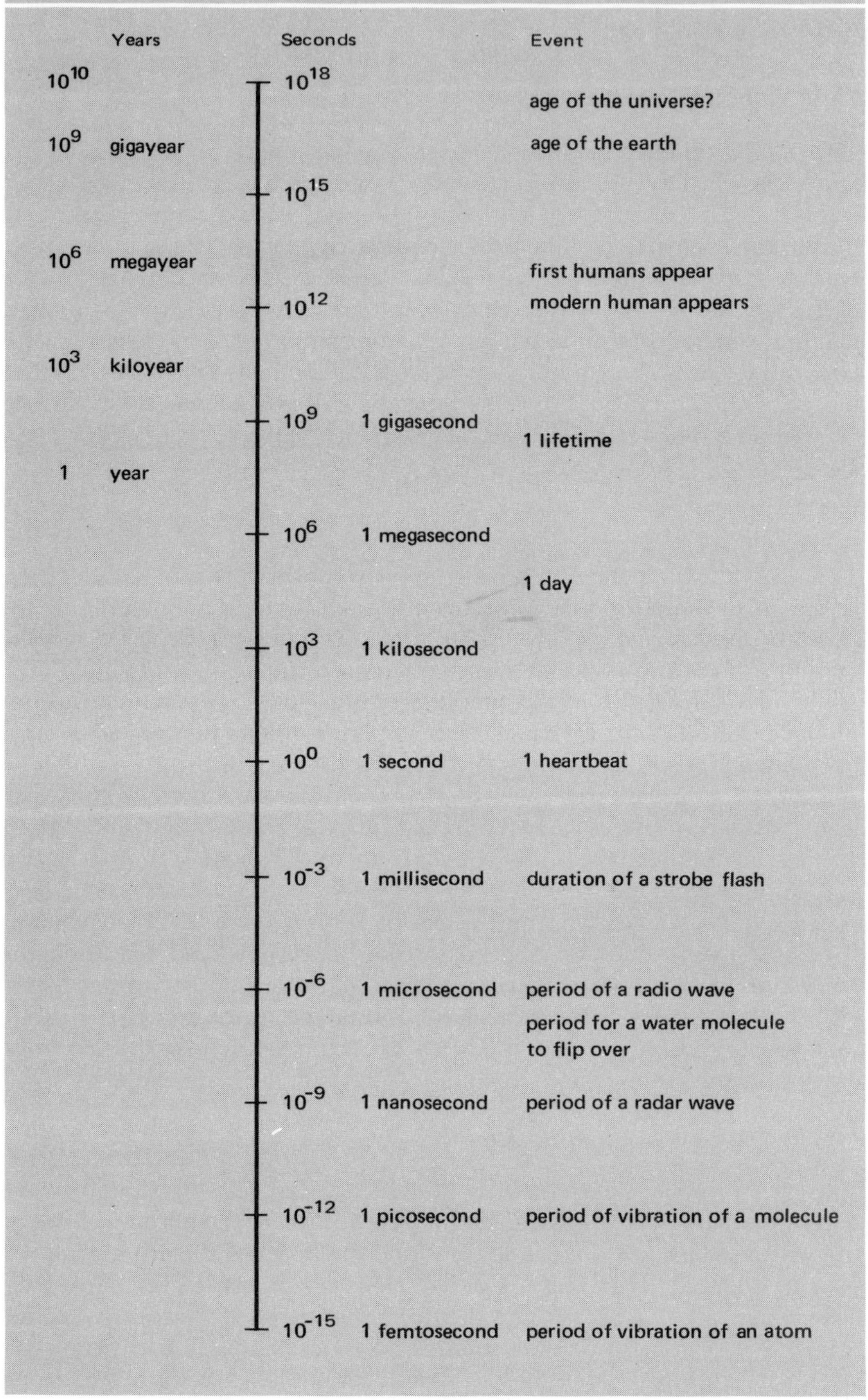

Orders of Magnitude of Time

100 milliliters of mercury at the freezing temperature of water we find that the volume of that same quantity of mercury will increase at the temperature of boiling water to 100.134 milliliters. A temperature change of 100 degrees has caused a volume increase of 0.134 milliliters per hundred milliliters of mercury, or 0.134 volume percent. On the assumption that the volume change corresponds in the same ratio to the temperature change anywhere on the scale we can say that a volume change of 0.00134 percent (compared to the volume at zero degrees) corresponds to a temperature change of one degree.

A number of other properties of substances change in proportion to temperature over a limited range. The electrical resistance of metals is one example; we will learn others as we continue. These properties can be used to make thermometers for temperatures at which mercury thermometers cannot function. Six reference points for calibrating such thermometers have been defined (Table 2.6).

Table 2.6

Temperature reference points

1063.6°C =	the freezing point of gold
960.8°C =	the freezing point of silver
444.6°C =	the boiling point of sulfur
100.0°C =	the boiling point of water
0.01°C =	water, ice, and water vapor (only) coexist
−183.97°C =	the boiling point of oxygen

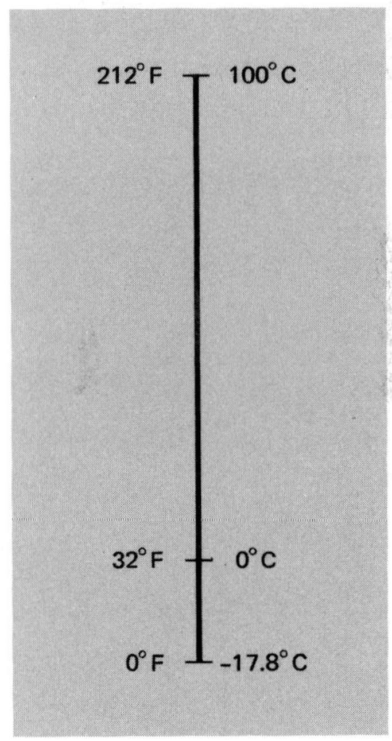

The Fahrenheit temperature scale marks off 180 degrees between the temperature of freezing and boiling water. A temperature interval of 5 Celsius degrees is therefore 9 Fahrenheit degrees. Then, to make things complicated, zero degrees on the Fahrenheit scale is placed at the lowest temperature available with a salt–ice mixture, which is 32 Fahrenheit degrees below pure freezing water.

To convert from one scale to the other, compute how many degrees the system is above or below freezing water, change the degrees into the other scale, and then adjust the new scale to the zero point, if necessary.

EXAMPLE 2.5

(a) Convert 98.6°F to degrees Celsius.
98.6°F is 66.6 Fahrenheit degrees above freezing (98.6 − 32).

$$66.6 \text{ deg F} \times \frac{5 \text{ deg C}}{9 \text{ deg F}} = 37 \text{ deg C above freezing, which is } 37°\text{C}.$$

(b) Convert 50°C to degrees Fahrenheit.
50°C is 50 Celsius degrees above freezing.

$$50 \text{ deg C} \times \frac{9 \text{ deg F}}{5 \text{ deg C}} = 90 \text{ deg F above freezing, which is } 90 \text{ deg F} + 32 \text{ deg F} = 122°\text{F above zero.}$$

Alternatively, you may wish to memorize the equations that describe the same operations.

$$\text{(a)}\quad °C = \frac{5}{9}(°F - 32)$$

$$\text{(b)}\quad °F = \frac{9}{5}°C + 32$$

EXAMPLE 2.6

(a) What is 14°F in degrees Celsius?

Solution:
$$°C = \frac{5}{9}(14 - 32) = \frac{5}{9}(-18) = -10°C$$

(b) What is 25°C in degrees Fahrenheit?

Solution:
$$°F = \frac{9}{5}(25) + 32 = 45 + 32 = 77°F$$

2.9 *Extensive properties*

In principle, a quantity of matter, space, or time can be measured by direct comparison with a set of weights, a measuring stick, or a clock. We can also count the number of items that make up the system to be described. Mass, space, time, and number are the basic measurements we can make on a system. As we shall see, all the other numerical properties that describe a system are defined by some relationship or combination of two or more of these basic measurements.

Mass, space, time, and number are called *extensive properties* of a system. Extensive properties depend on the amount of matter being considered. An extensive property can be added or subtracted; that is, the whole is the sum of its parts.

Mass is always additive. We can add one ton of sand to one ton of gravel, and the total will be two tons of mixed sand and gravel.

Volume is additive when identical materials are brought together, or when unlike materials do not mix. For example, 10 gallons of gasoline added to 10 gallons of gasoline will have a volume of 20 gallons. Likewise, 200 cubic inches of water plus a 100 cubic inch brick will occupy a volume of 300 cubic inches. But a cubic yard of sand mixed with a cubic yard of gravel will NOT fill a volume of two cubic yards.

We can add and subtract the numbers of members of the same

set. For example, 1 dozen eggs plus 1 dozen eggs is 2 dozen eggs, but 6 oranges subtracted from 12 grapefruit is nonsense.

There are only a few other properties of matter that have this characteristic of being additive. We will postpone examining them until we become more familiar with the basic quantities.

Because it is important to recognize which units of measurement specify a definite amount of material, we will use the word "quantity" only to refer to an extensive property.

EXAMPLE 2.7

Some examples of quantities (extensive properties) are:

(a) *one ton* of coal
(b) *one dozen* eggs
(c) *two pounds* of cheese
(d) *seven feet three inches* of height
(e) *seventeen gallons* of gasoline
(f) *two dollars* of money
(g) *one cubic yard* of earth
(h) *fifty minutes* of time

Related quantities 2.10

More often than not the quantity we want to find is not one we can measure conveniently. The number of atoms in the head of a pin, for example, would take 342 billion years to count at the rate of one atom per second. Since that amount of time is some thirty-five times longer than the universe has been in existence, we could say that it is impossible to count the number of atoms in the head of a pin. Nevertheless we can calculate the number if we measure the weight of the pinhead and know the weight per atom.

In everyday living we are constantly measuring one quantity and then mentally or mechanically converting it to another quantity. When you step on the bathroom scales, the length of extension of a spring is converted to a reading of pounds. The number of revolutions of the odometer cable in your car is converted to a reading of miles traveled. The number of turns in a gas meter is converted to a volume reading by an appropriate scale. Evidently a relationship exists between the two quantities, so that for every value of one there is a corresponding value of the other.

When there exists a set of variable quantities, x and y, such that for every value of x there is a definite value of y, we say that y is a function of x. If we *know* what the function is, we can compute one variable by measuring the other.

How to find a function

In order to find out how one quantity depends on the other, we must make a series of measurements on some quantity that is changing or variable, and, at the same time, look at the other quantity to see what changes take place. It is necessary, of course, to keep all the other conditions constant so we don't mix functions.

EXAMPLE 2.8 How does the weight of sea water depend on its volume?

Experiment: Make a series of measurements of different volumes of sea water and record the weight of each volume.

Data:

Volume	*Weight*
1.00 gal	8.55 lb
5.00 gal	42.8 lb
6.13 gal	52.4 lb
8.10 gal	69.3 lb
10.20 gal	87.2 lb

The easiest way to see whether one quantity depends on the other is to plot the results on a graph. Let all weights be plotted on the horizontal axis and all volumes on the vertical axis. The intersection of the two values of each ordered pair is a point on the graph. If you are reasonably sure that all the

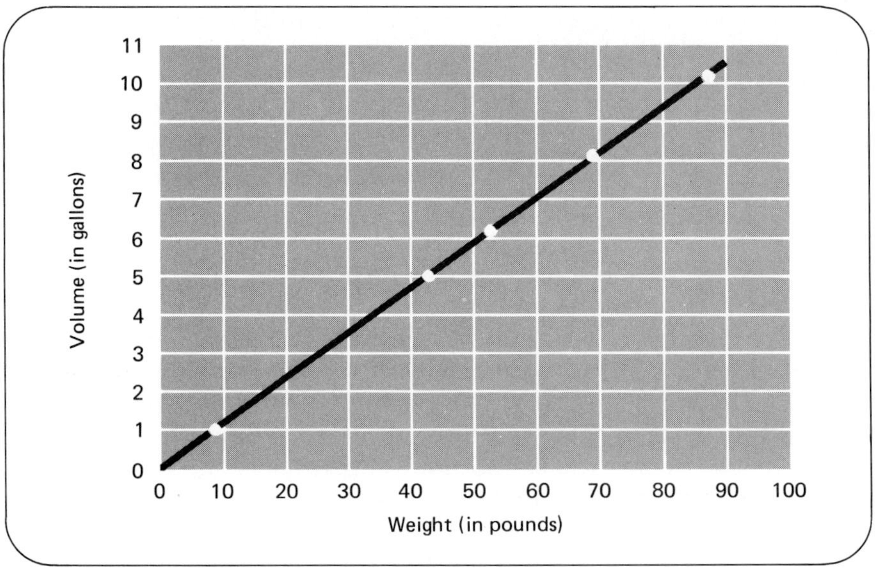

Figure 2.3

values between the points have a physical meaning, connect the points in a smooth curve. The curve (or straight line) tells how the one quantity changes for all values of the other, not just the measured ones.

Some of the points will not lie on the line because of errors in the measurements, but it is usually easy to see if the values line up in an orderly fashion. If they do not, the quantities are probably not related to each other, or something else is changing in an uncontrolled manner.

Graphs are important because it is easier to see the relationship between numbers as a line or curve than from the numbers themselves. If graphing was the only way to show a relationship, you can imagine that we would need a fantastic number of graphs, and some would be very bulky. There is a better way.

Looking at the graph of volumes versus weights of sea water, we see that the weight is increasing by a constant amount for each unit of increase of volume. In fact, if we mark off one unit of change of volume anywhere on the graph, we see there is a corresponding change of 8.55 units of weight. We also see that if we extend the line to zero volume the weight will also be zero. These facts can be expressed concisely in an algebraic equation as

$$w \text{ (lb of sea water)} = 8.55 \frac{\text{lb}}{\text{gal}} \times V \text{ (gal)}$$

where w is weight and V is volume.

The factor (8.55 lb/gal) is a property characteristic of sea water in general, and it relates the weight to the volume. Properties like these are not something we can "have," because they do not specify a particular amount of substance. They are something we "know." We use them to find the magnitude of one quantity when we have the other quantity.

EXAMPLE 2.9

A saltwater aquarium holds 20.0 gallons of sea water. How much will the water weigh?

Want: weight of salt water in pounds

Have: 20.0 gal of salt water

Know: the relation of weight to volume of salt water: 8.55 lb/gal

Solution: $20.0 \text{ gal} \times 8.55 \frac{\text{lb}}{\text{gal}} = 171 \text{ lb}$

EXAMPLE 2.10

1. The average weight of a penny can be found by *counting* and *weighing* several batches of pennies. In one such experiment the average weight was 0.115 ounces (oz) per penny. Having found the property of weight per penny, we can use the property to relate the weight and number of any amount of pennies.

$$w \text{ oz} \times \frac{1 \text{ penny}}{0.115 \text{ oz}} = n \text{ pennies}$$

or

$$n \text{ pennies} \times \frac{0.115 \text{ oz}}{1 \text{ penny}} = w \text{ oz}$$

2. The average gasoline consumption per mile can be found by measuring the *distance* traveled versus the *gallons* of gasoline used. (Note: This miles-per-gallon property decreases with speed.) Knowing the gasoline mileage enables one to compute the distance traveled given a volume of gasoline, or the volume needed to travel a given distance.

$$5.0\ \text{gal} \times 15\ \frac{\text{mi}}{\text{gal}} = 75\ \text{mi}$$

or $$75\ \text{mi} \times \frac{1\ \text{gal}}{15\ \text{mi}} = 5.0\ \text{gal}$$

When the factor that tells us how one quantity changes with the other is constant for a certain set of conditions, we say that the function or relationship between the quantities is linear. We get a straight line if we plot the two quantities on a graph. Most of the relationships of mass, number, and volume that we deal with in chemistry are of this kind.

Figure 2.4 shows why we can multiply a quantity by a factor to get a related quantity when the relationship is linear. When $\Delta y/\Delta x$ (read "the change in y per unit change in x," or "delta y over delta x") is a constant, y increases the same amount for each unit of x, and all we have to do to find y is to multiply $\Delta y/\Delta x$ by the number of units of x. Given a value for y, we may turn $\Delta y/\Delta x$ upside down and find x by multiplying y by $\Delta x/\Delta y$, because $\Delta x/\Delta y$ is the change in x per unit change in y.

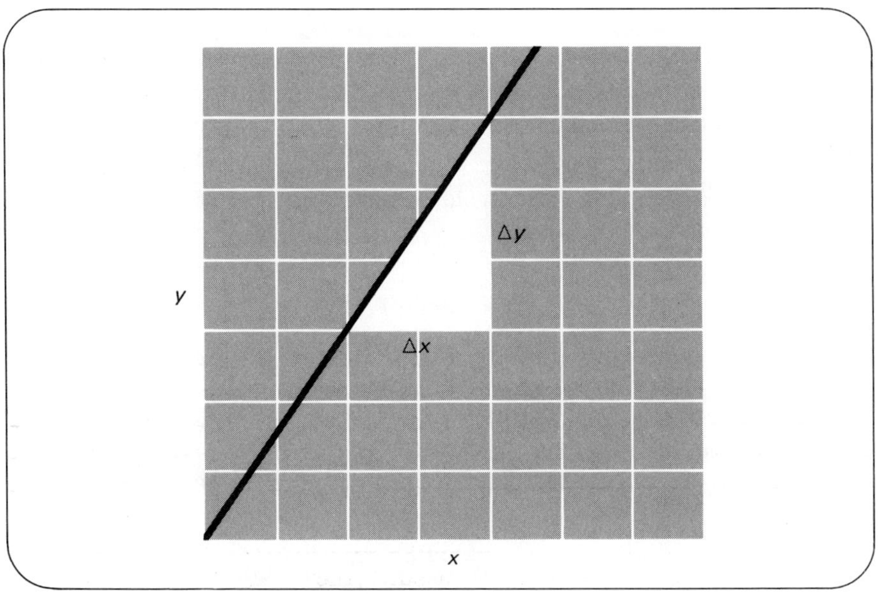

Figure 2.4

The factors $\Delta x/\Delta y$ are given many names depending on the situation in which they are used. They are called conversion factors, proportionality constants, derivatives (in math), relationship factors, or intensive properties (see next section). In all the problems in this book these factors will be constant for a given set of conditions, so we can find the value of y from the value of x by writing a linear equation:

$$y = \left(\frac{\Delta y}{\Delta x}\right)(x)$$

Intensive properties 2.11

The extensive properties of mass, volume, or number do not describe the kind of matter being considered. A pound of lead and a pound of feathers weigh the same, and the differences between lead and feathers are not disclosed by specifying one pound of each. The *volume* of one pound of lead is certainly different from the *volume* of one pound of feathers, and if we *compare* the weight to the volume of a sample of lead we will get a number different from the value obtained by comparing the weight to volume of feathers.

To compare two numbers means to form a ratio, or to divide one by the other. For example, we may measure the weight of a sample of lead as 57.0 grams, and the volume as 5.00 cubic centimeters. The ratio of the weight to the volume is

$$\frac{57.0 \text{ g}}{5.00 \text{ cm}^3} = 11.4 \frac{\text{g}}{\text{cm}^3}$$

This ratio of the weight to the volume is characteristic of lead at room temperature, and it is called its *density*. The ratio will remain constant at room temperature for any amount of lead. We can therefore use the density as a relationship factor to find the weight from the volume, and vice versa.

$$20 \text{ cm}^3 \text{ of lead} \times 11.4 \frac{\text{g}}{\text{cm}^3} = 228 \text{ g of lead}$$

$$\text{or} \qquad 114 \text{ g of lead} \times \frac{1 \text{ cm}^3}{11.4 \text{ g}} = 10.0 \text{ cm}^3 \text{ of lead}$$

In a similar manner we can compare weight to number, number to volume, weight of the part to weight of the whole, or *any measurable property* of a system to *any other measurable property* of the same system.

The property so obtained is called an *intensive property*, because it is independent of the amount of substance present, but it does depend on the kind, or condition, or relation of the substances present.

Because it is important to know the properties being compared by a particular intensive property, we will introduce them one at a time throughout this book. Of the properties discussed so far, temperature and density are intensive properties.

Problem solving may become easier to organize if you sort out those properties that specify an amount of material from those that do not. In this book quantities that specify an amount of material are "Have" quantities, and properties that are relationship factors or intensive properties are "Know" properties.

EXAMPLE 2.11

Write a linear equation that relates the "want" quantity in terms of the "have" quantity and the "know" property.

1. Find the number of tiles necessary to cover an area of 108 square feet (ft^2) if each tile is $\frac{9}{16}$ square feet in area.

Want: number of tiles

Have: 108 ft^2 of area

Know: the relationship of area to number of tiles is $\frac{9}{16}$ ft^2/tile

Equation: $108\ \text{ft}^2 \times \frac{16\ \text{tile}}{9\ \text{ft}^2} = 192\ \text{tiles}$

2. Find the volume displaced by an 8000 pound boat if the density of sea water is 64 lb/ft^3.

Want: volume in cubic feet

Have: 8000 lb of water displaced

Know: the relation between weight and volume of sea water is the density, 64 lb/ft^3

Equation: $8000\ \text{lb} \times \frac{1\ \text{ft}^3}{64\ \text{lb}} = 125\ \text{ft}^3$

2.12 *Percent*

Percent is a factor that relates the part of something to the whole of it. The quantities that describe the part and the whole have the same units. That is, we compare the weight of the part to the weight of the whole, or the volume of the part to the volume of the whole. The fraction obtained is multiplied by 100, because it is easier to say a number like 85% than it is to say 0.85.

Since we can compare any related quantities that are additive, we should specify what kind of quantities we are comparing.

EXAMPLE 2.12

1. The population of the U.S. is 52% women. We are evidently not comparing weight or volume here, but numbers.

$$\frac{\text{number of women}}{\text{total population}} \times 100 = 52\%$$

2. Sea water is 3% salt, by weight.

$$\frac{\text{weight of salt}}{\text{weight of sea water}} \times 100 = 3\%$$

Given the quantity of weight, volume, or number of either the part or the whole we can find the corresponding quantity of the other, if we know the right relationship as a percentage. Before we can do a calculation, though, we must write the percentage in functional form. That means that it must be written in arithmetical terms. A 10% salt solution means

$$\frac{10 \text{ g of salt}}{100 \text{ g of solution}}$$

This term can be multiplied and divided by another quantity, whereas there is no arithmetical operation associated with the % sign.

EXAMPLE 2.13

1. What weight of salt is in 200 grams of eyewash, if eyewash is 1% salt?

Want: weight of salt

Have: 200 g of eyewash

Know: the relation of weight of salt to weight of eyewash is

$$\frac{1 \text{ g of salt}}{100 \text{ g of eyewash}}$$

Equation: $200 \text{ g of eyewash} \times \frac{1 \text{ g of salt}}{100 \text{ g of eyewash}} = 2 \text{ g of salt}$

2. What volume of wine contains 500 milliliters of alcohol, if wine is 12.5% alcohol by volume?

Want: volume of wine, in milliliters

Have: 500 ml of alcohol

Know: the relation of volume of alcohol to volume of wine is

$$\frac{12.5 \text{ ml of alcohol}}{100 \text{ ml of wine}}$$

Equation: $500 \text{ ml of alcohol} \times \frac{100 \text{ ml of wine}}{12.5 \text{ ml of alcohol}} = 4000 \text{ ml of wine}$

If the ratio of the part to the whole is so small that multiplying by 100 still gives a ridiculously small number, one can multiply by a larger number, say a million or a billion. The art of chemical analysis has advanced to the point where extremely small fractions of substances can be detected. In these cases, the ratio is given as parts-per-million (ppm) or parts-per-billion (ppb).

EXAMPLE 2.14 The Clean Air Standards Act puts a limit of 275 ppm on hydrocarbon emissions from automobiles. This means

$$\frac{\text{275 pounds of hydrocarbons}}{10^6 \text{ pounds of exhaust}}$$

Expressed as a percent it would be

$$\frac{0.0275 \text{ lb}}{100 \text{ lb}} \quad \text{or} \quad 0.0275\%$$

2.13 *The chain rule*

Very often there is no direct relationship between the quantity you have and the one you want to find. The problem then becomes a matter of constructing a logic track, using the relationships that you do know, until the desired quantity is obtained.

EXAMPLE 2.15 Find the cost in dollars for gasoline to make a trip of 300 miles, if gasoline costs 50 cents per gallon and your car gets 15 miles per gallon.

(a) $300 \text{ mi} \times \frac{1 \text{ gal}}{15 \text{ mi}} = 20 \text{ gal}$

(b) $20 \text{ gal} \times 50 \frac{\text{cents}}{\text{gal}} = 1000 \text{ cents}$

(c) $1000 \text{ cents} \times \frac{1 \text{ dollar}}{100 \text{ cents}} = 10 \text{ dollars}$

Well, that was pretty easy, and you probably already knew how to do it. But can you do this one?

EXAMPLE 2.16 An average adult male has 7.5 quarts of blood in his body. He is considered too drunk to drive if his blood contains 0.15% alcohol by volume. One hundred proof whiskey contains 50% alcohol by volume. How many drinks containing one fluid ounce per drink will produce that concentration of blood alcohol, if it is absorbed right away and none is metabolized? One quart is 32 fluid ounces. Where do we start?

(a) $7.5 \text{ qt} \times 32 \dfrac{\text{fl oz}}{\text{qt}} = 240 \text{ fl oz of blood}$

(b) $240 \text{ fl oz} \times \dfrac{0.15 \text{ fl oz of alcohol}}{100 \text{ fl oz of blood}} = 0.36 \text{ fl oz of alcohol}$

(c) $0.36 \text{ fl oz} \times \dfrac{100 \text{ fl oz of whiskey}}{50 \text{ fl oz of alcohol}} = 0.72 \text{ fl oz of whiskey}$

(d) $0.72 \text{ fl oz} \times \dfrac{1 \text{ drink}}{1 \text{ fl oz}} = 0.72 \text{ drink}$

How did we know to start with 7.5 quarts of blood, and then convert to alcohol, whiskey, and drinks? I am tempted to say, "Lots of practice," but the answer is rather unsatisfying.

The first thing to observe is that a quantity multiplied by a relationship factor gives another quantity. Second, since a relationship factor or intensive property does not specify any amount of material, *if you want a quantity you must start with a quantity.* We wanted the *number* of drinks, a quantity, and the only other extensive quantity we had was *7.5 quarts* of blood. All the other terms are relationship factors. So we started with the quantity we had and multiplied it by the only factor that related it to another quantity.

The third observation will give us a short cut to solving problems that is no less rigorous or logical than the above, but it saves time and makes the whole process of problem solving almost automatic. Instead of solving the first equation, the left-hand side could be substituted for the 240 fl oz of blood in the second equation:

(b′) $7.5 \text{ qt} \times 32 \dfrac{\text{fl oz}}{\text{qt}} \times \dfrac{0.15 \text{ fl oz of alcohol}}{100 \text{ fl oz of blood}} = 0.36 \text{ fl oz of alcohol}$

And instead of solving (b′), we substitute the left side into (c′):

(c′) $7.5 \text{ qt} \times 32 \dfrac{\text{fl oz}}{\text{qt}} \times \dfrac{0.15 \text{ fl oz}}{100 \text{ fl oz}} \times \dfrac{100 \text{ fl oz of whiskey}}{50 \text{ fl oz of alcohol}}$
$= 0.72 \text{ fl oz whiskey}$

And finally:

(d′) $7.5 \text{ qt} \times 32 \dfrac{\text{fl oz}}{\text{qt}} \times \dfrac{0.15 \text{ fl oz}}{100 \text{ fl oz}} \times \dfrac{100 \text{ fl oz}}{50 \text{ fl oz}} \times \dfrac{1 \text{ drink}}{1 \text{ fl oz}} = 0.72 \text{ drink}$

The whole problem is solved in one equation!

In general, there is a rule in mathematics that says IF there is a relationship between A and B, between B and C, between C and D, and so on, you can find the relationship between A and any of the others by multiplying the appropriate relationship factors together.

Thus,

$$A \times \frac{\Delta B}{\Delta A} \times \frac{\Delta C}{\Delta B} \times \frac{\Delta D}{\Delta C} = D$$

It is called the "chain rule." Each of the links in the chain is one of the relationship factors (intensive property, conversion factor, etc.) that tells how B changes with A, C changes with B, and so on. Note that the links must be arranged so that quantities that refer to the same thing are diagonally next to each other so that the units cancel. The quantity that you *have* is first; the quantity that you *want* is named in the numerator of the last relationship factor.

Set up the previous problem in the following way. List the data given or measured in three categories.

Want: x drinks. The magnitude is unknown but you know the unit—1 drink, a quantity.

Have: 7.5 quarts of blood, a quantity

Know: **(a)** the relation of blood alcohol volume to blood volume:

$$\frac{0.15 \text{ fl oz of alcohol}}{100 \text{ fl oz of blood}}$$

(b) the relation of alcohol volume to whiskey volume:

$$\frac{50 \text{ fl oz alcohol}}{100 \text{ fl oz whiskey}}$$

(c) the relation of whiskey volume to drinks: $\frac{1 \text{ drink}}{1 \text{ fl oz}}$

(d) the relation of fluid ounces to quarts: $32 \frac{\text{fl oz}}{\text{qt}}$

We want a quantity; we must start with a quantity. Then look for a link (relationship factor) with the quantity you have. If you do not know it, you may have to look outside the problem and bring in more relationships. Link enough relationship factors together, being sure that every one is related (diagonally) to the one in front of it. When the quantity you want is named in the upper right corner, stop.

EXAMPLE 2.17

1. How many acres of peppers must Peter Piper pick to fill an order for 1000 cases if a case has 24 cans, each can has 8 peppers, the farm averages 12 peppers per square yard, and there are 4840 square yards per acre?

Want: x acres

Have: 1000 cases

Know: **(a)** the relationship of cans to cases: $24 \frac{\text{cans}}{\text{case}}$

(b) the relationship of peppers to cans: $8 \frac{\text{peppers}}{\text{can}}$

(c) the relationship of peppers to square yards: $12\ \frac{\text{peppers}}{\text{sq yd}}$

(d) the relationship of square yards to acres: $4840\ \frac{\text{yd}^2}{\text{acre}}$

Solution: $$1000 \text{ cases} \times 24\ \frac{\text{cans}}{\text{case}} \times 8\ \frac{\text{peppers}}{\text{can}} \times \frac{1 \text{ sq yd}}{12 \text{ peppers}} \times \frac{1 \text{ acre}}{4840 \text{ sq yd}} = 3.3 \text{ acres}$$

2. How much sugar is in a tank containing syrup that is 25% (by weight) sugar, if the tank holds 100 gallons per inch of depth, the syrup weighs 9.1 pounds per gallon, and the syrup is 75 inches deep?

Want: *x* pounds of sugar, a quantity

Have: 75 inches of depth, a quantity

Know: (a) the relation of depth to volume: $100\ \frac{\text{gal}}{\text{in.}}$

(b) the relation of volume to weight: $9.1\ \frac{\text{lb}}{\text{gal}}$

(c) the relation of weight of sugar to weight of syrup:

$$\frac{25 \text{ lb sugar}}{100 \text{ lb syrup}}$$

Solution: $$75 \text{ in.} \times 100\ \frac{\text{gal}}{\text{in.}} \times 9.1\ \frac{\text{lb}}{\text{gal}} \times \frac{25 \text{ lb}}{100 \text{ lb}} = 17{,}000 \text{ lb of sugar}$$

Summary 2.14

1. Matter and its interactions with other matter is described by its properties.
2. Any physical property has two necessary parts: a magnitude (a number) and a dimension (a unit).
3. The properties of matter can be divided into two classes:
   (a) Those that depend upon the *amount* of matter being considered: extensive properties; and,
   (b) Those that depend on the *kind* or *condition* or *relation* of matter being considered: intensive properties.
4. Properties that relate two or more measurable properties to each other can be used to calculate the magnitude of one property from the other.
5. Provided a functional relationship exists, extensive and intensive properties may be multiplied according to the chain rule to give equivalent properties.

$$A = D \times \frac{\Delta C}{\Delta D} \times \frac{\Delta B}{\Delta C} \times \frac{\Delta A}{\Delta B}$$

SUGGESTED READING

Allen V. Astin, "Standards of Measurement," *Scientific American* (June, 1968), p. 50.

EXERCISES

Solutions are at the end of the chapter.

I. Label the specific quantity or property on the right with the general concept on the left. Some are used more than once.

1. 5 lb/ft^3	**(a)** volume
2. 7 g	**(b)** area (surface)
3. 2 mi	**(c)** length (distance)
4. 5 cm	**(d)** mass
5. 4 cm^2	**(e)** speed
6. 100 km/hr	**(f)** density
7. 8 g/ml	**(g)** number
8. 2 ml	**(h)** time
9. 2 years	
10. 12 doz	

II. Convert the English units to the equivalent metric units using the given relationships.

1 m = 100 cm	1 kg = 2.2 lb
1 km = 1000 m	1 lb = 454 g
1 cm = 10 mm	1 liter = 1000 ml
1 in. = 2.54 cm	1 ml = 1 cm^3
1 ft = 12 in.	1 $in.^3$ = 16.4 ml
1 mi = 5280 ft	1 gal = 231 $in.^3$

1. 390 cubic inches to liters

2. 52.9 cents/gallon to cents/liter

3. 65 miles/hour to kilometers/hour

4. one-half inch to millimeters

5. 8 lb/gal to kilograms/liter

III. Write an equation that expresses one quantity in terms of the other.

1. The volume of a cylinder (V ml) depends on its height (h cm); relation is 5 ml/cm.

2. The weight of beans (w g) depends on the number of beans (n beans); relation is 0.5 g/bean.

3. The weight of water (w kg) depends on the volume of water (V liters); relation is 1.00 kg/liter.

4. **(a)** The cost of gasoline (x dollars) depends on the volume of gasoline (V gal); relation is 0.50 dollars/gal.

(b) The distance traveled (m mi) depends on the volume of gasoline used (V gal); relation is 15 mi/gal.

(c) Write an equation relating cost and distance using equations a and b.

SOLVED PROBLEMS

IV. **1.** 5.00 ml of mercury weighs 68.0 g. What is the density of mercury?

2. The density of kerosene is 0.80 g/ml. What is the weight of 500 ml?

3. The density of air at sea level is 1.29 g/liter. What is the volume of one pound (454 g) of air?

4. One quart of milk (946.4 ml) weighs 976 g. Its composition is 32 g protein, 48 g carbohydrate, 40 g butterfat, and the remainder is water and mineral salts.
Calculate **(a)** the density, **(b)** the percent composition of protein, **(c)** the percent carbohydrate, and **(d)** the percent butterfat in milk.

5. Molasses is 55% sugar. How much molasses is needed to have 100 kg of sugar?

6. Swiss cheese is 29% fat. Two ounces (56 g) contain how much fat?

7. A truck has a cylindrical gas tank 20 in. in diameter and 30 in. long. There are 231 $in.^3$ in a gallon. The truck gets 15 mi/gal. How far can it go on a full tank of gasoline?

8. The density of a 50% (by weight) alcohol solution in water is 0.95 g/ml. What volume of solution contains 28 g of alcohol?

9. Archimedes discovered that the weight *loss* of an object submerged in a liquid (called the buoyancy force) is equal to the weight of the liquid displaced.
An object that weighs 27 g in air weighs 17 g when suspended in water. The density of water is 1.0 g/ml. The same object weighs 19 g when suspended in kerosene.
(a) What is the volume of the object?
(b) What is the density of kerosene?

10. A wheel is 25 in. in diameter. If the car to which the wheel is attached is going 60 mph, how many revolutions per minute is the wheel turning?

ADDITIONAL PROBLEMS

I. Calculate the density of the following substances:

1. A 10.0 cm long aluminum wire that is 0.16 cm in diameter and weighs 0.540 g.

2. 50 liters of hydrogen gas weighing 4.46 g.

3. A cube of iron 2.0 cm on each edge weighing 57.6 g.

4. One gallon of gasoline weighing 6.0 lb. (Find density in g/ml.)

II. Calculate the percent composition of the following:

1. 25 liters of air composed of 19.5 liters of nitrogen, 5.0 liters of oxygen, and the rest smog.

2. 19.0 g of water and 1.0 g of salt.

3. 325 Democrats and 289 Republicans.

III. 1. What weight of plastic foam (density 1.2 lb/ft^3) is necessary to fill a rectangular cavity 8 ft x 16 ft x 0.25 ft?

2. Which is the better buy: 10 oz of dry cereal at \$.44 or 12 oz for \$.59? How much do you pay per pound for each one? (1 lb avdp = 16 oz)

3. A jet plane holds 1000 gal of fuel. Fuel weighs 6 lb/gal. The engine consumes 1200 lb/hr. The speed of the plane is 550 mph. How far can the plane go on 1000 gal of fuel?

4. A metric ton is 1000 kg. A U.S. ton is 2000 lb. If you ordered a U.S. ton of quadrified gilfinkels from Germany at \$1.50 per pound, and received instead a metric ton, how much more (or less) would they cost than you expected?

5. The average distance to the moon from the surface of the earth is 386,000 km. If radio waves travel 3×10^5 km/sec, how much time is necessary for a radio signal to go from the earth to the moon and back?

6. Fruit juice that has not fermented is called "a wine must." A typical must contains 22.7% sugar and 0.70% acid as tartaric acid. How much sugar and how much acid are in 20.0 liters of must of density 1.095 kg/liter?

7. A piece of ore weighs 8.00 g in air. Submerged in oil it weighs 7.00 g. The density of the oil is 0.81 g/ml.
   (a) What is the volume of the ore?
   (b) What is the density of the ore?

8. The density of sea water is 0.037 lb/in.^{3}. One standard atmosphere of pressure is 14.7 lb/in.^{2}. What is the height in feet of a column of sea water that exerts a pressure of one standard atmosphere on its base?

   Note: $\text{height} = \dfrac{\text{volume}}{\text{area}}$; $\text{pressure} = \dfrac{\text{force (weight)}}{\text{area}}$

9. Premium gasoline uses 4.00 ml of tetraethyl lead (density 1.66 g/ml) per gallon. Tetraethyl lead is 64% lead. How much lead (in kilograms) is exhausted into the air each day by one million automobiles burning an average of 7.0 gal per day?

10. How many cigarettes containing 30 mg of tar per cigarette must a person smoke to inhale one pound (454 g) of tar? If a person smokes 20 cigarettes per day, what period of time is necessary to ingest one pound of tar?

SOLUTIONS

I.
1. density
2. mass
3. distance (length)
4. length
5. area
6. speed
7. density
8. volume
9. time
10. number

II. 1. $390 \text{ in.}^3 \times 16.4 \dfrac{\text{ml}}{\text{in.}^3} \times \dfrac{1 \text{ liter}}{1000 \text{ ml}} = 6.4 \text{ liters}$

2. $\frac{32.9 \text{ cents}}{\text{gal}} \times \frac{1 \text{ gal}}{231 \text{ in.}^3} \times \frac{1 \text{ in.}^3}{16.4 \text{ ml}} \times \frac{1000 \text{ ml}}{\text{liter}} = \frac{8.7 \text{ cents}}{\text{liter}}$

Note: If we want an intensive property we can start with a ratio of two quantities or another relationship factor.

3. $\frac{65 \text{ mi}}{\text{hr}} \times \frac{5280 \text{ ft}}{\text{mi}} \times \frac{12 \text{ in.}}{\text{ft}} + \frac{2.54 \text{ cm}}{\text{in.}} \times \frac{1 \text{ meter}}{100 \text{ cm}} \times \frac{1 \text{ km}}{1000 \text{ m}} = \frac{105 \text{ km}}{\text{hr}}$

4. $0.5 \text{ in.} \times \frac{2.54 \text{ cm}}{\text{in.}} \times \frac{10 \text{ mm}}{\text{cm}} = 12.7 \text{ mm}$

5. $\frac{8.3 \text{ lb}}{\text{gal}} \times \frac{454 \text{ g}}{\text{lb}} \times \frac{1 \text{ kg}}{1000 \text{ g}} \times \frac{1 \text{ gal}}{231 \text{ in.}^3} \times \frac{1 \text{ in.}^3}{16.4 \text{ ml}} \times \frac{1000 \text{ ml}}{\text{liter}} = \frac{0.99 \text{ kg}}{\text{liter}}$

Note: We want a property (kg/liter) in which both the numerator and denominator are different from the property given. The first two factors change the numerator to kilograms and the next three change the denominator to liters.

III. **1.** $V \text{ ml} = 5 \frac{\text{ml}}{\text{cm}} \times h \text{ cm}$

2. $W \text{ g} = 0.5 \frac{\text{g}}{\text{bean}} \times n \text{ beans}$

3. $W \text{ kg} = 1.00 \frac{\text{kg}}{\text{liter}} \times V \text{ liter}$

4. **(a)** $x \text{ dollars} = 0.50 \frac{\text{dollars}}{\text{gal}} \times V \text{ gal}$

(b) $m \text{ mi} = 15 \frac{\text{mi}}{\text{gal}} \times V \text{ gal}$

(c) $x \text{ dollars} = m \text{ mi} \times \frac{1 \text{ gal}}{15 \text{ mi}} \times 0.50 \frac{\text{dollars}}{\text{gal}}$

or $m \text{ mi} = x \text{ dollars} \times \frac{1 \text{ gal}}{0.50 \text{ dollar}} \times 15 \frac{\text{mi}}{\text{gal}}$

or $\frac{x \text{ dollars}}{m \text{ mi}} = 0.50 \frac{\text{dollars}}{\text{gal}} \times \frac{1 \text{ gal}}{15 \text{ mi}}$

IV. **1.** $\text{density} = \frac{\text{weight of mercury}}{\text{volume of mercury}} = \frac{68.0 \text{ g}}{5.00 \text{ ml}} = 13.6 \frac{\text{g}}{\text{ml}}$

2. **Want:** x g of kerosene

Have: 500 ml

Know: $\frac{0.80 \text{ g}}{\text{ml}}$

Solution: $500 \text{ ml} \times \frac{0.80 \text{ g}}{\text{ml}} = 400$ g of kerosene

3. **Want:** x liters of air

Have: 454 g of air

Know: $\frac{1.29 \text{ g of air}}{\text{liter}}$

Solution: $454 \text{ g} \times \frac{1 \text{ liter}}{1.29 \text{ g}} = 352$ liters of air

4. (a) **Want:** $\frac{\text{weight of milk}}{\text{volume of milk}}$

Have: 976 g; 946.4 ml

Solution: $\frac{976 \text{ g}}{946.4 \text{ ml}} = \frac{1.032 \text{ g of milk}}{\text{ml of milk}}$

(b) **Want:** $\frac{\text{weight of protein}}{\text{weight of milk}} \times 100$

Have: 32 g protein; 976 g of milk

Solution: $\frac{32 \text{ g}}{976 \text{ g}} \times 100 = 3.3\%$ protein

(c) $\frac{48 \text{ g carbohydrate}}{976 \text{ g milk}} \times 100 = 4.9\%$ carbohydrate

(d) $\frac{40 \text{ g fat}}{976 \text{ g milk}} \times 100 = 4.1\%$ butterfat

5. **Want:** kg of molasses

Have: 100 kg of sugar

Know: $\frac{55 \text{ kg of sugar}}{100 \text{ kg of molasses}}$

Solution: 100 kg of sugar $\times \frac{100 \text{ kg of molasses}}{55 \text{ kg of sugar}}$
= 182 kg of molasses

6. 56 g of cheese $\times \frac{29 \text{ g of fat}}{100 \text{ g of cheese}} = 16$ g of fat

7. **Want:** x miles

Have: dimensions of tank

Know: volume of tank from dimensions; $\frac{15 \text{ mi}}{\text{gal}}$; $\frac{231 \text{ in.}^3}{\text{gal}}$

$V = \frac{\pi}{4} d^2h$

Solution: $\frac{3.14(20 \text{ in.})^2(30 \text{ in.})}{4} \times \frac{1 \text{ gal}}{231 \text{ in.}^3} \times \frac{15 \text{ mi}}{\text{gal}} = 610$ mi

8. **Want:** x ml of solution

Have: 28 g of alcohol

Know: $\frac{50 \text{ g of alcohol}}{100 \text{ g of solution}}$; $\frac{0.95 \text{ g of solution}}{\text{ml}}$

Solution: 28 g of alc. $\times \frac{100 \text{ g of soln.}}{50 \text{ g of alc.}} \times \frac{1 \text{ ml}}{0.95 \text{ g of soln.}}$
= 59 ml of solution

9. (a) **Want:** volume of object in ml

Have: buoyancy force = 27 g − 17 g = 10 g of water

Know: $\frac{1.0 \text{ g of water}}{\text{ml}}$, vol of object = vol of water displaced

Solution: 10 g $\times \frac{1.0 \text{ ml}}{\text{g}} = 10$ ml

(b) **Want:** $\frac{\text{weight of kerosene}}{\text{volume of kerosene}}$

Have: buoyancy force = 27 g − 19 g = 8 g of kerosene
volume of kerosene = vol of object = 10 ml

Solution: $\dfrac{8\text{ g}}{10\text{ ml}} = \dfrac{0.8\text{ g of kerosene}}{\text{ml}}$

10. Want: $x\,\dfrac{\text{revolutions}}{\text{minute}}$

Know: $\dfrac{1\text{ rev}}{\pi(25)\text{ in.}};\ \dfrac{60\text{ mi}}{\text{hr}};\ \dfrac{5280\text{ ft}}{\text{mi}};\ \dfrac{60\text{ min}}{\text{hr}};\ \dfrac{12\text{ in.}}{\text{ft}}$

circumference = πd

Solution: $\dfrac{1\text{ rev}}{3.14(25)\text{ in.}} \times \dfrac{12\text{ in.}}{\text{ft}} \times \dfrac{5280\text{ ft}}{\text{mi}} \times \dfrac{60\text{ mi}}{\text{hr}} \times \dfrac{1\text{ hr}}{60\text{ min}}$

$= 8.1 \times 10^2\,\dfrac{\text{revolutions}}{\text{minute}}$

The task of understanding the physical world begins with understanding things we can see, touch, and manipulate.

After reading this chapter you should be able to

1. Describe the phases of matter;
2. Recognize heterogeneous and homogeneous mixtures;
3. Describe several methods of separating and purifying substances;
4. Name the requirements that establish the purity of a substance;
5. Predict the effects of temperature and pressure on the properties of a substance;
6. Formulate a hypothesis about the microstructure of matter that partially explains what we see;
7. Distinguish between kinetic and potential energy.

3 form and substance

The world that we perceive with our senses is the macroscopic world. This is the arena in which we can experiment directly, study causes and effects, and make measurements. The lower limit of "vision" of an electron microscope is about the size of a virus particle. The practical lower limit of the amount of substance that can be isolated and identified is about one microgram.

The upper limit is the world. We are certainly causing direct changes in the ecology of the earth, but our control of the experiment is in doubt.

We infer that what goes on in the microscopic world beyond our vision causes observable changes in the macroscopic world if a sufficient number of things happen. Our picture of the details of what is happening must take into account the great numbers involved and treat each event as a statistical average. There is no way we can cause or observe a single microscopic event. We have no molecular tweezers.

That part of the universe that is at cosmic distances is beyond our capacity to experiment with at the moment. We can observe what happens and relate what we see to what we know about the earth, but there is no way we can affect the events.

COSMIC	Universe Galactic clusters Galaxies Stars Solar systems
MACRO	Planets Ecosystems Communities Species Individuals Cells
MICRO	Organelles Molecules Atoms Subatomic particles ?

3.1 *Forms of matter*

The idea that the form and behavior of macroscopic things depends on microscopic structure is very old. The ancients thought that things were composed of four "elements": air, earth, fire, and water. Only when speculation was combined with experiment was some measure of understanding and control achieved.

Matter in nature usually occurs as a mixture of various substances. The first problem, then, is the separation of the components from one another. Obviously some things are grossly different and therefore are easily separable: air, earth, and water, for example. It is not so obvious that a given substance can occur as a solid, liquid, or gas under different conditions of temperature and pressure without changing its identity. These facts can be used not only to effect the separation of mixtures, but also to help explain the microstructure of the substance.

Under ordinary temperature and pressure matter can exist in three states or phases. A *phase* is any volume of space containing matter which has a visible surface and which has no abrupt changes of properties within it. The word "phase" derives from the Greek word *phasis* meaning "appearance." If a sample of matter looks uniform throughout and has no internal surface large enough to scatter light it has one phase.

Matter containing only one phase is said to be *homogeneous.* This does not mean that only one substance is present, but rather that whatever is in the sample is distributed evenly enough so that no differences in appearance or composition are visible.

Matter with more than one phase is *heterogeneous.* A heterogeneous sample contains regions larger than the wavelength of light (400–700 nanometers) where abrupt changes of properties do occur so that boundary surfaces are visible. The change of properties may be purely physical as in a mixture of snow and water, or due to composition differences as in an oil and vinegar salad dressing.

3.2 *Kinds of phases*

Four characteristics of external form give us clues about the microstructure of matter:

1. the regularity or orderliness of shape;
2. the degree of self-cohesion and resistance to deformation;
3. the amount of unoccupied space within a substance inferred from the effects of temperature and pressure on it; and,

4. the presence or absence of a surface.

Depending on which of these attributes a substance has, it may be classified as a crystalline solid, an amorphous or glassy solid, a liquid, or a gas.

Symmetry of form: the crystalline phase 3.3

There is beauty in symmetrical forms, and people have always been fascinated by objects that have resisted the universal drift into disorder and decay. The beauty of diamonds, rubies, and sapphires is due not only to their brilliance and fire, but also to the permanent underlying order of structure which manifests itself in a regular array of facets and angles (Figure 3.1).

Figure 3.1 Crystalline structures of diamond and garnet.

A solid substance that under a sharp blow shatters into pieces with plane (not curved) faces with characteristic angles between the faces is a *crystal.* Most solids seem to be made of crystals bonded together on their faces. A grain of sand or a grain of salt would look like a gemstone to a cockroach.[1] The orderliness of form is not limited to saltlike or rocklike substances either. Metals show the same kind of structure (Figure 3.2), although there must be some differences in the way the building blocks are held together to account for the ability to be drawn into wire (ductile), and forged with a hammer (malleable).

[1] It is not known whether a cockroach appreciates the beauty.

Figure 3.2 Crystalline structure of aluminum.

The tendency to form crystals in the solid state extends even to water (Figure 3.3). Even though it is said that no two snowflakes are alike, all of them show the same hexagonal arrangement of the faces.

If we suppose that the external form is the sum of all the microstructures, then the form of a crystal must be the result of the shape of the fundamental building blocks, the way they are packed together, and the forces acting between them.

Crystals resist mechanical forces that tend to change their shape. Large pressures have very little effect on the volume of crystals. Temperature has little effect until the melting point is reached and the crystal collapses and becomes a liquid. Some crystalline substances (solid carbon dioxide, Dry Ice, for example) do not melt under ordinary pressures, but pass directly into a gas at a particular temperature.

3.4 *The amorphous or glassy phase*

Some substances do not freeze and form crystals. Instead they congeal into a solid that resists deformation by mechanical forces but shows no evidence of microscopic order. These substances tend to soften slowly rather than melt sharply as the temperature increases over a broad range. Broken pieces of these substances have curved faces rather than flat planes.

Evidently these substances are composed of molecules that are not packed together in orderly arrays, but have fairly strong forces holding them together. Such substances are known as *amorphous* or *glassy solids.* Asphalt, polystyrene, borosilicate glass, and many plastics are examples of this phase of matter.

Figure 3.3 Crystalline structure of water in the solid state.

The liquid phase 3.5

A substance with a definite surface boundary but no definite shape is a *liquid.* A force acting on a liquid will cause the liquid to accommodate itself to the shape of the container. If no external force acts on a liquid, it will take the shape with the smallest surface-to-volume ratio, a sphere. There is no long-range microscopic order to be found in liquids (Figure 3.4). Liquids are slightly more compressible than crystals, but the volume change is very small when pressure is applied.

Figure 3.4 Molecular structure of water.

The mobility of liquids can be explained by the presence of holes or voids in the material into which neighboring molecules can jump. As soon as enough holes exist the possibility of gross movement of molecules occurs, which we perceive as a flow of the liquid.

3.6 *The gaseous phase*

Substances that have no surface boundary and that spontaneously fill any container are gases. Gases press against the walls of a container in all directions and will expand rapidly if the container is opened. This behavior is described aptly enough as the "escaping tendency."

Temperature and external pressure affect the density of any substance, but their effect on gases is extraordinary. The volume of a balloon containing some gas is halved when the pressure on the balloon is doubled, whereas the same change in pressure on a liquid or solid would scarcely be noticed.

When a number of different gases are mixed, the only boundary surface is that of the container itself, so gases are always homogeneous.

It was once thought that the high escaping tendency of gases was due to the springiness of the molecules. All gases, however, can be condensed to liquids or solids if the temperature is low enough, and it is difficult to imagine springs that suddenly contract to one-thousandth of their length at a particular temperature.

It would be more in accord with the facts to suppose that the particles in a gas are quite far apart and moving at high speeds in all directions (Figure 3.5). Then whatever attractive force exists between them would be diluted by distance and overcome by the motion. Perhaps the temperature is some measure of the average motion. At lower temperatures the particles would slow down enough to stick together and become liquids or solids.

3.7 *Physical separations*

The methods used to separate the parts of a mixture rely on differences in physical properties of the substances in the mixture. Much chemical and chemical engineering research has been devoted to finding new and better ways of isolating and purifying substances from mixtures. Today it is sometimes possible to analyze components that comprise only a few parts per million of the mixture.

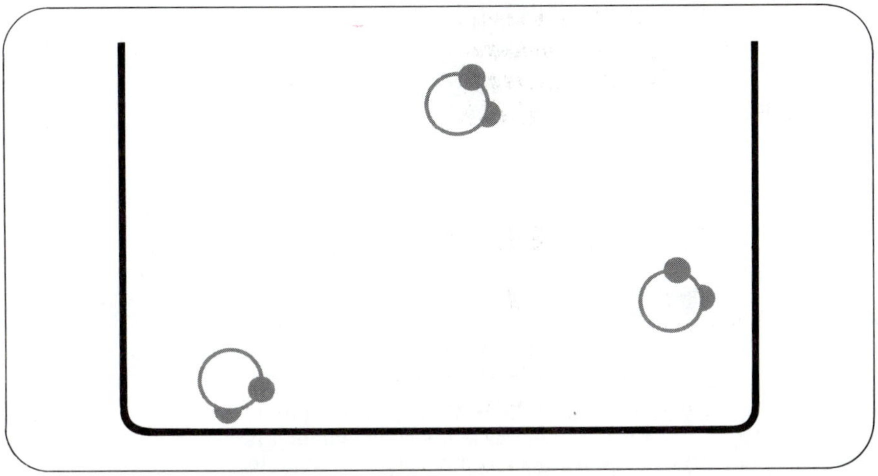

Figure 3.5 Water in the gaseous state.

Heterogeneous Mixtures

Solids suspended in liquids and gases can be separated by sedimentation (that is, allowing them to settle out) or filtration if the particle size is not too small (Figure 3.6). Very fine particles tend to remain in suspension and either clog filters or go right through them. Dust particles in a gas are sometimes removed with a cyclone separator. This machine whirls gases rapidly so that centrifugal force drives the denser dust particles to the outer rim. Particles suspended in liquids can be centrifuged at ultrahigh speeds to collect the solids. The denser particles sediment faster (Figure 3.7 A and B).

Particles in smoke are too small for either of these methods. Electrostatic precipitators pass smoke through a highly charged screen where the particles pick up an electric charge. On their way out of the stack they pass through a region with an opposite charge. This region attracts the smoke particles and holds them for later collection.

Sometimes physical properties other than size or density can be exploited. The iron ore in taconite is magnetic and can be pulled from the finely crushed rock by a magnet. Some other ores are wetted more by oil than water so that they can be floated away from the rock in a froth.

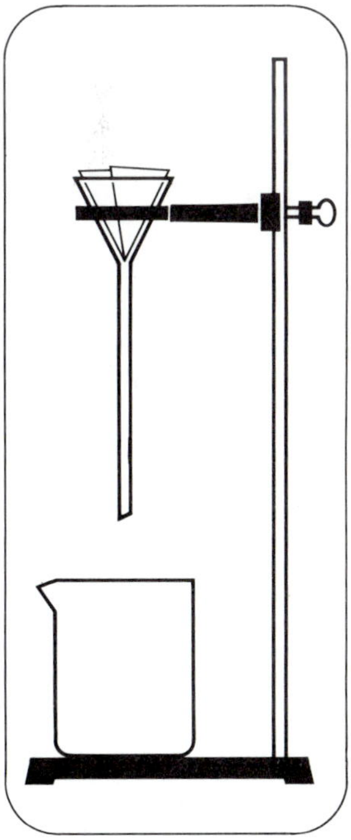

Figure 3.6 Filtration apparatus.

Liquid mixtures that will not blend with each other to form a single phase are said to be *immiscible*. Immiscible liquids such as oil and water can be separated by pouring off the top layer or drawing off the bottom layer. A laboratory device for that purpose is called a

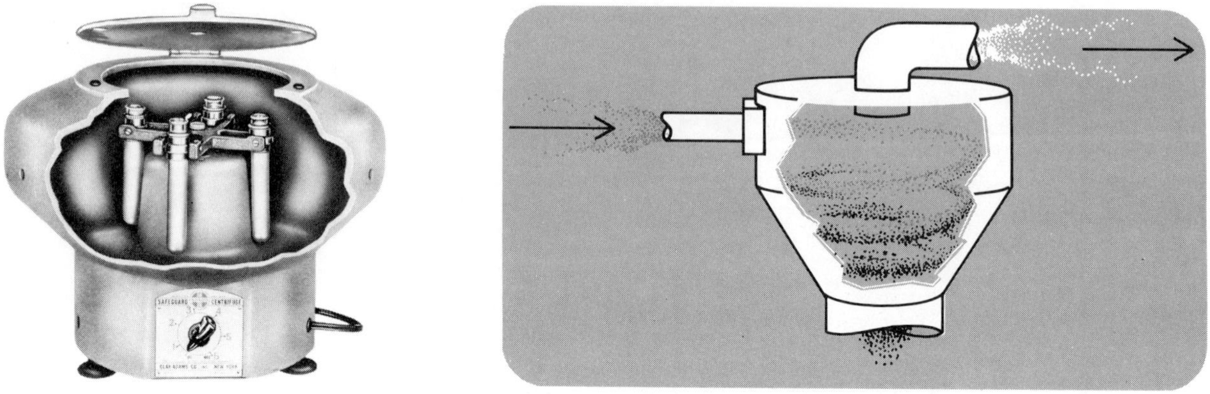

Figure 3.7 **(A)** Laboratory centrifuge. This machine whirls liquid suspensions at 3000–4000 rpm in order to increase the sedimentation rate of the suspended solids. **(B)** Cyclone separator. This machine separates dust particles from air using the centrifugal force of whirling gas.

separatory funnel (Figure 3.8). If the liquids are so thoroughly mixed that they do not separate by themselves (an emulsion), then centrifugation will usually cause them to separate.

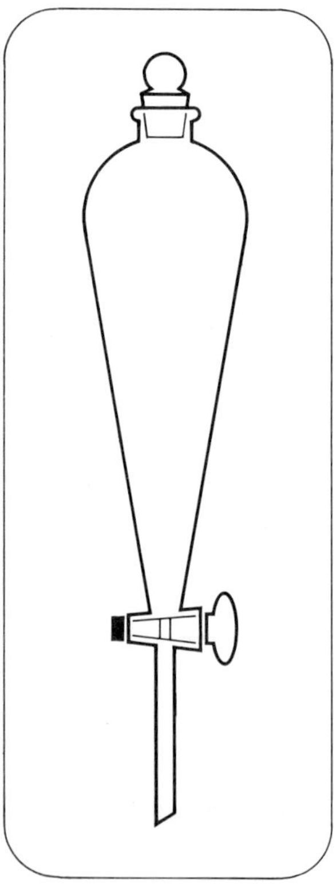

Figure 3.8
Separatory funnel.

Homogeneous Mixtures

A homogeneous mixture is called a *solution.* Two or more components are so thoroughly dispersed that even with a microscope separate phases cannot be seen. To separate a solution it is necessary to change the temperature or pressure so that usable physical differences will appear.

DISTILLATION. One way to separate a homogeneous mixture is to cause one or more of the components to change its phase. The Arabs discovered long ago that if they heated wine in a retort and caught the wine vapors in a cool vessel they got a liquid that was just what the doctor ordered. The alcohol and much of the flavor and aroma of the wine boiled at a lower temperature than the water and were therefore richer in the gas phase. The technique is called *distillation* and its use depends upon differences in volatility of the components in the mixture (Figure 3.9). A highly volatile substance has a great tendency to escape the surface (evaporate) and therefore has a low boiling temperature. Other substances, like salts and most metals, have little tendency to escape into the gas phase until the temperature is very high.

CRYSTALLIZATION. Solutions of solid crystalline substances in a liquid can often be separated by partially evaporating the liquid or by

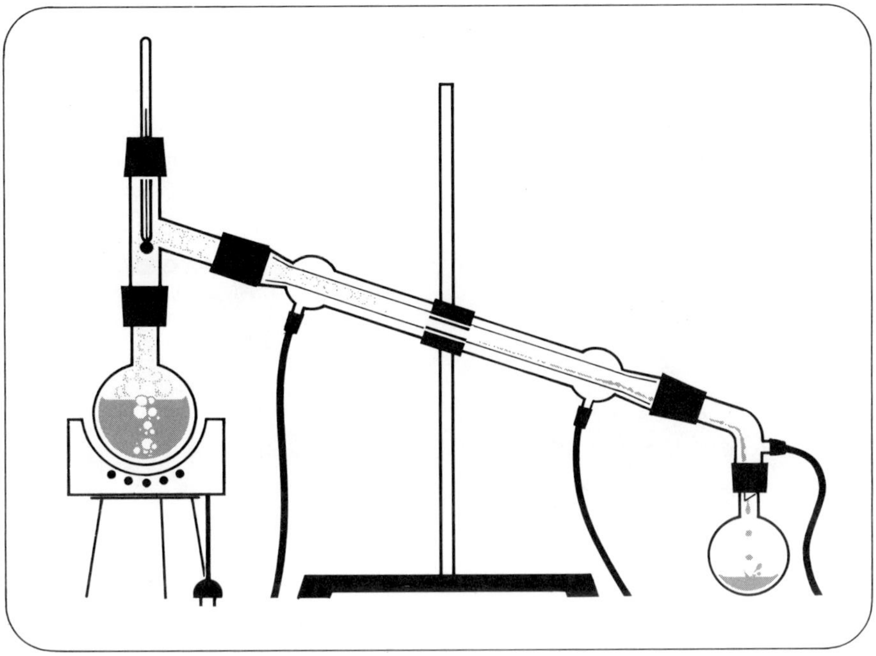

Figure 3.9 Distillation apparatus.

cooling the solution. Under ideal circumstances, much of the dissolved solid component will come out of solution or "recrystallize" so that it can be separated from the liquid by filtration.

The amount of solid that dissolves in a given amount of the liquid depends in part on how strongly the dispersing medium (solvent) attracts the substance being dispersed (solute). If the solvent and solute particles are strongly attracted to each other, compared to the attraction of each for its own kind, relatively large amounts of solute will be dissolved. The solute is said to be very soluble. Conversely, if each is strongly attracted to its own kind and interacts weakly with the other, the solid will be sparingly soluble or insoluble.

Higher temperatures usually increase the solubility of a substance. At higher temperatures solvents have more space between molecules to accommodate solute particles, and the particles of the solid can escape the crystal more easily.

If a liquid solvent has as much solute as it can hold the solution is *saturated.* At that point a loss of solvent by evaporation or a lowering of the temperature will result in the growth of crystals from the solution.

This process is a useful method of purification because the growing crystal tends to exclude molecular units that differ. The odd-shaped or unlike molecules do not fit in the framework, so crystals tend to be made of units of the same size and shape.

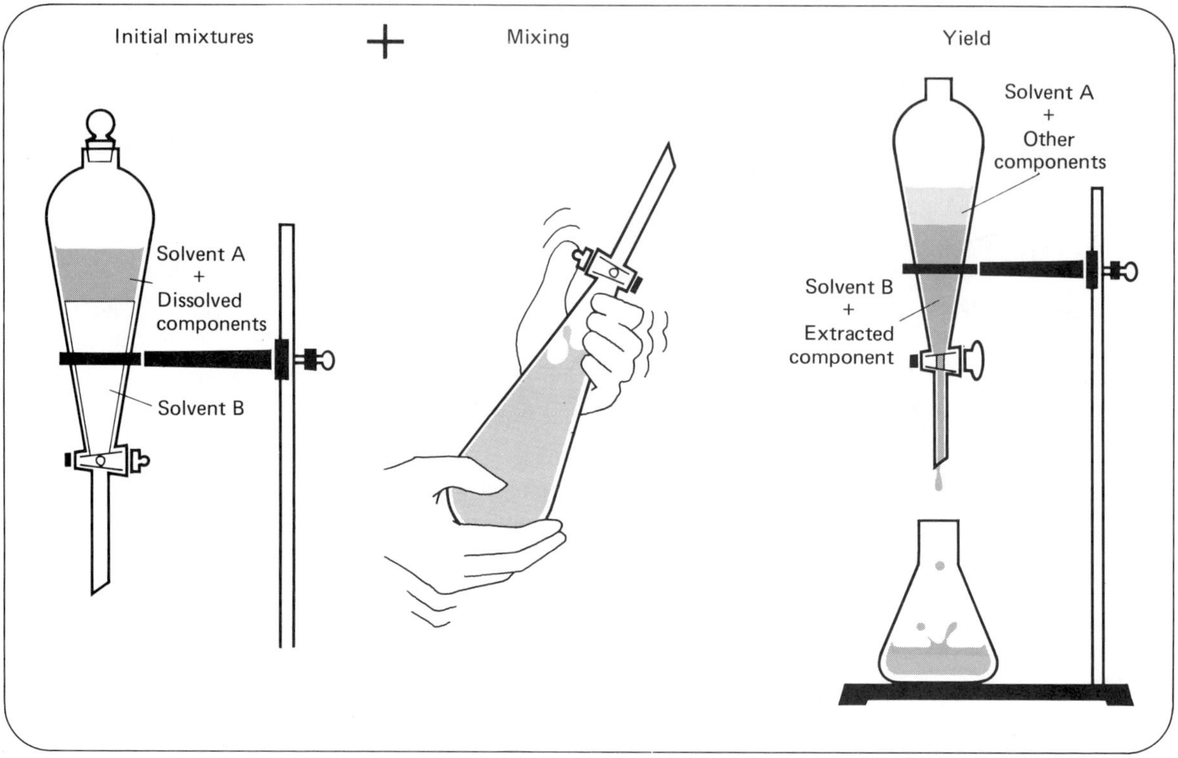

Figure 3.10 Liquid extraction.

The extremely pure crystals of silicon and germanium that are necessary to make transistors for your TV, radio, and electronic calculator are purified by recrystallization without a solvent. A long slug of the impure crystal is melted starting with a narrow zone at the bottom. The heat source is then moved slowly up the slug as the metal crystallizes below it. The impurities tend to concentrate in the melted zone and are swept to the top, which is cut off.

EXTRACTION. Given the opportunity to move to a place where there is more economic opportunity or more elbow room many people will emigrate from one country to another. Molecules do the same thing, not of course by volition, but purely on the basis of statistical considerations. It is more likely that a molecule will move to a phase that closely matches its nature than to stay in a phase that is a poor solvent for it.

This principle is used in the technique of *extraction.* A familiar example is the extraction of ground coffee beans with hot water. The desirable flavor components in the coffee bean migrate into the hot water. The bitter components are less soluble in water, so they stay in the solid phase. The solid grounds and the liquid are separated by filtration, and the undesirable part is discarded.

Components in liquid mixtures can be selectively extracted with solvents that are immiscible with each other. The two liquids are agitated to bring them in intimate contact with each other so that the desired component can migrate rapidly into the better solvent. On standing, the two liquids form two layers, and the lower layer can be drawn off through a separatory funnel (Figure 3.10).

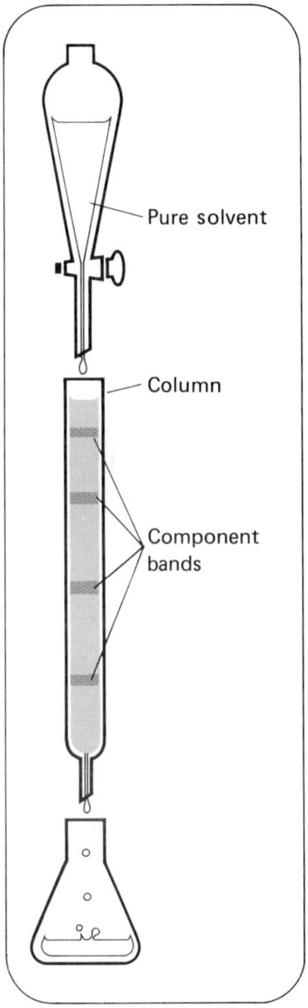

Figure 3.11
Column chromatography.

CHROMATOGRAPHY. The tendency for a substance to distribute itself between two phases is the basis for some very efficient methods of separation grouped under the term *chromatography*. Chromatography differs from extraction in that one phase is moving and one is stationary.

The technique was discovered when it was observed that a solution of plant pigments could be separated into many colored bands by a column of a finely powdered solid. The solution was placed at the top of the column and, as the solvent trickled out the bottom, more pure solvent was added at the top.

The pigments moved through the column at different rates and appeared as different colored bands depending on the degree to which each pigment was attracted to the solid surface or the moving solvent (Figure 3.11).

A technique useful for separating extremely minute amounts of substances is called *thin-layer chromatography* (TLC). This method uses a stationary phase of a powdered solid coated with a very thin layer of a liquid, usually water. The solid is spread on a glass plate. The mobile phase in TLC is a liquid that moves through the finely powdered medium by capillary action. The samples are "spotted" at the bottom of the plate and each component is carried up the medium at a rate that depends on its relative affinity for the stationary phase or the mobile phase.

After the solvent front has reached the top of the plate, the plate is removed and dried. If the spots are not colored, some chemical or optical means must be used to make them visible (Figure 3.12).

Gas-liquid chromatography (GLC), sometimes called simply gas chromatography, uses a stream of gas that flows over a thin layer of a non-volatile liquid. The liquid is coated on an inert solid support packed in a long tube. The sample components are introduced at one end of the tube and are alternately swept along by the gas or dissolved in the liquid (Figure 3.13). Since each component in the sample has a characteristic tendency to dissolve or evaporate, it is swept along at a different rate and arrives at the end of the tube at a different time.

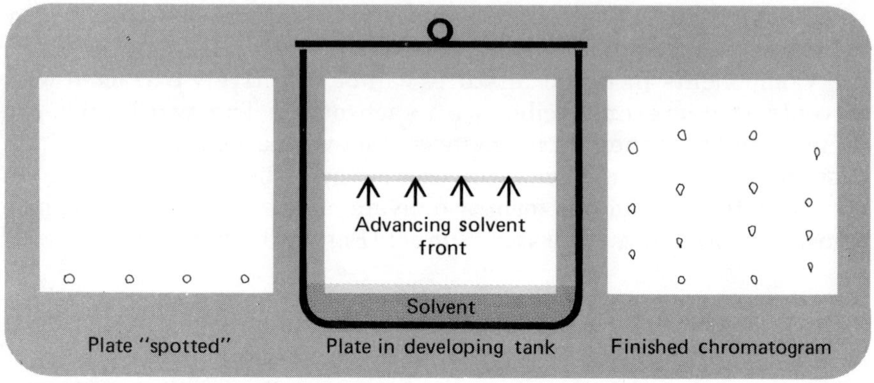

Figure 3.12 Thin-layer chromatography.

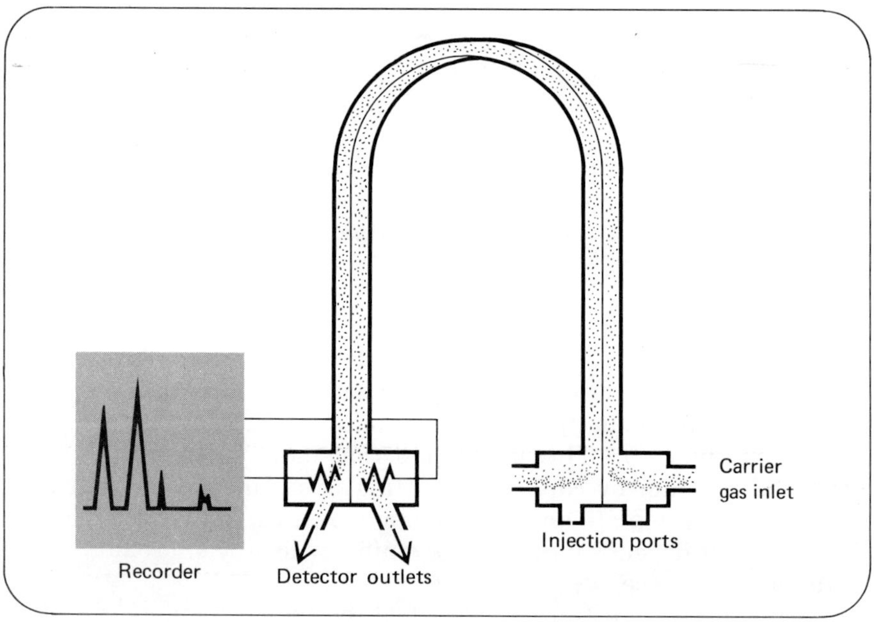

Figure 3.13 Gas chromatography.

3.8 *Pure substances and chemical change*

The result of all these physical separations is intended to be the isolation of one or more pure substances. A *pure substance* is unchanged in any of its intensive properties by repeated physical separations. Pure water, for example, will have the same density (at some reference temperature), the same boiling point (at some reference pressure), and the same melting point regardless of its source and how often it is distilled.

Physical separations do not change the nature and identity of pure substances. If the substances obtained from a physical separation are put together again, the original mixture will be obtained, except for possible differences in mechanical subdivisions.

It is not always possible to use differences in physical properties to separate mixtures. The chemist must sometimes resort to chemical changes in order to produce new substances with appropriate physical differences.

A chemical change results in a new substance, not originally present in the mixture, so the mixture cannot be restored by simply restoring the original conditions.

Energy I[2] 3.9

All physical and chemical changes occur as a result of interactions between matter. *Energy* is a measure of the capacity of a system to change the state or condition of something else by interacting with it. Since there are a number of ways one thing can interact with another, there are many ways of measuring the energy changes that take place.

Kinetic energy

We all know that an object in motion relative to something else has a certain capacity to cause changes in the motion and/or state of organization of the other if they should collide. Anyone who drives a car should know that this capacity increases with the *square* of the velocity. The capacity of an object to cause changes due to its mass and velocity is called *kinetic energy*. It is a quantity that must be calculated as the product of one-half the mass times the square of the velocity.

$$\text{kinetic energy} = \tfrac{1}{2}\ \text{mass} \times (\text{velocity})^2$$

When the mass is in kilograms and the velocity is in meters per second the units of energy are kg–m^2/sec^2 or *joules* (pronounced jools).

EXAMPLE 3.1

A 1000 kilogram car with a velocity of 15 meters per second has

$$\tfrac{1}{2} \times 1000\ \text{kg} \times \left(15\ \frac{\text{m}}{\text{sec}}\right)^2 = 112{,}000\ \text{joules}$$

kinetic energy with respect to a stationary object. If its speed is doubled to 30 meters per second (about 67 miles per hour) the kinetic energy is

$$\tfrac{1}{2} \times 1000\ \text{kg} \times \left(30\ \frac{\text{m}}{\text{sec}}\right)^2 = 450{,}000\ \text{joules},$$

or four times as much.

[2] This topic is discussed in more detail in Chapter 14.

Potential energy

The owner of a house at the bottom of a cliff keeps a wary eye out for loose boulders above the house knowing that they have the *potential* of disorganizing the living room. Potential energy comes from *position* and the existence of a *force,* in this case gravity, which would cause the boulder to move if not restrained. Gravity is not the only force that acts on matter. The electromagnetic force acts on charged bodies, and the nuclear force acts on subatomic particles. A body may have potential energy by virtue of its position in any of these force fields.

Thermal energy

The energy change that accompanies a phase change (melting, boiling, dissolving, etc.) or a temperature change is thermal energy or *heat.*

Heat energy units are measured in terms of the amount of physical change that a reference substance will undergo when it absorbs or gives up heat. The quantity of heat given up by a system can be measured by the quantity of ice melted, or by the temperature change of a given quantity of water. Calculations of heat changes are given in Chapter 14.

Energy is conserved!

Energy is an extensive property. The energy change associated with a physical or chemical change depends on the amount of material present.

It is our experience that the total energy of the universe is constant. The energy lost or gained by a system is exactly matched by a corresponding gain or loss in energy by the surroundings. This is a statement of the Law of Conservation of Energy. We have never found an exception to this law, so we believe that it is one of the fundamental laws of nature.

3.10 *A first hypothesis about the microstructure of matter*

Let us suppose that matter is made of different kinds of building blocks, which we will call "atoms," and that these atoms are in continuous motion, vibrating, tumbling, and ricochetting about. These atoms differ in their attraction for each other, depending on their nature, but all of them repel each other strongly when they get too close.

Some stack closely together in orderly arrays, with strong binding forces holding them together.

Some have strong binding forces, but do not fit together evenly, so that the packing is random like peanut brittle.

Some have weak or moderate forces holding them together or are moving about so violently that the order is destroyed and empty spaces appear in the structure.

Some have very weak forces of interaction so that they fly about chaotically with lots of empty space in which to move.

Phase	*Attractive Force*	*Order*	*Free Volume*
Crystalline	High	High	Low
Glass	High	Low	Low
Liquid	Intermediate	Low	Low
Gas	Low	None	High

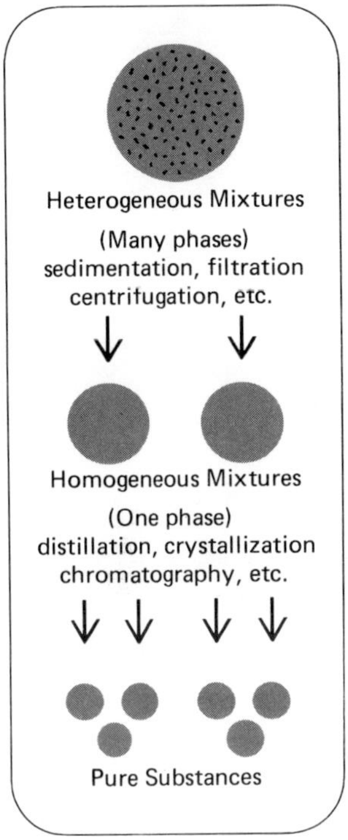

The observation that an increase in temperature has the effect of converting a highly ordered compact substance like ice into a disordered diffuse substance like steam fits our hypothesis if we think of an increase in temperature being associated with an increase in the average kinetic energy of the atoms. The increased speed of the atoms overcomes their attraction for each other and causes them to spread out.

Summary 3.11

1. Under ordinary conditions of temperature and pressure matter can exist in three phases: solid (crystalline or amorphous), liquid, and gaseous.
2. Matter is usually found in nature as a mixture of different substances. Mixtures can be separated into a number of pure substances by using differences in physical properties.
3. Pure substances are uniform throughout and do not change intensive physical properties upon repeated purification processes.
4. Each pure substance has a characteristic range of temperature and pressure in which it is a solid, liquid, or a gas. Lower temperatures and high pressures enable a substance to exist as a solid or a liquid. Higher temperatures promote the formation of the gaseous phase.
5. We suppose that matter is made of different kinds of building blocks, called atoms, with different degrees of attractive force

between them. We suppose that temperature is a measure of the average motion of the atoms, because all substances become more disordered and spread out as the temperature is increased.

6. Energy is a measure of the capacity of a system to change the state or condition of something else by interacting with it. The energy something possesses because of its mass and motion is kinetic energy. The energy something possesses because of its position in a force field is potential energy.

Energy is never lost or gained, but may be transformed into other forms of energy.

SUGGESTED READING

Cyril Stanley Smith, "Materials," *Scientific American,* (September, 1967), p. 68.

EXERCISES

I. Define the following terms:

(a) phase
(b) heterogeneous
(c) homogeneous
(d) solution
(e) crystalline
(f) liquid
(g) gas
(h) kinetic energy
(i) potential energy
(j) temperature
(k) physical change
(l) chemical change
(m) pure substance

II. 1. Classify the following as heterogeneous or homogeneous:

(a) blood
(b) milk
(c) beer (include the foam)
(d) tap water
(e) table salt

2. Classify the following as pure substances or mixtures:

(a) diamond
(b) table salt
(c) peanut butter
(d) sterling silver
(e) cotton

III. Suggest a means for separating

1. mud from water
2. cream from milk
3. oil from peanut butter
4. salt from sea water
5. sulfur dioxide (a gas) from air

To develop our ideas about the microstructure of matter we study the simplest form of matter first.

After reading this chapter you should be able to

1. Define the intensive property pressure, and describe how it is measured;
2. Change Celsius temperature to absolute temperature;
3. Compute the volume change in a fixed amount of gas for any change of temperature and pressure;
4. Define Avogadro's number and a mole;
5. Write the ideal-gas law equation, and know the value(s) of the ideal-gas constant;
6. Use the relationship between moles and volume, the molar volume, to convert volume to moles, and vice versa;
7. Use the relationship between weight and moles, the molar weight, to change weight to moles and vice versa;
8. Use the three-way relationships of the intensive properties molar volume, molar weight, and density to convert each to the others;
9. Compute the partial pressure of a gas from the total pressure and the other partial pressures;
10. (optional) Explain what happens on a microscopic scale when the temperature, pressure, or number of molecules of a gas is changed;
11. (optional) Compute the relative rates of diffusion of two gases from the molar weights.

gases

4

The first real progress toward understanding matter was made when the early chemists studied gases. They found that the effects of temperature and pressure, though large, followed simple laws. The simplicity of the laws led first to their envisioning models of what gases must be like and then to other predictions. Chemistry was off and running.

Pressure 4.1

Everybody has some intuitive notions about pressure from experiences with tires (flat and otherwise), lawn sprinklers, eardrum-sensations when changing altitude, and vague references to highs and lows by the weather reporter on TV. What is pressure, exactly, and how is it measured?

Pressure is the amount of force directed perpendicular to a surface per unit area of surface. A *force* is any interaction between matter that causes a change in the magnitude or direction of motion. The actual motion of a body depends on the net amount of force applied after the directions of all the forces acting on the body have been taken into account. The gravitational force between you and the earth always exists, but when you are seated that force is balanced by the force of the chair on your body in the opposite direction.

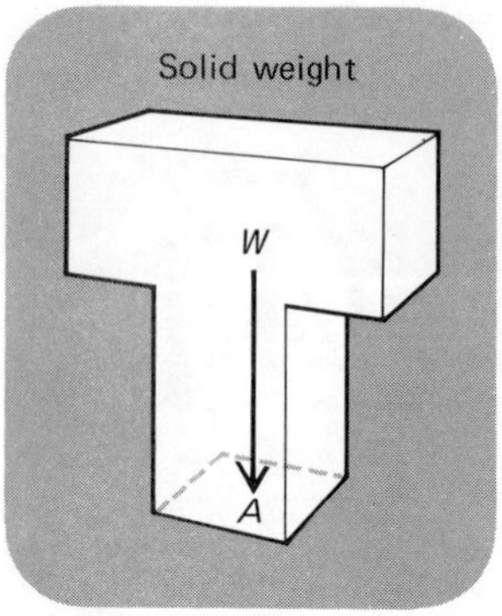

Figure 4.1 Pressure on the base of a solid equals total weight divided by area of the base $\left(P = \frac{W}{A}\right)$.

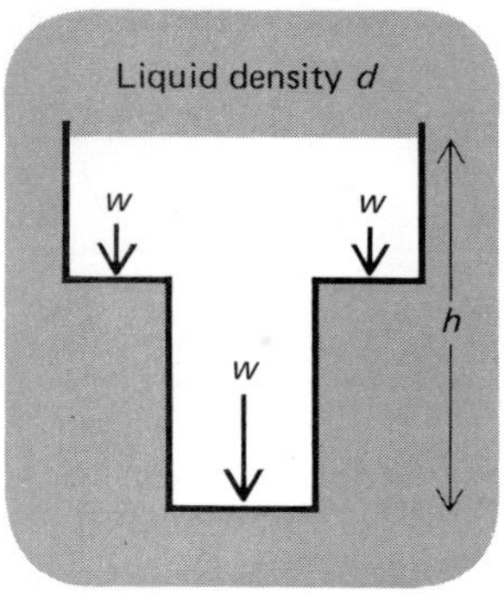

Figure 4.2 Pressure of a liquid equals height times density ($P = h \times d$).

The force on your body due to gravity is simply your weight, but how you *feel* your weight is due to pressure. We normally try to spread the force over as many square inches of our bodies as possible. Walking barefoot on sharp pebbles is uncomfortable because the area supporting body weight has decreased so that the pressure on the skin is increased.

Pressures are measured by balancing the unknown force acting on a surface with a known force in the opposite direction. Liquids are often used to supply the known force because they will flow to cancel any unbalanced forces. For this reason the pressure exerted by a liquid due to gravity is proportional to its depth and independent of the amount present.

The pressure on the base of a solid is its total weight divided by the area of the base, because a solid can transmit off-center (shear) forces within its structure (Figure 4.1). The pressure on A from a liquid is unaffected by any liquid not directly above A because a liquid cannot transmit shear force (Figure 4.2). The pressure of the liquid is therefore the weight of a uniform column A square centimeters in cross section and h centimeters high, divided by A square centimeters. The weight of the liquid is the volume above A times its density.[1]

$$
\begin{aligned}
\text{pressure} &= \frac{\text{weight}}{\text{area}} = \frac{\text{volume} \times \text{density}}{\text{area}} \\
&= \frac{\text{height} \times \text{area} \times \text{density}}{\text{area}} \\
&= \text{height} \times \text{density}
\end{aligned}
$$

EXAMPLE 4.1 What is the pressure at the bottom of a full water tank 30 meters high, if the density of water is 1.0×10^{-3} kilograms per cubic centimeter?

Solution: The pressure is equal to the *weight* of a column of water 30 meters high and 1 square centimeter in cross section, divided by 1 square centimeter. The weight of water is its volume times its density.

$$\text{pressure} = 30\ \text{m} \times 100\ \frac{\text{cm}}{\text{m}} \times 1.0 \times 10^{-3}\ \frac{\text{kg}}{\text{cm}^3} = 3.0\ \frac{\text{kg}}{\text{cm}^2}$$

[1] In the English System the unit of force is the pound, so we can properly speak of pressure in pounds per square inch. The unit of force in the International System is the *newton,* and pressure should be in newtons per square meter. A newton is the force which will accelerate one kilogram one meter per second per second. One kilogram of mass weighs 9.81 newtons at the earth's surface. Nevertheless, pressure is often given in kilograms (force) per square meter, where it is understood that it is the *weight* of a kilogram that is being measured, not the mass. For more discussion consult a freshman physics text, for example, *The Principles of Physics and Chemistry,* Brackenridge and Rosenberg, McGraw-Hill, 1970.

Atmospheric pressure 4.2

We live at the bottom of a rather thin ocean of air that has weight and exerts pressure like any other fluid. The pressure of the atmosphere at sea level is nominally equal to the pressure exerted by a column of mercury 76.0 centimeters high at 0°C. This pressure is defined as *one standard atmosphere.* A *torr* is the pressure exerted by a column of mercury one millimeter high. There are 760 torr per atmosphere.

Meteorologists use the unit of pressure called a *bar,* which is defined as 10^5 newtons/meter^2. A standard atmosphere is 1013 millibars.

The relationship of volume and pressure of gases 4.3

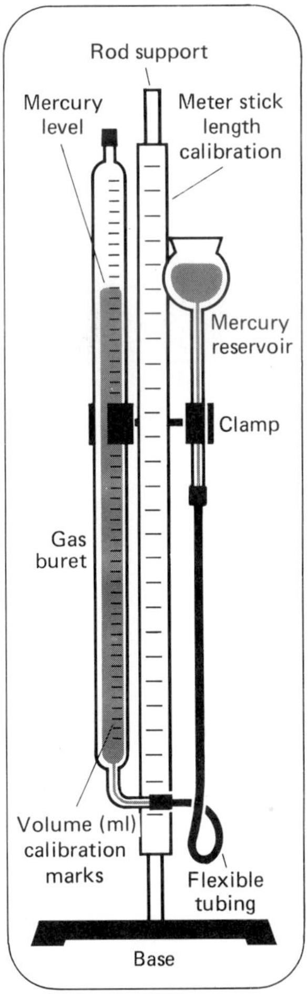

Figure 4.3

Sir Robert Boyle used an apparatus like that in Figure 4.3 to find out how the volume and pressure of a gas were related, if at all. A sample of gas (e.g., air, nitrogen, carbon dioxide) was trapped in a closed volume measuring device over mercury. The mercury acted as a piston to change the volume of the gas when the mercury reservoir was raised or lowered. The pressure exerted by the mercury piston was related to the difference in height between the upper and lower surfaces of the mercury.

Boyle found that all gases behaved very much alike. When the pressure was increased, the volume was decreased by a proportional amount. With a constant amount of gas and a constant temperature, he got numbers similar to those in Table 4.1.

Table 4.1
Pressure versus volume at constant temperature

Pressure (torr)	Volume (ml)	PV Product
950	30.3	2.88×10^4
750	38.4	2.88×10^4
500	57.6	2.88×10^4
350	82.3	2.88×10^4
300	96.0	2.88×10^4

A graph of these data would look like the solid line in Figure 4.4. A sample with more gas would result in a similar curve (dashed line) further from the axis. Higher temperatures would also displace the curve from the axis.

This kind of curve is characteristic of an *inverse* relationship. The equation for the curve can be written as

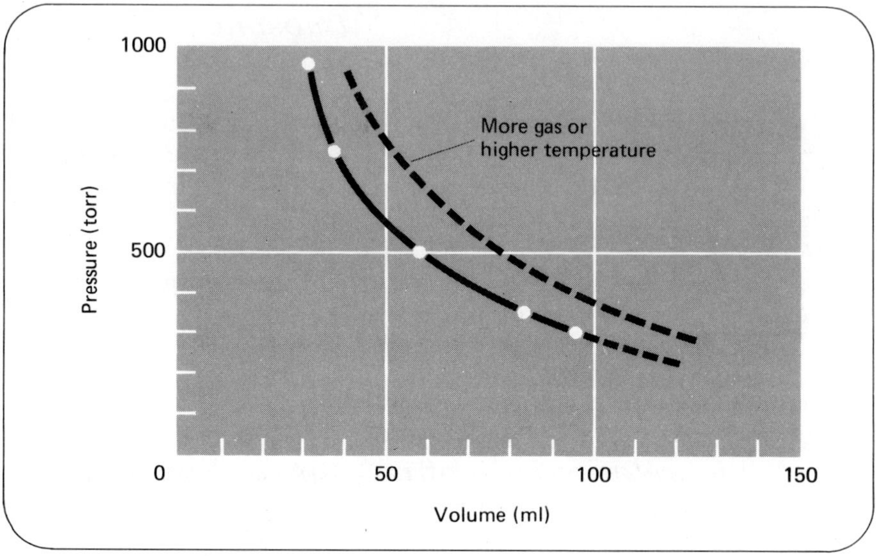

Figure 4.4

$$V = \frac{k}{P} \tag{4.1}$$

where k is a proportionality constant that evidently depends on the temperature and the amount of gas. As the pressure increases, the fraction k/P gets smaller but never becomes zero or negative. As the pressure decreases, the volume increases to become very large as the pressure approaches zero.

The experiment shows that at constant temperature, the volume of a constant amount of gas will change with pressure in such a way that the product of the new pressure and volume will equal the product of the initial pressure and volume.

$$P_f V_f = P_i V_i \tag{4.2}$$

where the subscript i means initial and f means final.

To compute the final volume from the initial volume, we multiply the initial volume by the ratio of pressures:

$$V_f = V_i \times \frac{P_i}{P_f} \tag{4.3}$$

EXAMPLE 4.2 A particular sample of a gas at room temperature had a volume of 35.0 milliliters at a pressure of 725 torr. What would its volume be at 760 torr?

Solution: If the product of 35.0 ml and 725 torr is a constant at a fixed temperature and amount of gas, then the product of the new volume and 760 torr is the same constant. Two things equal to the same thing are equal to each other, so

$$35.0 \text{ ml} \times 725 \text{ torr} = V_f \text{ ml} \times 760 \text{ torr}$$

$$V_f \text{ ml} = 35.0 \text{ ml} \times \frac{725 \text{ torr}}{760 \text{ torr}} = 33.4 \text{ ml}$$

Because the volume of a given amount of gas is inversely proportional to the pressure, we can compute volume changes due to pressure changes by "common sense" rather than algebra. An increase of pressure squeezes a gas; the final volume will be the initial volume times a ratio of pressures less than one. A decrease of pressure allows a gas to "press out" or expand; the final volume will be the initial volume times a factor greater than one.

EXAMPLE 4.3

1. The pressure on 10 liters of gas is doubled. What is the new volume?

Solution: Increasing the pressure squeezes a gas; multiply the initial volume by a pressure ratio smaller than one.

$$V_f = 10 \text{ liters} \times \frac{1}{2} = 5 \text{ liters}$$

2. The initial pressure on 10 liters of gas is 760 torr. What is the volume when the pressure is decreased to 380 torr?

Solution: Decreasing the pressure allows a gas to expand; multiply the initial volume by a factor greater than one.

$$V_f = 10 \text{ liters} \times \frac{760}{380} = 20 \text{ liters}$$

The relationship of volume and temperature: a gas thermometer 4.4

In Chapter 2 we observed that the relative volume change of a quantity of mercury could be used as a thermometer. Figure 4.5 shows a similar device using the relationship between the volume and temperature of a gas.

A quantity of gas is trapped in the bulb over mercury. The pressure is kept constant by raising or lowering the mercury reservoir so that the difference between the two levels is constant.

To calibrate the thermometer it is necessary to measure the volume of the gas at two reference temperatures. The freezing point of water (0°C) and the boiling point of water (100°C) are convenient.

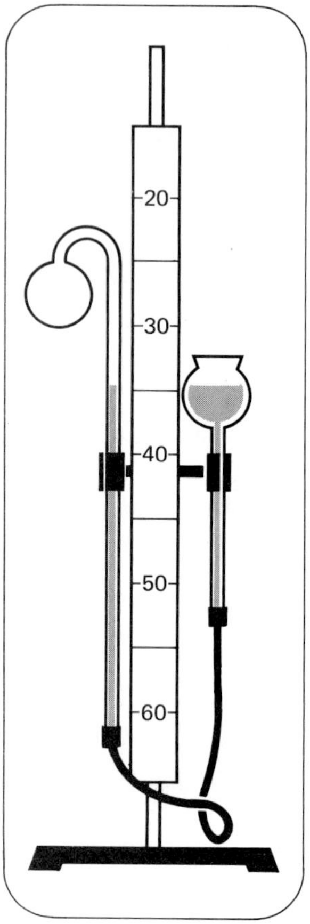

Figure 4.5 A gas thermometer.

The results of one experiment might be

temperature	*volume*
0°C	35.0 ml
100°C	47.8 ml

It is hard to see what relationship holds here, except that the volume increases with temperature, so let's plot the data on a graph (Figure 4.6). The volumes between 0°C and 100°C do not *necessarily* lie on the straight line, but if the intermediate points are checked with a mercury thermometer, we find that they do agree fairly well.

The equation for this linear relationship can be written as:

$$V = V_0 \alpha t + V_0 \tag{4.4}$$

where V is the volume at the temperature t°C and V_0 is the volume at 0°C. The constant α is the fractional change in volume per degree, also known as the thermal coefficient of expansion. Knowing the other values, we can solve the equation for α:

$$\alpha = \frac{V - V_0}{V_0 t}$$

$$\alpha = \frac{47.8 \text{ ml} - 35.0 \text{ ml}}{(35.0 \text{ ml})(100 \text{ deg})}$$

$$\alpha = 0.00366 \text{ deg}^{-1}$$

This means that the volume change is 0.00366 ml per milliliter of volume at 0°C per degree of temperature change, or 0.366% per degree. Compare that to mercury which only changes 0.00134% per degree (Chapter 2).

If we repeat this experiment with different gases, at other pressures, and with other amounts of gas, our results will be much the same. All gases have about the same coefficient of expansion (α).

We may now use the equation to find the temperature by measuring the volume change between 0°C and any other temperature.

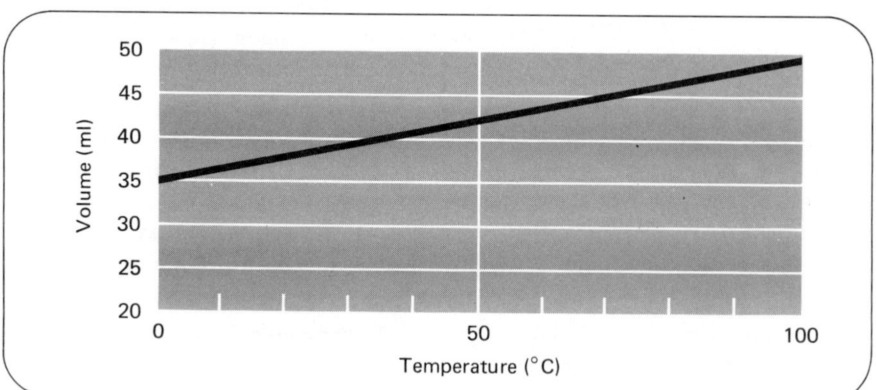

Figure 4.6 Volume vs. temperature in the gas thermometer.

EXAMPLE 4.4

1. Use equation 4.4 to calculate the temperature at which the volume is 38.2 milliliters, if V_0 is 35.0 milliliters.

Solution: We must first solve for t.

$$V = V_0\alpha t + V_0$$
$$V - V_0 = V_0\alpha t$$
$$t = \frac{V - V_0}{V_0\alpha}$$

Substituting the values of V, V_0, and α, we find

$$t = \frac{38.2 \text{ ml} - 35.0 \text{ ml}}{(35.0 \text{ ml})(0.00366 \text{ deg}^{-1})}$$
$$t = 25°\text{C}$$

4.5 *Absolute temperature*

Section 4.4 was intended to convey three impressions about the volume of a gas and the temperature.

1. Temperature is measured by comparing some property of a substance to the same property at a reference temperature.
2. A linear relationship exists between volume and temperature which is the same for all gases.
3. The equation that relates temperature and volume is a pain and an aggravation when Celsius temperature is used, because it is necessary to measure or compute the volume at 0°C for every computation. If we had a temperature scale such that the volume is zero when the temperature is zero we would not have to find the volume at 0°C in order to find the values at some other temperature.

What would the temperature be if we could cool the gas so much that the volume approached zero? We ask this question of the equation, not reality, because if we cool a real gas enough it becomes a liquid and does not follow the equation. To find out, we let the volume in equation 4.4 be zero, and solve for t.

$$V_0\alpha t + V_0 = 0$$
$$V_0\alpha t = -V_0$$
$$t = \frac{-V_0}{V_0\alpha} = -\frac{1}{\alpha}$$
$$t = \frac{-1}{0.00366}$$
$$t = -273°\text{C}$$

The temperature where the volume hypothetically goes to zero is −273°C. If we choose a new temperature scale with the new zero at the point where the volume goes to zero, the equation relating gas volume and temperature becomes a simple linear equation with no additive constant.

$$V = cT \tag{4.5}$$

where T is now $t + 273$, and c is a constant that depends on the pressure and the amount of gas.

The zero point on the new scale is absolute zero; nothing can be colder than that. It is called the Kelvin (K) temperature scale. To convert from Celsius to Kelvin simply add 273 to the Celsius temperature.

Equation 4.5 states that the volume is directly proportional to the absolute temperature when the pressure and amount of gas are held constant. If the Kelvin scale of temperature is used, volumes can be compared by direct comparison of temperatures.

$$V_i = cT_i$$

and $$V_f = cT_f$$

Therefore,

$$\frac{V_i}{V_f} = \frac{T_i}{T_f} \tag{4.6}$$

EXAMPLE 4.5

1. Calculate the volume of a gas at 50°C if the volume at 25°C is 38.2 milliliters.

   Solution: First convert the temperature to Kelvin.

   25°C + 273 = 298°K
   50°C + 273 = 323°K

   Then,

   $$V_f = 38.2 \text{ ml} \times \frac{323}{298} = 41.4 \text{ ml}$$

2. Calculate the temperature when the volume is 32.0 milliliters, if the volume of the same amount of gas is 23.5 milliliters at 25°C.

   Solution: First convert 25°C to Kelvin.

   25°C + 273 = 298°K

   Then,

   $$T_f = 298°\text{K} \times \frac{32.0 \text{ ml}}{23.5 \text{ ml}} = 406°\text{K} = 133°\text{C}$$

Because the volume of a particular amount of gas at constant pressure is directly proportional to the absolute temperature, we can apply common sense to volume changes due to temperature changes. The final volume will be the initial volume times a temperature correction factor. Volumes increase with a temperature increase, so the temperature correction factor (the ratio of absolute temperatures) will be greater than one. A decrease of temperature will shrink the volume, so the temperature correction factor will be less than one.

EXAMPLE 4.6

1. One liter of air at 0°C is taken up in a jet engine and exhausted at the same pressure at 1100°C. What volume does it occupy in the exhaust?

Solution: The temperatures in Kelvin degrees are

$$T_i = 0 + 273 = 273°K$$
$$T_f = 1100 + 273 = 1373°K$$

The gas is being expanded; therefore,

$$1 \text{ liter} \times \frac{1373}{273} = 5.0 \text{ liters}$$

The combined gas law 4.6

The results of the two experiments above can be summarized by the statement that the volume of a gas is directly proportional to the absolute temperature and inversely proportional to the pressure. Since volume is an extensive property, it also depends on the amount of gas present.

It is sometimes necessary to know what would happen to the volume of a gas if both the temperature and pressure changed. We can derive the right relationship using the laws we have already discovered.

Because the volume is inversely proportional to the pressure, we can apply a pressure correction factor to the initial volume to find a new volume:

$$V = V_i \times \frac{P_i}{P_f} \tag{4.3}$$

Then, because the volume is directly proportional to the absolute temperature, we can apply a temperature correction factor to the volume (V) calculated from equation 4.3:

$$V_f = V \times \frac{T_f}{T_i} \tag{4.6}$$

It is not really necessary to carry out the calculation in two steps. The right side of equation 4.3 can be substituted for V in equation 4.6 so that we have

$$V_f = V_i \times \frac{P_i}{P_f} \times \frac{T_f}{T_i} \tag{4.7}$$

EXAMPLE 4.7 Let us suppose we fill a 10-liter balloon with air on a high mountain like Aspen or Mammoth. The pressure there is about 0.6 atmosphere and the temperature is about 0°C. What will the volume of the balloon be at sea level, where the pressure is 1.0 atmosphere and the temperature is 25°C?

Solution:

$$V_f = 10 \text{ liters} \times \frac{0.6}{1.0} \times \frac{298}{273} = 6.5 \text{ liters}$$

Note that the pressure factor decreases the volume and the temperature factor increases the volume.

Equation 4.7 has six variables: P_i, V_i, T_i, P_f, V_f, and T_f. With a little algebraic rearranging we can find any one of them if we know the other five. Equation 4.7 is usually written in this easily remembered form:

$$\frac{P_i V_i}{T_i} = \frac{P_f V_f}{T_f} \tag{4.8}$$

EXAMPLE 4.8 Suppose a scuba tank is filled with air at 2250 psi at 25°C and then thrown in the trunk of a car where the temperature can reach 75°C on a hot day. Will it explode? In this situation the volume and the amount of gas are constant (unless it does blow up).

Solution: Rearranging equation 4.8 to get P_f, we have

$$P_f = P_i \times \frac{V_i}{V_f} \times \frac{T_f}{T_i}$$

The volume is unchanged in the example, so the ratio of V_i/V_f is one. Changing the temperatures to absolute temperatures and solving for P_f, we find

$$P_f = 2250 \text{ psi} \times \frac{(75 + 273)}{(25 + 273)} = 2630 \text{ psi}$$

4.7 *The mole*

The fact that the volume of a gas depends not only on the pressure and temperature but also on the amount of gas present should not go unnoticed. It would be nice to know *how* the volume depends on the quantity of gas so we could calculate one from the other.

Early in the last century Avogadro suggested that the factor PV/T depended only on the *number* of gas molecules present. If the size of the factor PV/T depends on the number of molecules we can write

$$\frac{PV}{T} = Nk \qquad \text{or} \qquad PV = NkT \tag{4.9}$$

where N is the total number of molecules and k is a proportionality constant *that is the same for all gases.*

The number (N) of molecules must be very large in order to have an appreciable amount of gas. It is so large that it is convenient to define a group number (like a dozen or a billion) so that we can calculate using the name of the group number rather than units. The number chosen is the number of molecules in 22.4 liters of a gas at 0°C and under one atmosphere pressure. It is equal to 6.02×10^{23} molecules and is called a *mole* (mol).

If the number of units of something happens to be known, the number of moles can be found by dividing the number (N) by 6.02×10^{23}.

EXAMPLE 4.9

1. A sample of gas has 3.01×10^{24} molecules. How many moles are there?

Solution:

$$3.01 \times 10^{24} \text{ units} \times \frac{1 \text{ mol}}{6.02 \times 10^{23} \text{ units}} = 5.00 \text{ mol}$$

2. Approximately three billion people live on the earth today. Express that number in terms of a mole.

Solution:

$$3.0 \times 10^{9} \text{ units} \times \frac{1 \text{ mol}}{6.02 \times 10^{23} \text{ units}} = 5.0 \times 10^{-15} \text{ mol}$$

The example should make it evident that nobody has ever counted a mole of anything. The number is simply too big. Instead some quantity related to number is measured, and the number of moles is calculated from that.

4.8 *The equation of state for gases: the ideal gas law*

Equation 4.9 relates the pressure, volume, and temperature to a very large number (N) through a very small constant (k). If (N) is divided by 6.02×10^{23}, the number of molecules is expressed in moles. When

(k) is multiplied by 6.02×10^{23}, a constant of reasonable size is obtained. This constant is given the symbol R and called the "universal gas constant." Equation 4.9 may now be written as

$$PV = nRT \tag{4.10}$$

where n is the number of moles of a gas, and R is the proportionality constant that relates PV units per mole to the absolute temperature. The numerical value of R depends on the PV units being used (see Table 4.2).

Table 4.2

Values and units of R

P	V	n	T	R
atm	liter	mol	Kelvin	$0.0821 \frac{\text{liter-atm}}{\text{mol-deg}}$
torr	liter	mol	Kelvin	$62.4 \frac{\text{liter-torr}}{\text{mol-deg}}$
$\frac{\text{newtons}}{\text{sq meter}}$	cu meters	mol	Kelvin	$8.314 \frac{\text{joules}}{\text{mol-deg}}$

4.9 *The relationship between moles and volume: the molar volume of gases*

The Ideal Gas Law relates two quantities, the volume and the number of moles, by means of two intensive factors, the pressure and temperature, and a proportionality constant.

$$V = \frac{RT}{P}\, n \tag{4.11}$$

and

$$n = \frac{P}{RT}\, V \tag{4.12}$$

At 273°K (0°C) and one atmosphere pressure,

$$\frac{RT}{P} = 22.4\ \frac{\text{liters}}{\text{mol}}$$

Since the value of R is the same for all gases, equal volumes of all gases under the same conditions of temperature and pressure must contain the same number of molecules. Now we have a way to count molecules by measuring the volume of a gas. This is going to come in handy later on.

EXAMPLE 4.10

1. The molar volume, RT/P, is 22.4 liters per mole at standard temperature and pressure, 0°C and 1 atm pressure. What is the molar volume at 298°K and 0.95 atm?

$$\text{molar volume in } \frac{\text{liters}}{\text{mol}} = \frac{0.0821 \dfrac{\text{liter-atm}}{\text{mol-deg}} (298°\text{K})}{0.95 \text{ atm}} = 25.8 \frac{\text{liters}}{\text{mol}}$$

2. How many moles of gas are in 350 ml of gas at 25°C and 725 torr?

Want: moles of gas

Have: 350 ml of gas

Know: **(a)** the relationship of volume to moles of a gas $= \dfrac{RT}{P} \dfrac{\text{liter}}{\text{mol}}$

(b) the relationship of milliliters to liters $= 1000 \dfrac{\text{ml}}{\text{liter}}$

Solution:

$$350 \text{ ml} \times \frac{1 \text{ liter}}{1000 \text{ ml}} \times \frac{725}{(62.4)(298)} \frac{\text{mol}}{\text{liter}} = 0.0136 \text{ mol}$$

The relationship between mass and number: the molar weight 4.10

A balloon filled with hydrogen or helium will rise in the air because it weighs less than the air it displaces. In other words, it is less dense than the surrounding air. We know that equal volumes of gases contain equal numbers of molecules when RT/P is the same. It follows then that some molecules must be heavier than others. One mole of hydrogen weighs 2 grams. One mole of helium weighs 4 grams. One mole of oxygen weighs 32 grams. *The weight in grams of one mole of a pure substance is its molar weight.*[2] It is the property that fixes the relationship between the mass of a substance and the number of molecules of that substance. Molar weight (MW) has the dimensions *grams per mole,* and if the weight and the number of moles of a substance is

[2] The relationship between mass and number is a fixed characteristic of each pure substance. The molar weight defined above is more formally called the gram-molar weight because it is the weight in grams of one gram-mole, Avogadro's number (6.02×10^{23} units). The pound-molar weight is the weight of one pound-mole of substance. A pound-mole has 454 times as many molecules as a gram-mole, just as a pound has 454 times as much mass as a gram, so the numerical value of the molar weight does not change. The word "mole" in this book means gram-mole.

known, the molar weight is found by dividing the weight by the moles.

EXAMPLE 4.11

1. 0.050 mole of a substance weighs 2.2 grams. What is its molar weight?

Solution:

$$\text{molar weight} = \frac{\text{weight in grams}}{\text{number of moles}} = \frac{2.2\ \text{g}}{0.050\ \text{mol}} = 44\ \frac{\text{g}}{\text{mol}}$$

2. 1.36 liters of a gas at 298°K and 0.90 atm weighed 2.2 grams. What is its molar weight?

Now what do we do? We want grams per mole, but we have grams and volume. Luckily we know a relationship that changes volume to moles for a gas.

Want: MW in $\frac{\text{g}}{\text{mol}}$

Have: 2.2 g; 1.36 liters

Know: $\frac{RT}{P} = \frac{(0.0821)(298)}{0.90}\ \frac{\text{liter}}{\text{mol}}$

Solution: $\text{MW} = \frac{2.2\ \text{g}}{1.36\ \text{liters}} \times \frac{(0.0821)(298)}{0.90}\ \frac{\text{liter}}{\text{mol}} = 44\ \frac{\text{g}}{\text{mol}}$

4.11 *The relationship of density, molar weight, and molar volume*

Mass, volume, and number (moles) are related to each other by the three intensive properties density, molar weight, and molar volume $\left(\frac{RT}{P}\text{ for gases}\right)$.

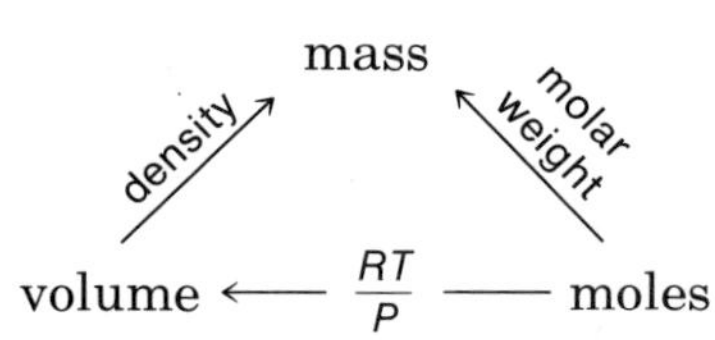

Density is the relationship of mass to volume.

Molar weight is the relationship of mass to moles.

Molar volume is the relationship of volume to moles.

Observing that the product of two related intensive properties is another intensive property, we see an interesting three-sided relationship appear.

$$\text{density}\left(\frac{\text{g}}{\text{liter}}\right) \times \text{molar volume}\left(\frac{\text{liter}}{\text{mol}}\right) = \text{molar weight}\left(\frac{\text{g}}{\text{mol}}\right)$$

$$\frac{dRT}{P} = \text{MW} \qquad (4.13)$$

$$\text{molar weight}\left(\frac{\text{g}}{\text{mol}}\right) \times \frac{1}{\text{molar volume}}\left(\frac{\text{mol}}{\text{liter}}\right) = \text{density}\left(\frac{\text{g}}{\text{liter}}\right)$$

$$(\text{MW})\left(\frac{P}{RT}\right) = d \qquad (4.14)$$

$$\text{molar weight}\left(\frac{\text{g}}{\text{mol}}\right) \times \frac{1}{\text{density}}\left(\frac{\text{liter}}{\text{g}}\right) = \text{molar volume}\left(\frac{\text{liter}}{\text{mol}}\right)$$

$$(\text{MW})\left(\frac{1}{d}\right) = \frac{RT}{P} \qquad (4.15)$$

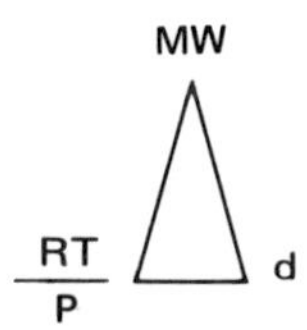

It would be foolish to memorize these equations. It is of far more value *to know exactly* which quantities are related by each intensive property and to know that one can be converted to the other when the related property is known.

EXAMPLE 4.12

1. The density of butane gas at 745 torr and 298°K is 2.32 grams per liter. What is the molar weight?

   Want: molar weight in grams per mole

   Know: **(a)** the relationship of weight to volume: $2.32\ \frac{\text{g}}{\text{liter}}$

   (b) the relationship of volume to moles:

   $$\frac{RT}{P} = \frac{(62.4)(298)}{745}\ \frac{\text{liter}}{\text{mol}}$$

   Solution:

   $$\text{MW}\ \frac{\text{g}}{\text{mol}} = 2.32\ \frac{\text{g}}{\text{liter}} \times \frac{(62.4)(298)}{745}\ \frac{\text{liter}}{\text{mol}} = 57.9\ \frac{\text{g}}{\text{mol}}$$

   Note: R is $62.4\ \frac{\text{liter-torr}}{\text{mol-deg}}$ when P is in torr.

2. What is the density of oxygen gas, molar weight 32 g/mol, at standard temperature and pressure (0°C and 1 atm)?

   Want: density in grams per liter

Know: **(a)** the relationship of grams to moles: $32 \frac{\text{g}}{\text{mol}}$

(b) the relationship of volume to moles: $\frac{RT}{P} = \frac{(0.0821)(273)}{1} \frac{\text{liter}}{\text{mol}}$

Solution: $\text{density} = 32 \frac{\text{g}}{\text{mol}} \times \frac{1}{(0.0821)(273)} \frac{\text{mol}}{\text{liter}}$

$= 1.43 \frac{\text{g}}{\text{liter}}$

Note: $R = 0.0821 \frac{\text{liter-atm}}{\text{mol-deg}}$ when P is in atmospheres; T is *always* in Kelvin.

3. What volume is occupied by 0.352 g of a gas at 373 K and 730 torr? The molar weight is 154 g/mol.

Want: volume in liters

Have: 0.352 g of gas

Know: **(a)** the relationship of grams to moles: $154 \frac{\text{g}}{\text{mol}}$

(b) the relationship of volume to moles: $\frac{RT}{P} = \frac{(62.4)(373)}{730} \frac{\text{liter}}{\text{mol}}$

Solution: $\text{volume} = 0.352 \text{ g} \times \frac{1 \text{ mol}}{154 \text{ g}} \times \frac{(62.4)(373)}{730} \frac{\text{liter}}{\text{mol}}$

$= 0.0729 \text{ liter} = 72.9 \text{ ml}$

4.12 *The law of partial pressures*

Gases being mostly empty space, nothing prevents mixing more than one gas in the same volume. The total pressure will go up because there will be more molecules hitting the walls, but each gas will contribute pressure in proportion to its numbers as though the other gases were not there. (The possibility of chemical reactions that lead to explosions is another subject.)

This means that the total pressure in a container is the sum of the pressures of each of the components in the mixture. This law is useful when computing the amount of one component in a gas mixture when the total pressure and the partial pressure of the other components are known.

EXAMPLE 4.13 Exactly one liter of oxygen was collected over water at 25°C. The total pressure was 750 torr and the partial pressure of water vapor was 25 torr. How many moles of oxygen were collected?

Solution: We know a property that relates volume to moles, RT/P, but we must be sure that the property only includes the pressure of the gas we are interested in. The total pressure includes pressure due to oxygen and pressure due to water vapor. The partial pressure of oxygen is therefore 750 torr − 25 torr = 725 torr, and the number of moles is

$$1.0\ \text{liter} \times \frac{725}{(62.4)(298)} \frac{\text{mol}}{\text{liter}} = 0.039\ \text{mol}$$

Kinetic molecular theory *4.13

The qualitative behavior of gases has been explained by the assumption that gases are composed of atoms or molecules in rapid motion, with little attraction for each other, and with a lot of free volume in which to move. Now that we have some quantitative laws describing gas behavior, let us contrive a model that will explain it.

All gases exert pressure on the sides of their container. What causes that pressure? The most reasonable explanation is that the molecules are striking the walls with an enormous number of collisions per second. We do not perceive the individual impacts because they are so frequent and numerous, but each impact contributes a small impulse to the wall that must be resisted by a force in the opposite direction.

A handful of BB shot in a blender going at high speed is analogous to the situation in a microscopic region of the gas. The impact of particles in a blender will lift the top if it is not held down. The impact of gas molecules will move the walls of the container if the walls are not restrained in some way. The atoms in the wall do move, of course, but they are bound to the other atoms in the wall so that they are all jiggling together.

The size of an impulse is measured by the change in momentum that occurs during the collision. The momentum of a particle is the product of its mass and velocity (mv). Momentum is changed whenever a particle changes its speed or direction of motion. The wall must exert a force on the molecule for an instant of time in order to reverse the direction of motion of the molecule.

Each impact on the wall delivers an impulse of twice the momentum of the molecule; one unit of momentum to stop the molecule and one more to send it flying back the way it came.

$$\frac{\text{impulse}}{\text{impact}} = \frac{2mv}{\text{impact}} \tag{4.16}$$

Force is the rate of delivering impulse, so if the number of

* Optional topic.

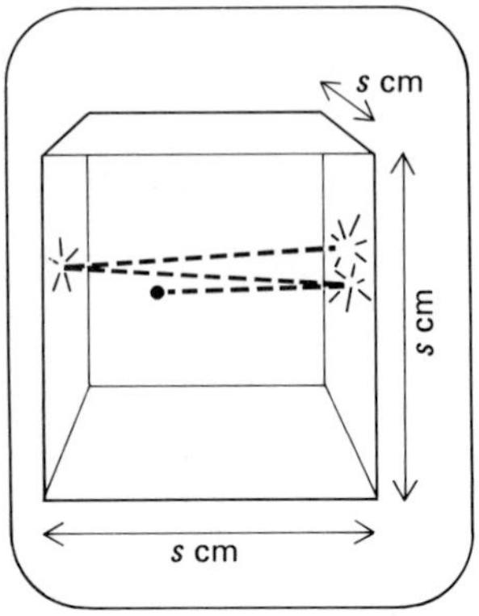

Figure 4.7 The molecule travels 2*s* centimeters between impacts on the same wall.

impacts per second can be found we can find the force. The number of impacts per second must be related to the number of molecules and the frequency of impact of each molecule. If we imagine a cubic box with only one molecule in it we can see that a molecule must travel twice the length of the box between impacts on the same wall (Figure 4.7). There is an average of

$$\frac{1 \text{ impact}}{2s \text{ centimeters}} \tag{4.17}$$

The distance ($2s$) is related to the time between impacts by the velocity of the molecule, v cm/sec, so the frequency of impacts, the number per second, is the product of those two factors.

$$\frac{1 \text{ impact}}{2s \text{ cm}} \times \frac{v \text{ cm}}{\text{sec}} = \frac{v}{2s} \frac{\text{impacts}}{\text{sec}} \tag{4.18}$$

The force due to one molecule hitting the wall is the product of equations 4.16 and 4.18.

$$\text{force per molecule} = \frac{\text{impulse}}{\text{sec}}$$

$$= \frac{2\ mv}{\text{impact}} \times \frac{v}{2s} \frac{\text{impacts}}{\text{sec}}$$

$$\frac{mv^2}{s} = \text{units of momentum change per second per molecule}$$

A large number, say N, of molecules put in the box will soon move randomly in all directions. Because there are three directions in space, only one third of the velocities will be directed perpendicular to one wall. The contribution of N molecules to the force is $N/3$ times the force of one molecule.

$$\text{force on one wall} = \frac{N}{3} \times \frac{mv^2}{s}$$

where v is now the *average velocity* of the molecules.

The pressure on the wall is the force divided by the area, and the area of the wall is s^2.

$$\text{pressure} = \frac{\text{force}}{\text{area}} = \frac{N}{3} \times \frac{mv^2}{s} \times \frac{1}{s^2} = \frac{Nmv^2}{3s^3} \tag{4.19}$$

But s^3 is the volume of the box; thus we can substitute volume for s^3 and write

$$\text{pressure} = \frac{Nmv^2}{3(\text{volume})} \tag{4.20}$$

or

$$\text{pressure} \times \text{volume} = \frac{Nmv^2}{3}$$

The factor mv^2 should look familiar; it is twice the average kinetic energy of the molecules.

$$mv^2 = 2(KE)$$

Substituting $2(KE)$ for mv^2 in equation 4.20 we have

$$PV = \frac{2}{3} N(KE) \qquad (4.21)$$

Now we have discovered the nature of the constant in the equation $PV =$ a constant. It depends on the number of molecules and on the average kinetic energy per molecule!

Average kinetic energy and temperature *4.14

The average kinetic energy of the molecules is a difficult property to measure directly. The measurement requires complicated apparatus and lots of figuring. Some reflection on the nature of temperature might provide an alternative means for indirectly measuring that property.

When a hot object is placed in contact with a cool one, changes take place until both objects are at the same temperature. In fact we recognize that they *are* at the same temperature because no more changes (in volume, pressure, or whatever) occur. What was transferred from the hot object to the cold one to effect that change? It was not mass, because no detectable change of mass is observed. The only thing that can be transferred is the motion of the molecules. Fast molecules collide with slow ones, thereby increasing the speed of the slow and decreasing the speed of the fast. The transfer of kinetic energy by collision occurs not only at the interface of the two objects but throughout each material until the *average kinetic energy* is the same.

This must mean that average kinetic energy and temperature are related. The relationship is found by comparing equations 4.9 and 4.21. We found by experiment that

$$PV = NkT \qquad (4.9)$$

From kinetic molecular theory we found that

$$PV = \frac{2}{3} N(KE) \qquad (4.21)$$

Two things equal to the same thing must be equal to each other; therefore,

$$(KE) = \frac{3}{2} kT \qquad (4.22)$$

* Optional topic.

The average kinetic energy per molecule is directly proportional to the absolute temperature. In fact they are the same thing in the same way that 1 kilogram/liter is the same as 62.4 pounds/cubic foot. They just have different units.

*4.15 *The law of diffusion*

If a bottle of perfume and a bottle of skunk oil were opened simultaneously some distance from you, which odor would you smell first? The ability to predict diffusion rates and other transport properties of gases like viscosity and the speed of sound is a direct result of kinetic molecular theory.

If all other conditions are the same, these properties depend only on the speed of the molecules. The speed of the molecules is related to their average kinetic energy, and kinetic molecular theory relates the average kinetic energy to the absolute temperature. At the same temperature all substances have the same average kinetic energy. It follows that the ratio of average molecular speeds is inversely proportional to the square root of the ratio of their masses.

$$KE_1 = KE_2$$

$$\frac{1}{2} m_1 v_1^2 = \frac{1}{2} m_2 v_2^2$$

$$\frac{v_1^2}{v_2^2} = \frac{m_2}{m_1}$$

$$\frac{v_1}{v_2} = \sqrt{\frac{m_2}{m_1}}$$

The masses of the molecules are directly proportional to their molar weights, so

$$\frac{v_1}{v_2} = \sqrt{\frac{\text{MW}_2}{\text{MW}_1}} \qquad (4.23)$$

This equation says that if the molar weight of the perfume is two times that of the skunk oil, the skunk oil will diffuse $\sqrt{2}$ or 1.41 times as fast as the perfume.

EXAMPLE 4.14 Each of two balloons at the same temperature and pressure has a small pinhole in it. One contains hydrogen gas; the other contains oxygen. Which balloon will deflate more quickly, and at what comparative rate?

* Optional topic.

Solution: The rate that a gas diffuses through a hole evidently depends on how often a molecule hits the hole. That depends on the number of collisions per second on the wall, and we know that collisions per second depend on the velocity of the molecules. From equation 4.23 and the data that the molar weight of hydrogen is 2 grams per mole and that of oxygen is 32 grams per mole, we find

$$\frac{v_{\text{hyd}}}{v_{\text{oxy}}} = \sqrt{\frac{32}{2}} = 4$$

Hydrogen will initially diffuse 4 times faster than oxygen.

The fact that light molecules have correspondingly greater average speeds than heavy molecules has been used in such diverse applications as rocket fuel mixtures and divers' breathing mixtures. Divers who go deep for extended periods of time dissolve large quantities of the gas they breathe into their tissues and blood. As they return to the surface, they must allow time for the dissolved gas to diffuse out of their bodies through the lungs. If they come up too fast, bubbles of gas form in their bodies and cause a painful and sometimes fatal condition. The nitrogen in air, besides being toxic at high pressures, is a very slow diffusing gas. It is replaced by helium which leaves the body much faster.

Helium does cause a communications problem, however. The speed of sound is faster in helium, which causes the divers to sound like Donald Duck.

Real gases 4.16

The Ideal Gas Law describes the ideal behavior of gases. If gas molecules really had no attractive forces between them, and if the volume occupied really approached zero when the temperature got near absolute zero, then the Ideal Gas Law would hold at any pressure and temperature.

But real gases do occupy space and there are attractive forces between them. These facts cause deviations from ideal behavior when the pressure is very high or the temperature is very low.

At high pressures the molecules are crowding each other, and the free volume in which they move is much less than the volume of the container. The actual pressure will go higher than predicted because the actual volume is lower than the calculated volume.

At temperatures near the boiling point of the gas the molecules are slowed down and the attractive forces have more effect. Molecules begin to clump together so that the gas behaves as though there were fewer molecules present. The pressure is less than that calculated under these conditions.

At temperatures well above the boiling point and at moderate pressures the Ideal Gas Law predicts the behavior of gases with good accuracy and can be used with confidence.

4.17 *Summary*

1. The force exerted perpendicular to a surface per unit area of surface is called *pressure.* The pressure exerted by a liquid is the height of the column of liquid times its density.

2. The unit of atmospheric pressure is the *atmosphere.* One atmosphere is defined to be equal to the pressure exerted by a column of mercury 76.0 cm high, or 760 torr (millimeters of mercury).

3. At constant temperature the volume of a given amount of gas is inversely proportional to the pressure.

4. At constant pressure the volume of a given amount of gas is directly proportional to the *absolute temperature.* The absolute temperature (Kelvin scale) is the Celsius temperature plus 273 degrees.

5. The above two laws may be combined to give

$$\frac{P_iV_i}{T_i} = \frac{P_fV_f}{T_f}$$

6. The value of PV/T depends only on the number of molecules present. The number of molecules in 22.4 liters of a gas at standard temperature and pressure (0°C and 1 atm) is called a *mole.*

7. When the amount of gas is expressed in moles and the absolute temperature scale is used

$$PV = nRT$$

The proportionality constant R is called the *ideal gas constant.*

8. The relationship of mass to moles is the *molar weight (MW).*

9. The pressure exerted by a gas is independent of other gases which may be present in a mixture. The total pressure is the sum of the partial pressures of each gas present.

10. (optional) The pressure-volume product of an ideal gas equals

$$PV = \frac{Nmv^2}{3} = \frac{2}{3}N(KE)$$

The average kinetic energy (KE) of a gas is directly proportional to the absolute temperature; therefore

$$PV = NkT$$

11. (optional) The ratio of average speeds of gas molecules at the same temperature is inversely proportional to the square root of the ratio of their molecular weights.

$$\frac{v_1}{v_2} = \sqrt{\frac{MW_2}{MW_1}}$$

12. Real gases do not follow the Ideal Gas Law at temperatures near their boiling points or at very high pressures.

EXERCISES

I. Define the following terms.

1. pressure
2. temperature
3. kinetic energy
4. potential energy
5. mole (gram-mole)
6. molar volume
7. molar weight
8. density
9. Celsius temperature
10. Kelvin temperature
11. ideal gas constant

***II.** Use the Kinetic Molecular Theory to explain why

1. At constant volume and moles, the pressure increases when the temperature increases.
2. At constant temperature and volume, the pressure increases when the number of moles increases.
3. Equal volumes of gases at the same temperature and pressure have the same number of moles.
4. The velocity of diffusion of helium (MW = 4) is 2.65 times faster than nitrogen (MW = 28) under the same conditions.
5. A vacuum is an excellent insulator for heat and sound.
6. The *PV* product is less than *nRT* for a real gas when the temperature is very low.
7. The *PV* product is more than *nRT* for a real gas when the pressure is very high.

SOLVED PROBLEMS

III. Calculate the missing value. The quantity of gas in each case is constant.

	P_i	V_i	T_i	P_f	V_f	T_f
1.	745 torr	10.0 liters	298°K	700 torr	?	273°K
2.	1.0 atm	5.0 liters	27°C	2.0 atm	?	−3°C
3.	720 torr	75 cu ft	35°C	?	3 cu ft	25°C
4.	0.6 atm	1.0 liter	273°K	1.0 atm	0.5 liter	?
5.	14.7 psi	10.0 liters	273°K	44.1 psi	?	373°K

* Optional topic.

IV. Calculate the missing value.

	P	V	T	n (mol)
1.	1.0 atm	20.0 liters	298°K	?
2.	745 torr	200 ml	280°K	?
3.	715 torr	? liter	273°K	0.03
4.	2.00 atm	? ml	298°K	0.25
5.	700 torr	125 ml	20°C	?

V. Calculate the missing values.

	P	V (liter)	T	n	MW (g/mol)	wt (g)	d (g/liter)
1.	800 torr	0.82	308°K	?	?	2.46	?
2.	300 torr	?	348°K	?	30	1.5	?
3.	712 torr	1.47	303°K	?	?	1.98	?
4.	2.00 atm		303°K		?		2.07
5.	1.00 atm	1.5	300°K	?	44	?	?

VI. Word Problems

1. A steel cylinder has a diameter of 2.0 cm and is 20 cm high. The density of steel is 7.9 g/cm^3. What pressure does the cylinder exert when it is standing on its base? Use units of grams (weight) per square centimeter.

2. What volume will 75 cubic feet of air occupy if the external pressure is increased from 14.7 psi to 2250 psi?

3. The volume of gas in the cylinder of an engine when the piston is all the way down is 375 cm^3. When the piston is at the top of the compression stroke there remains 50 cm^3 of gas volume. If the pressure is 0.80 atm in the cylinder at the beginning of the compression stroke, what is the pressure at the end?

4. A gas thermometer is more convenient to use if the volume of the gas is held constant and the pressure is measured. The fractional pressure change per degree is the same as the volume coefficient, 0.00366 degree^{-1}. If a gas at 0°C exerts 650 torr pressure, what is the temperature at which the pressure is 750 torr?

5. What volume will one liter of oxygen gas at its boiling point, −182°C, occupy at room temperature, 25°C?

6. What volume will 50 liters of helium at 20°C and 1.0 atm pressure occupy at 80,000 ft, where the pressure is 0.10 atm and the temperature is −50°C?

7. What will be the pressure in a tire inflated to 28 psi at 25°C if its temperature is increased to 50°C? Assume the volume of the tire does not change.

8. Express 1.00×10^{24} units in

(a) dozens (12 units per dozen)

(b) reams (500 units per ream)

(c) billions (10^9 units per billion)
(d) moles (6.02×10^{23} units per mole)

9. What is the molar volume of a gas at 700 torr and 300°K?
10. A chemical reaction produces 0.050 mol of oxygen at 700 torr and 300°K. What volume will the oxygen occupy?
11. 1.500 mol of a substance weighs 66.0 g. What is its molar weight?
12. The molar weight of a gas is 56 g/mol. How many moles are in 100 g of the gas?
13. What volume will one pound of Dry Ice (454 g) occupy when it evaporates at 1 atm pressure and warms to room temperature (20°C)? Dry Ice is solid carbon dioxide, molar weight 44 g/mole.
14. At 21°C and 760 torr the density of air is 1.20 g/liter. What is the effective "molar weight" of air, calculated as if it were a pure substance?
15. A sample of a new compound in a bulb of capacity 268 ml weighed 0.0516 g. The pressure was 17.9 torr and the temperature was 69°C. Calculate the molar weight of the gas.

ADDITIONAL PROBLEMS

VII. Compute the missing values:

	P_i	V_i (liter)	T_i	P_f	V_f (liter)	T_f
1.	720 torr	1.0	288°K	760 torr	?	273°K
2.	0.5 atm	5.0	298°K	?	2.5	298°K
3.	5.0 psi	1.3	310°K	14.7 psi	?	310°K
4.	3.0 atm	1.5	273°K	1.0 atm	?	298°K
5.	1.0 atm	10.0	273°K	1.0 atm	30.0	?

VIII. Compute the missing values:

	P	V	T	n	wt(g)	MW (g/mol)	d (g/liter)
1.	700 torr	35.0 ml	373°K	?	0.162	?	?
2.	720 torr	?	323°K	?	5.00	58	?
3.	600 torr	20.0 liters	310°K	?	?	46	?
4.	1.0 atm		298°K			?	1.80
5.	745 torr	?	27°C	0.01 mol	?	2.0	?

IX. Word Problems

1. What pressure above atmospheric pressure is felt by a diver 60 ft down, if the density of sea water is 64 lb/ft^3? Express in units of pounds per square inch.
2. What volume will 10 liters of air occupy if the external pressure is increased from 0.7 to 0.95 atm?
3. If 75 cubic feet at 1 atm is compressed to 3 cubic feet at constant temperature, what is the new pressure?

4. A strip of brass is 10.000 cm long at 0°C. The coefficient of length increase for brass is 1.8×10^{-5} deg^{-1}. What is the temperature when the brass strip is 10.010 cm long?
5. The pressure of hydrogen is 100 torr in a vessel of constant volume at the temperature of melting ice. What is the pressure at the temperature of boiling water?
6. An aerosol can is pressurized with a gas to 2.0 atm above atmospheric pressure at 20°C. At what temperature will the pressure be 3.0 atm?
7. The molar weight of hydrogen sulfide is 34 g/mol. A quantity of 1.7 g of hydrogen sulfide is equal to how many moles?
8. A vessel of 100 ml volume holds 0.050 g of a gas at 127°C and 1.0 atm pressure. What is the molar weight of the gas?
9. The volume of gas collected by displacement of water in a graduated cylinder is 225 ml. Atmospheric pressure is 720 torr and the temperature is 22°C. The vapor pressure of water at that temperature is 20 torr. What would be the volume of the dry gas at standard conditions (0°C and 760 torr)?
10. A 100-liter vessel contains 8 g of hydrogen, 28 g of carbon dioxide, and 8 g of helium. The molar weights are: hydrogen, 2 g/mol; carbon dioxide, 44 g/mol; and helium, 4 g/mol. The temperature is 25°C. What is the total pressure?

SOLUTIONS

III.

1. $V_f = 10.0 \text{ liters} \times \frac{745 \text{ torr}}{700 \text{ torr}} \times \frac{273°\text{K}}{298°\text{K}} = 9.75 \text{ liters}$

2. $V_f = 5.0 \text{ liters} \times \frac{1.0 \text{ atm}}{2.0 \text{ atm}} \times \frac{270°\text{K}}{300°\text{K}} = 2.3 \text{ liters}$

3. $P_f = 720 \text{ torr} \times \frac{75 \text{ cu ft}}{3 \text{ cu ft}} \times \frac{298°\text{K}}{308°\text{K}} = 17{,}000 \text{ torr} (1.7 \times 10^4)$

4. $T_f = 273°\text{K} \times \frac{1.0 \text{ atm}}{0.6 \text{ atm}} \times \frac{0.5 \text{ liter}}{1.0 \text{ liter}} = 230°\text{K}$

5. $V_f = 10.0 \text{ liters} \times \frac{14.7 \text{ psi}}{44.1 \text{ psi}} \times \frac{373°\text{K}}{273°\text{K}} = 4.55 \text{ liters}$

IV.

1. $n = 20.0 \text{ liters} \times \frac{1.0 \quad \text{mol}}{(0.0821)(298) \text{ liter}} = 0.82 \text{ mol}$

2. $n = 0.200 \text{ liter} \times \frac{745 \quad \text{mol}}{(62.4)(280) \text{ liter}} = 0.00853 \text{ mol}$

3. $V = 0.03 \text{ mol} \times \frac{(62.4)(273) \text{ liter}}{715 \quad \text{mol}} = 0.7 \text{ liter}$

4. $V = 0.25 \text{ mol} \times \frac{(0.0821)(298) \text{ liter}}{2.0 \quad \text{mol}} \times 1000 \frac{\text{ml}}{1} = 3100 \text{ ml}$

5. $n = 125 \text{ ml} \times \frac{1 \text{ liter}}{1000 \text{ ml}} \times \frac{700}{(62.4)(293)} = 0.00479 \text{ mol}$

V. **1.** $n = 0.82 \text{ liter} \times \frac{800}{(62.4)(308)} \frac{\text{mol}}{\text{liter}} = 0.034 \text{ mol}$

$\text{MW} = \frac{2.46 \text{ g}}{0.034 \text{ mol}} = 72 \frac{\text{g}}{\text{mol}}$

$d = \frac{2.46 \text{ g}}{0.82 \text{ liter}} = 3.0 \frac{\text{g}}{\text{liter}}$

2. $n = 1.5 \text{ g} \times \frac{1 \text{ mol}}{30 \text{ g}} = 0.050 \text{ mol}$

$V = 0.050 \text{ mol} \times \frac{(62.4)(348)}{300} \frac{\text{liter}}{\text{mol}} = 3.6 \text{ liters}$

$d = \frac{1.5 \text{ g}}{3.6 \text{ liters}} = 0.42 \frac{\text{g}}{\text{liter}}$

3. $n = 1.47 \text{ liters} \times \frac{712}{(62.4)(303)} \frac{\text{mol}}{\text{liter}} = 0.0554 \text{ mol}$

$\text{MW} = \frac{1.98 \text{ g}}{0.0553 \text{ mol}} = 35.8 \frac{\text{g}}{\text{mol}}$

$d = \frac{1.98 \text{ g}}{1.47 \text{ liters}} = 1.35 \frac{\text{g}}{\text{liter}}$

4. $\text{MW} = 2.07 \frac{\text{g}}{\text{liter}} \times \frac{(0.0821)(303)}{2.00} \frac{\text{liter}}{\text{mol}} = 25.7 \frac{\text{g}}{\text{mol}}$

5. $n = 1.5 \text{ liters} \times \frac{1.0}{(0.0821)(300)} \frac{\text{mol}}{\text{liter}} = 0.061 \text{ mole}$

$\text{weight} = 0.061 \text{ mol} \times 44 \frac{\text{g}}{\text{mol}} = 2.7 \text{ g}$

$d = \frac{2.7 \text{ g}}{1.5 \text{ liters}} = 1.8 \frac{\text{g}}{\text{liter}}$

VI. **1. Want:** $\text{pressure} = \frac{\text{weight}}{\text{area of base}} \frac{\text{g}}{\text{cm}^2}$

Have: dimensions of cylinder; 2.0 cm diameter, 20 cm high

Know: area of base $= \frac{\pi d^2}{4}$; volume $= \frac{\pi d^2 h}{4}$; density $= 7.9 \frac{\text{g}}{\text{cm}^3}$

Solution: $P = \frac{\pi(2.0)^2(20)}{4} \text{ cm}^3 \times 7.9 \frac{\text{g}}{\text{cm}^3} \times \frac{4}{\pi(2.0)^2\text{cm}^2} = 158 \frac{\text{g}}{\text{cm}^2}$

Note: Since a cylinder is a uniform column, the pressure on the base is simply the height times density.

2. Want: final volume (ft^3) after compression

Have: $V_i = 75 \text{ ft}^3$

Know: $P_i = 14.7$ psi; $P_f = 2250$ psi

Solution: $V_f = 75 \text{ ft}^3 \times \frac{14.7 \text{ psi}}{2250 \text{ psi}} = 0.49 \text{ ft}^3$

3. Want: final pressure (atm) when gas is compressed

Have: $V_i = 375$ cc; $V_f = 50$ cc

Know: $P_i = 0.80$ atm

Solution: $P_f = 0.80 \text{ atm} \times \dfrac{375 \text{ cc}}{50 \text{ cc}} = 6.0 \text{ atm}$

4. **Want:** temperature in Celsius scale
Have: constant volume
Know: $T_o = 0°\text{C}$; $P_o = 650$ torr; $P = 750$ torr; $\alpha = 0.00366 \text{ deg}^{-1}$

Solution: $T°\text{C} = \dfrac{750 \text{ torr} - 650 \text{ torr}}{(0.00366)(650 \text{ torr})} \text{ deg} = 42°\text{C}$

In Kelvin: $T = 273°\text{K} \times \dfrac{750}{650} = 315°\text{K} = 42°\text{C}$

5. **Want:** final volume when gas is warmed
Have: 1.0 liter
Know: $T_i = -182°\text{C} = 91°\text{K}$; $T_f = 25°\text{C} = 298°\text{K}$

Solution: $V_f = 1.0 \text{ liter} \times \dfrac{298°\text{K}}{91°\text{K}} = 3.3 \text{ liters}$

6. **Want:** final volume (liters)
Have: $V_i = 50$ liters
Know: $P_i = 1.0$ atm; $T_i = 293$ K; $P_f = 0.10$ atm; $T_f = 223°$K

Solution: $V_f = 50 \text{ liters} \times \dfrac{1.0 \text{ atm}}{0.1 \text{ atm}} \times \dfrac{223°\text{K}}{293°\text{K}} = 380 \text{ liters}$

7. **Want:** final pressure when gas has been warmed
Know: $P_i = 28$ psi, $T_i = 298°$K; $T_f = 323°$K

Solution: $P_f = 28 \text{ psi} \times \dfrac{323}{298} = 30.3 \text{ psi}$

8. **(a)** $1.00 \times 10^{24} \text{ units} \times \dfrac{1 \text{ dozen}}{12 \text{ units}} = 8.33 \times 10^{22} \text{ dozen}$

(b) $1.00 \times 10^{24} \text{ units} \times \dfrac{1 \text{ ream}}{500 \text{ units}} = 2.0 \times 10^{21} \text{ reams}$

(c) $1.00 \times 10^{24} \text{ units} \times \dfrac{1 \text{ billion}}{10^9 \text{ units}} = 1.0 \times 10^{15} \text{ billions}$

(d) $1.00 \times 10^{24} \text{ units} \times \dfrac{1 \text{ mole}}{6.02 \times 10^{23} \text{ units}} = 1.66 \text{ moles}$

9. **Want:** molar volume in liters per mole
Know: $P = 700$ torr; $T = 300°$K

Solution: $\text{molar volume} = \dfrac{62.4 \dfrac{\text{liter-torr}}{\text{mol-deg}} \times 300 \text{ deg}}{700 \text{ torr}} = 26.7 \dfrac{\text{liters}}{\text{mol}}$

10. **Want:** volume in liters
Have: 0.050 mol of gas
Know: the relation between volume and moles of a gas, $\dfrac{RT}{P} \dfrac{\text{liter}}{\text{mol}}$

$P = 700$ torr; $T = 300°$K

Solution: $V = 0.050 \text{ mol} \times \dfrac{(62.4)(300)}{700} \dfrac{\text{liter}}{\text{mol}} = 1.34 \text{ liters}$

11. Want: molar weight in grams per mole

Have: weight in grams = 66.0 g; number of moles = 1.500 mol

Solution: $\text{MW} = \dfrac{66.0\text{ g}}{1.500\text{ mol}} = 44.0\ \dfrac{\text{g}}{\text{mol}}$

12. Want: number of moles

Have: 100 g

Know: the relation between weight and moles = $56\ \dfrac{\text{g}}{\text{mol}}$

Solution: $n = 100\text{ g} \times \dfrac{1\text{ mol}}{56\text{ g}} = 1.8\text{ mol}$

13. Want: volume in liters

Have: 1 lb (454 g)

Know: the relation of volume to moles, RT/P; $T = 293$, $P = 1$ atm
the relation of weight to moles, 44 g/mol.

Solution: $V = 454\text{ g} \times \dfrac{1\text{ mol}}{44\text{ g}} \times \dfrac{(.0821)(293)}{1.0}\,\dfrac{\text{liter}}{\text{mol}} = 248\text{ liters}$

14. Want: molar weight in grams per mole

Know: density = 1.20 g/liter

molar volume $= \dfrac{RT}{P}\,\dfrac{\text{liter}}{\text{mol}}$; $P = 760$ torr; $T = 294°\text{K}$

Solution: $\text{MW} = 1.20\ \dfrac{\text{g}}{\text{liter}} \times \dfrac{(62.4)(294)}{760}\,\dfrac{\text{liter}}{\text{mol}} = 29.0\ \dfrac{\text{g}}{\text{mol}}$

15. Want: molar weight in grams per mole

Have: 0.0516 g; 268 ml

Know: molar volume $= \dfrac{RT}{P}\,\dfrac{\text{liter}}{\text{mol}}$; $P = 17.9$ torr; $T = 342°\text{K}$

Solution: $\text{MW} = \dfrac{0.0516\text{ g}}{0.268\text{ liter}} \times \dfrac{(62.4)(342)}{17.9}\,\dfrac{\text{liter}}{\text{mol}} = 230\ \dfrac{\text{g}}{\text{mol}}$

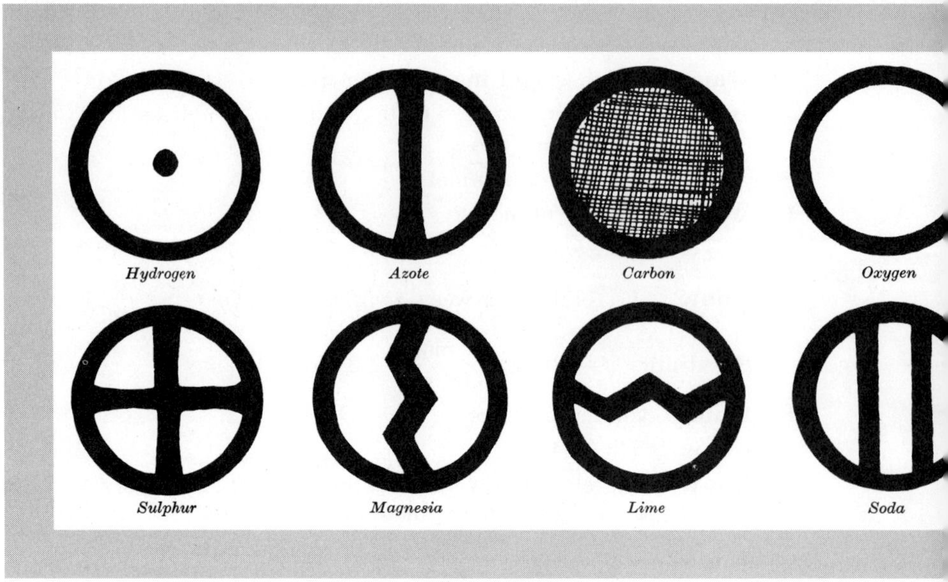

The innumerable pure substances in the world are made of a relatively small number of elemental substances.

After reading this chapter you should be able to

1. Describe a practical way to distinguish elements from compounds;
2. State the main points of (Dalton's) Atomic Theory;
3. Use the relationship of weight to number, atomic weight (amu/atom) and molar weight (grams/mole), to compute moles from weight and vice versa;
4. Name the elements that occur as diatomic molecules;
5. Define atomic number and the mass number of atoms;
6. Define isotope;
7. Write the symbols and names of the more common elements;
8. Recognize the chemical groups and periods in the periodic table;
9. State the information given by a chemical empirical formula;
10. Calculate the molar weight of a compound from the formula;
11. Calculate the weight percent composition of the elements in a compound;
12. Calculate the amount of an element in a given amount of a compound;
13. Calculate the empirical formula of a compound from its weight composition.

5 composition

In the seventeenth century an empiricist school of thought founded by Francis Bacon held that knowledge could only be gained by the collection, comparison, and ordering of facts. According to Bacon, the form (substance) of a thing was the sum of its natures (physical properties). As the nature changed so did the substance. The way to knowledge was to collect all the facts, arrange them in tables, and compare those that in some degree possessed like natures. In this manner relationships should become apparent and lead to greater understanding.

The sheer numbers of substances in the world makes this plan impractical if not impossible. Bacon knew this and allowed for intermediate levels of understanding. The death blow to this grand plan, however, is the changeableness of matter. What happens to the substance when the nature changes? When gold dissolves in aqua regia the nature of gold disappears and something else takes its place. The plan of Bacon had no means for relating the substance of gold and its salt.

The explanation of this apparently hopeless situation was put together only one hundred fifty years ago by a few men, using imprecise balances and materials of doubtful purity, but with great faith in the natural order of things and in their own reasoning powers.

5.1 *Compounds and elements*

What evidence formed the basis for the concepts of these men? Qualitatively it had been known for some time that some pure substances could be decomposed into their parts under conditions such that simply mixing the parts did not restore the original substance. Water, for example, could be decomposed into two components by passing an electric current through it. A mixture of the two components was simply a mixture of gases that did not in the least resemble water (Figure 5.1). Similarly, red mercury oxide decomposed into a gas and liquid metallic mercury (Figure 5.2). The components did not form the red powder no matter how well they were mixed.

A much smaller number of pure substances could not be decomposed. Metals such as copper, silver, and lead and nonmetals such as sulfur and oxygen could be changed into something else, but all the known chemical changes of these substances resulted in a weight gain. That is, they would combine with something else to form a new substance that weighed more than the original substance.

A pure substance that can be decomposed into two or more components by chemical means is a *compound.* A substance that cannot be decomposed by any known chemical means is an *element.* This

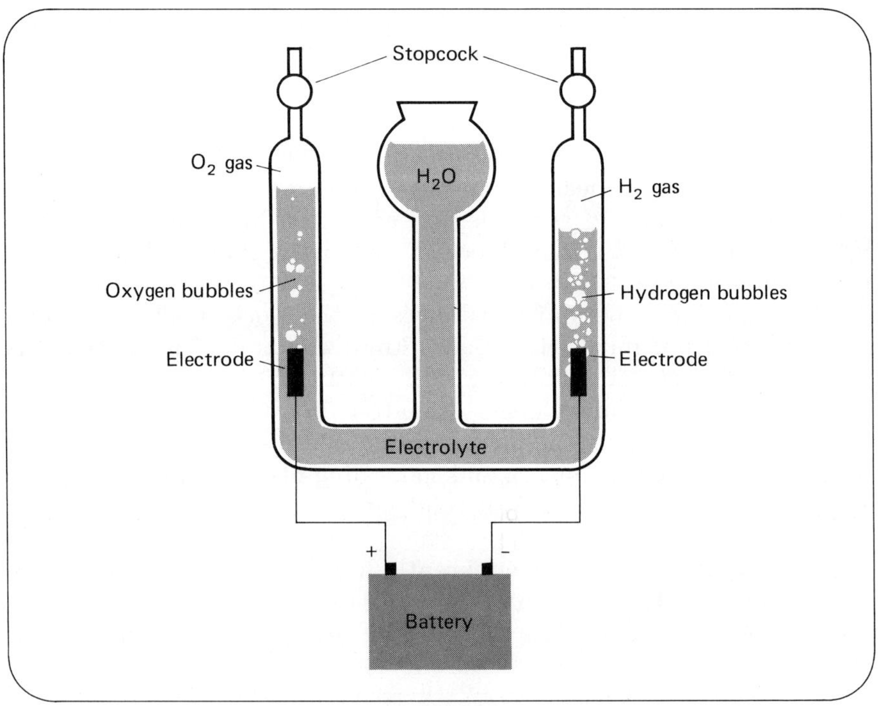

Figure 5.1 Decomposition of water.

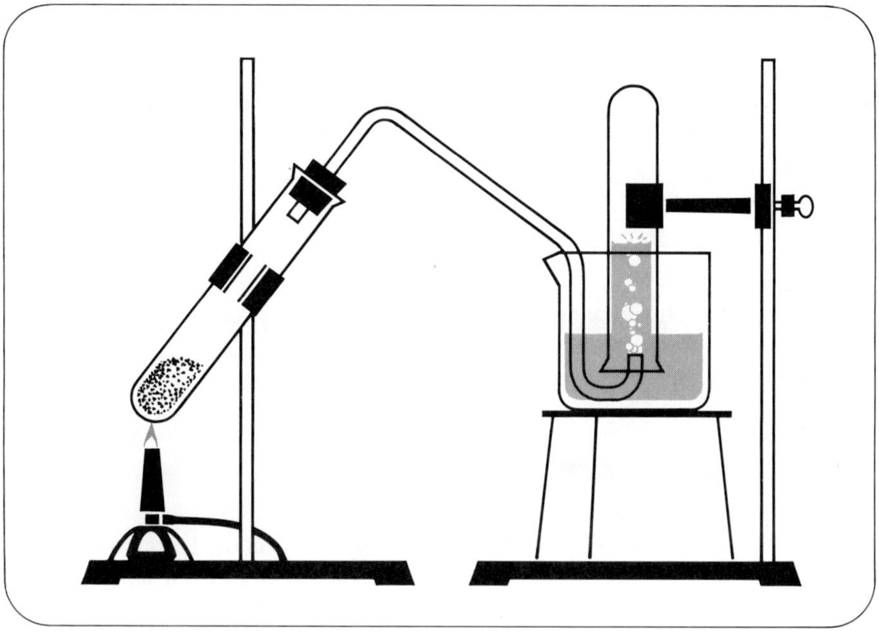

Figure 5.2 Decomposition of mercury oxide.

negative definition can and has led to mistakes, but it was the only one available until the twentieth century. A better definition is given in the next chapter.

Elements must consist of only one kind of atom since they cannot be decomposed into two or more different substances. Elements combine chemically to form compounds. Compounds are not only composed of two or more kinds of atoms, but they have a definite fixed composition. *The weight ratio of elements forming a compound is always the same for a given compound, no matter how it is made or what its source is.*

This law was disputed for some time. And it was established only after means were developed to purify substances from mixtures, and instruments were invented to measure the properties accurately enough to be sure that the substance being analyzed was pure.

EXAMPLE 5.1

1. One gram of hydrogen combines with 8 grams of oxygen to give 9 grams of water together with much heat and light. Conversely, 9 grams of water can be decomposed electrically into 1 gram of hydrogen and 8 grams of oxygen.

$$1\text{ g hydrogen} + 8\text{ g oxygen} = 9\text{ g water}$$

The information about the composition of water is usually expressed as a weight percent. One-ninth (11.1%) of the weight of the compound is hydrogen. Eight-ninths (88.9%) of the weight of the compound is oxygen.

If there happens to be 10 grams of oxygen with 1 gram of hydrogen before reaction, there will still be 9 grams of water formed and 2 grams of oxygen will be left.

2. 2.04 grams of zinc dust and 1.0 gram of sulfur combine vigorously to form 3.04 grams of zinc sulfide.

$$2.04 \text{ g zinc} + 1.0 \text{ g sulfur} = 3.04 \text{ g zinc sulfide}$$

The composition of zinc sulfide is $\frac{2.04}{3.04} \times 100 = 67.1\%$ zinc and 32.9% sulfur.

If excess sulfur is present it remains as sulfur and can be removed from the zinc sulfide by physical means.

5.2 *The atomic theory*

The foregoing evidence might lead to the conclusion that compounds are not simply mixtures of elements. Laws must govern the way elements combine and take into account the extreme specificity of the weight relationships.

Early in the nineteenth century John Dalton proposed an explanation that, with some modifications, has proved to be the only one consistent with all known facts. The main points of his hypothesis are

1. Matter is not continuous but consists of discrete indivisible particles, henceforth called *atoms.*
2. All atoms of a particular element are identical.
3. The atoms of different elements differ in mass and other properties.
4. Compounds are formed by atoms combined in small whole number ratios. Each compound is composed of units that are identical with respect to the kinds and numbers of atoms.
5. The same elements can combine with each other in more than one way to form different compounds.

The consequences of Dalton's theory are these. If the relative masses of the atoms are known, the relative numbers of atoms that combine can be found. If the numbers of atoms that combine is known, perhaps the rules governing the combinations can be found, and then we will be able to predict chemical changes.

As things turned out, Dalton's contribution was limited to outlining the nature of the problem. The problem can be summarized like this:

1A. Suppose that each element is composed of extremely small identical atoms.

1B. Then the weight of a particular amount is equal to the number of atoms times the weight per atom.

$$\text{Total weight} = \text{number of atoms} \times \text{weight per atom}$$

2A. Suppose that compounds are formed by the clumping together of a particular number of different kinds of atoms.

2B. Then the weight ratio of the elements in the compound (which is known by experiment) is equal to the number ratio of the elements in the compound (which we want to find) multiplied by the weight ratio of each atom to the other (which is not known either!).

$$\frac{\text{Weight of A}}{\text{Weight of B}} = \frac{\text{no. of A's}}{\text{no. of B's}} \times \frac{\text{weight per A}}{\text{weight per B}} \qquad (5.1)$$

EXAMPLE 5.2

Water is composed of 8 parts oxygen and 1 part hydrogen by weight. What is the number ratio of hydrogen to oxygen?

$$\frac{\text{8 parts oxygen}}{\text{1 part hydrogen}} = \boxed{\frac{\text{no. of O's}}{\text{no. of H's}}} \times \frac{\text{weight per O}}{\text{weight per H}}$$

To find the answer we need to know the weight of the oxygen atom compared to the hydrogen atom. We cannot solve one equation with two unknowns!

Figure 5.3 Dalton's hypothesis.

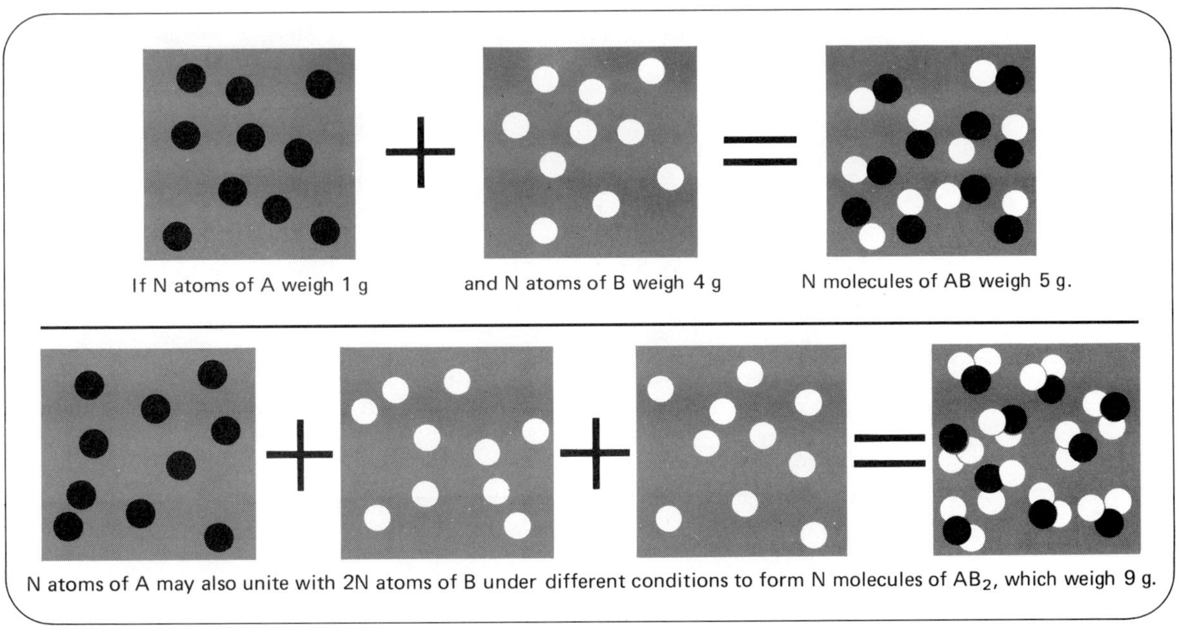

5.3 *Avogadro's Hypothesis*

In order to solve equation 5.1 a means had to be found to compare either the weights of atoms or the numbers of atoms that reacted with each other. Inspired by his knowledge of the gas laws, Avogadro proposed that equal volumes of all gases at the same temperature and pressure contained equal numbers of molecules. If the number of molecules is proportional to the volume of the gas, the numbers that react can be compared by comparing the volumes.

If $\quad\text{Number of A atoms} \propto \text{Volume of A}$

and $\quad\text{Number of B atoms} \propto \text{Volume of B}$

then
$$\frac{\text{Number of A atoms}}{\text{Number of B atoms}} = \frac{\text{Volume of A}}{\text{Volume of B}} \tag{5.2}$$

This suggestion was not immediately accepted as a stroke of genius because the volumes of products formed (and therefore the number of molecules) were twice as much as expected. Some elemental gases combined in the ratio of one volume each to give two volumes of products!

Dalton held that the elements consisted of individual atoms. In order for two molecules to be formed by combination of two atoms it is necessary to divide the atoms in some way, and that is contrary to his hypothesis that atoms are indivisible.

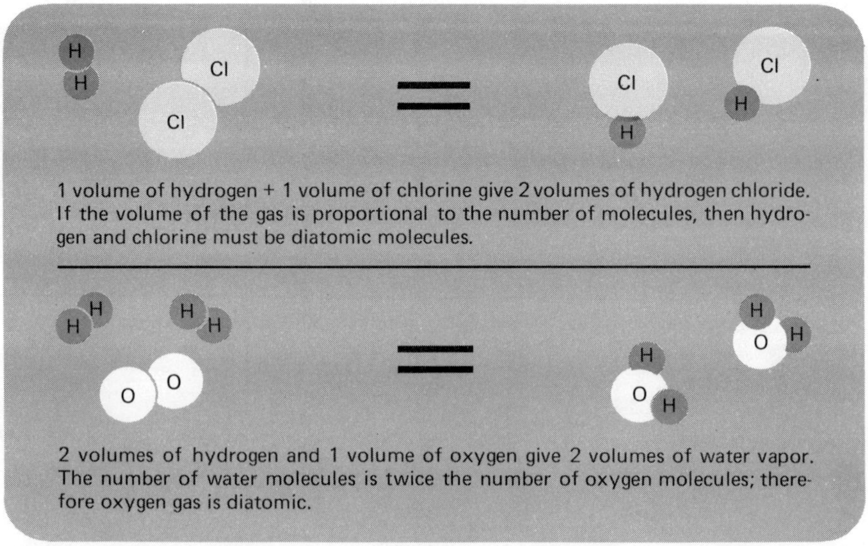

1 volume of hydrogen + 1 volume of chlorine give 2 volumes of hydrogen chloride. If the volume of the gas is proportional to the number of molecules, then hydrogen and chlorine must be diatomic molecules.

2 volumes of hydrogen and 1 volume of oxygen give 2 volumes of water vapor. The number of water molecules is twice the number of oxygen molecules; therefore oxygen gas is diatomic.

Figure 5.4

EXAMPLE 5.3

1. One volume of hydrogen and one volume of chlorine combine to form *two* volumes of hydrogen chloride. If N atoms of hydrogen combine with N atoms of chlorine the expected number is N hydrogen chloride molecules (2 atoms form 1 molecule), but the actual number calculated from the volume is $2N$. (See Figure 5.4.)
2. Two volumes of hydrogen combine with one volume of oxygen to give *two* volumes of water vapor. If $2N$ hydrogen atoms combine with N oxygen atoms the expected number is N water molecules (3 atoms become 1 molecule), but instead the number calculated from the volume is $2N$.

Stanislao Cannizzarro, a student of Amedeo Avogadro, explained this complication. The correct number of product molecules could be predicted if the elemental gases were not atoms, but diatomic molecules, two identical atoms bound together.

EXERCISE 5.1

From the facts that

1 volume of hydrogen and 1 volume of chlorine give
2 volumes of hydrogen chloride;

2 volumes of hydrogen and 1 volume of oxygen give
2 volumes of water vapor;

1 volume of oxygen and 1 volume of nitrogen give
2 volumes of nitric oxide;

3 volumes of hydrogen and 1 volume of nitrogen give
2 volumes of ammonia;

prove that hydrogen, oxygen, nitrogen, and chlorine are all diatomic.

From these facts it is now possible to determine the relative weight of a hydrogen atom to an oxygen atom. Hydrogen combines with oxygen in a *number ratio* of two hydrogens to one oxygen. The ratio of combining weights is one part hydrogen to eight parts oxygen. Therefore an oxygen atom weighs 16 times as much as a hydrogen atom.

$$\frac{8 \text{ parts oxygen}}{1 \text{ part hydrogen}} = \frac{1 \text{ O atom}}{2 \text{ H atoms}} \times \boxed{\frac{16 \text{ mass units per O}}{1 \text{ mass unit per H}}}$$

Atomic weights 5.4

For chemical purposes it is not really necessary to know the absolute weight of one atom. If one oxygen atom is 16 times as heavy as one hydrogen atom, a large number (say N) oxygen atoms will be 16 times as heavy as N hydrogen atoms. One gram of hydrogen therefore will have the same number of atoms as 16 grams of oxygen, and that is all we really need to know.

If we are going to talk about the weight of an atom, though, we need a unit that expresses the *idea* of the weight per atom. So without knowing size, we will speak of weight per atom in atomic mass units (amu). We can say, then, that an oxygen atom weighs 16 amu and a hydrogen atom weighs 1 amu.

The atomic weights of many of the other elements can also be found by comparing weights and volumes. A compound of hydrogen and chlorine establishes that chlorine is 35.5 times as heavy as hydrogen, so we will assign chlorine 35.5 amu per atom.

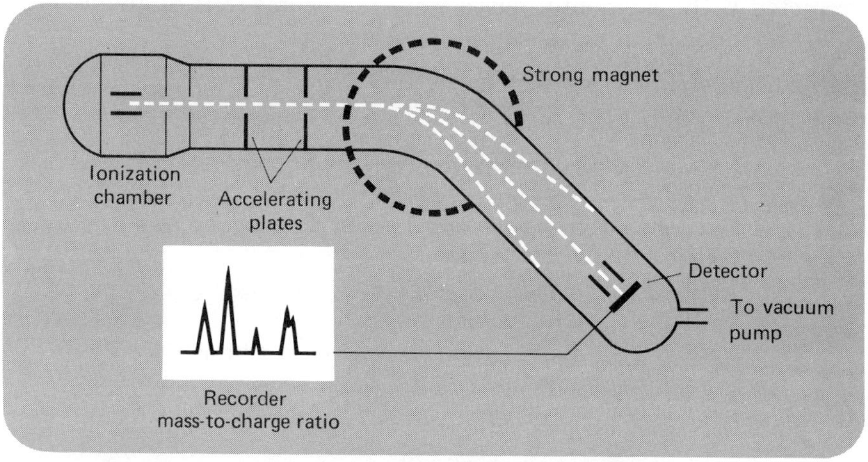

Figure 5.5 Schematic diagram of a mass spectrograph. Atoms and molecules are charged in the ionization chamber and formed into a beam by slits and accelerating plates. The magnet bends the beam in proportion to the mass/charge ratio of the particles.

Carbon is not a gas at any reasonable temperature, so we cannot measure the gas volume of carbon. We can examine gaseous compounds of carbon with oxygen, hydrogen, and chlorine, and we find that carbon always occurs in multiples of 12 amu in its compounds. Therefore carbon must weigh 12 amu per atom. In a similar fashion it was determined that nitrogen weighs 14 amu per atom, sulfur weighs 32 amu per atom, and so on.

A list of the relative weights of the elements has been established by assigning an amu value to one of them and then comparing the rest to that one. We used hydrogen as an example, but for physical reasons it was more convenient to use carbon as the reference atom. As presently *defined*, one atom of the most common isotope of carbon weighs precisely 12.0000 amu, and the rest of the elements are compared to that.

The relative masses of the atoms can be found by the stepwise process described, but a more modern technique compares the masses of the atoms directly. A *mass spectrograph* is a machine that separates and spreads out atoms and molecules according to their masses and the electrical charge on them. If the charge on them is known, the relative masses may be determined. The masses of the elements range from 1 amu for hydrogen to over 240 amu for the heaviest element.

The mole 5.5

Knowing that a carbon atom weighs 12 amu and an oxygen atom weighs 16 amu, we can confidently predict that 16 g of oxygen will have the same number (N) of oxygen atoms as there are carbon atoms in 12 g of carbon. The number (N) has been measured in a variety of experiments, and, as you might expect, it is very large. To three significant figures, there are 6.02×10^{23} atoms in 12.0 g of carbon. This is nearly a trillion trillion carbon atoms. The number is so big that it has a name to characterize it: it is Avogadro's Number or a *mole.*

The size of an atomic mass unit can now be calculated. We know that

$$\text{carbon} = 12\ \frac{\text{amu}}{\text{atom}}$$

$$\text{carbon} = 12\ \frac{\text{g}}{\text{mol}}$$

$$1\ \text{mol} = 6.02 \times 10^{23}\ \text{atoms}$$

$$12\ \frac{\text{amu}}{\text{atom}} = 12\ \frac{\text{g}}{\text{mol}} \times \frac{1\ \text{mol}}{6.02 \times 10^{23}\ \text{atoms}} = 1.99 \times 10^{-23}\ \frac{\text{g}}{\text{atom}}$$

$$1\ \text{amu} = \frac{1}{12}\ \text{the mass of a carbon atom} = 1.66 \times 10^{-24}\ \text{g}$$

Avogadro's Number and the mass of an amu are too big and too small, respectively, to work with, so we compare numbers of atoms in *moles.* A mole is the number of atoms that weigh an amount in grams equal to the atomic weight in amu. The molar atomic weight (in $\frac{\text{grams}}{\text{mole}}$) is the same number as the atomic weight in $\frac{\text{amu}}{\text{atom}}$.

Since the number is the same, it is usually not necessary to distinguish between molar atomic weight and the atomic weight. Since we will be dealing with weighable amounts of matter, we will use the units of grams per mole.

The atomic weight of an element is a characteristic property of the element. It can be used to convert weight to number (in moles) and vice versa. A table of atomic weights is on the inside cover of this book.

EXAMPLE 5.4

1. The atomic weight of magnesium is 24.3 g/mol.

   (a) How many moles are in 50.0 grams of magnesium?

$$50.0 \text{ g} \times \frac{1 \text{ mol}}{24.3 \text{ g}} = 2.06 \text{ mol of magnesium}$$

   (b) How many atoms are in 50.0 grams of magnesium?

$$2.06 \text{ mol} \times 6.02 \times 10^{23} \frac{\text{atoms}}{\text{mol}} = 1.24 \times 10^{24} \text{ atoms}$$

2. The atomic weight of nitrogen is 14 g/mol, but elemental nitrogen gas is diatomic. The molar weight of nitrogen gas is therefore 28 g/mol. How much does 2.0 moles of nitrogen gas weigh?

$$2.0 \text{ mol} \times 28 \frac{\text{g}}{\text{mol}} = 56 \text{ g}$$

5.6 *The parts of an atom*

Dalton thought of an atom as an indestructible sphere, possibly with hooks so it could hook onto other atoms. We have since found that atoms are not indestructible, not identical for the same element, and are not held together with hooks. In fact, the innermost workings of an atom are so complicated that the laws governing them are not all known. The outside workings of atoms are pretty well known and since that is what chemistry is concerned with, let's see what a typical atom is made of.

An atom is mostly empty space. Its mass is almost completely concentrated in a small structure in the center called the *nucleus.* The size of a nucleus is of the order of a few femtometers (10^{-15} meters) across, whereas the atom itself is some ten thousand times bigger. The nucleus is constructed of particles of almost equal mass called *protons* and *neutrons.* The mass of a single proton or neutron is almost exactly 1 atomic mass unit.[1] Protons have a unit positive electrical charge and the neutrons are not charged. Particles in the nucleus are held together by a poorly understood interaction called the strong force or nuclear force.

The number of protons in the nucleus of an atom determines the

[1] Collected together into nuclei the average mass may be more or less than 1 amu for reasons that pertain to nuclear energy.

identity of the element. An atom with one proton is hydrogen. An atom with six is carbon; with 92 uranium. The number of protons in the nucleus is called the *atomic number* and is symbolized by *Z*.

The number of neutrons in the nucleus is variable within a small range. That is, the ratio of protons to neutrons in the nucleus cannot exceed a certain range without having the nucleus fall apart. The nuclei of some elements do fall apart sometimes, and those elements are radioactive.

The sum of the protons and neutrons for a given atom is the *mass number,* symbolized by *A*. Elements with the same atomic number but different mass numbers are called *isotopes.* When it is important to specify which isotope of an element is being considered, the mass number is given after the name of the element or is written as a superscript to the left of the symbol.

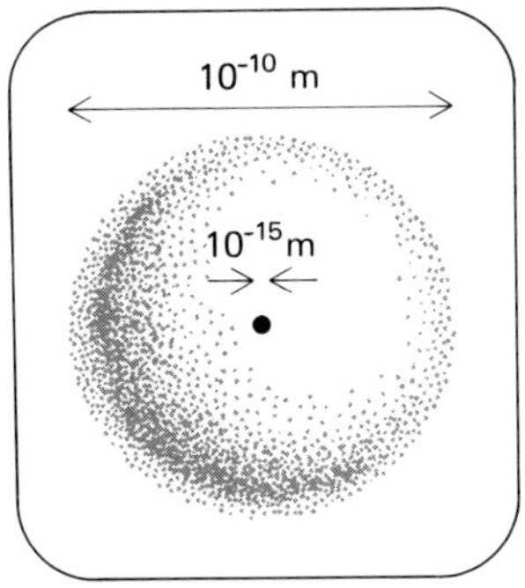

Figure 5.6 A way of visualizing a hydrogen atom.

EXAMPLE 5.5

carbon-12	^{12}C	the most common isotope of carbon
carbon-14	^{14}C	a radioactive isotope of carbon
cobalt-60	^{60}Co	a source of gamma radiation
uranium-235	^{235}U	the principal "fuel" in atomic reactors

The naturally occurring ratio of the isotopes of the elements is fairly uniform over the earth, because isotopes of the same element are almost identical chemically, and the process of mixing has been going on a long time. The *chemical atomic weight* is defined as the weight of one mole of atoms of an element, as its naturally occurring mixture of isotopes, compared to the weight of one mole of carbon atoms with the mass number of 12. The atomic weight is therefore a weighted average.

EXAMPLE 5.6

Suppose you had 500 one dollar bills and 100 five dollar bills. What is the average value of one of your bills? The average is simply the total value divided by the number of bills.

$$500 \text{ bills} \times \frac{\$1}{\text{bill}} = \$500$$

$$100 \text{ bills} \times \frac{\$5}{\text{bill}} = \$500$$

$$\text{Total:} \quad 600 \text{ bills} \qquad \$1000$$

$$\text{Average} = \frac{\$1000}{600 \text{ bills}} = \$1.67 \text{ per bill}$$

Obviously you have no bills with the value of \$1.67. The average value means that if you chose at random a large number, say 100, of bills from a bushel basket of ones and fives with that 5:1 ratio, you would more likely have \$167 than any other particular value.

EXERCISE 5.2 If 989 out of 1000 carbon atoms have a mass number of 12 and the remainder have a mass number of 13, calculate the average atomic weight of carbon.

Outside the nucleus and moving around it in a way that cannot be completely described are the *electrons.* Electrons have very little mass but they have a unit negative electrical charge. A neutral atom has equal numbers of electrons and protons.

Electrons and protons attract each other by means of the "coulombic force" or "electromagnetic force." This force can also be repulsive; particles with a charge of the same sign repel each other.

Table 5.1 summarizes the principal atomic particles.

5.7 *Symbols*

The names of the elements were conferred by their discoverers subject only to the rule that the names of the metals and semimetals have an -ium ending. The few exceptions go back to antiquity. The symbol for the element is the first letter of its name if that letter is not already used. If necessary the initial letter plus the second or subsequent letter is used. The initial letter is always capitalized; the second is always lower case.

A large part of learning any new branch of knowledge is becoming familiar with the symbolism and notation of that subject. A good notation has two objectives: 1. to communicate a complex idea in a concise and meaningful way; 2. to improve one's economy of thought. In order to communicate in chemistry you must learn to convert words and symbols into each other quickly and easily.

5.8 *The periodic table of the elements*

The periodic table is an array of the elements in order of their atomic numbers arranged in rows and columns so that elements in any column have similar chemical properties.

The rows are *chemical periods,* and they are numbered in arabic numerals from 1 through 7. The last member of a period is always a noble gas. Noble gases (helium, neon, argon, krypton, and xenon) are monatomic (single atom) gases with remarkably little chemical reactivity. The number of elements in a period increases in the order 2, 8, 8, 18, 18, 32, 17 . . . (the last period is incomplete). The most

Table 5.1

Principal atomic particles

Name	Symbol	Mass No.	Charge
proton	*p* or H^+	1	+1
neutron	*n*	1	0
electron	*e*	5×10^{-4}	−1

Table 5.2

Common elements and their symbols

Elements with single letter symbols					
B	Boron	I	Iodine	P	Phosphorus
C	Carbon	K*	Potassium	S	Sulfur
F	Fluorine	N	Nitrogen	U	Uranium
H	Hydrogen	O	Oxygen	V	Vanadium
Elements with double letter symbols					
Al	Aluminum	Ca	Calcium	Mn	Manganese
Ar	Argon	Cl	Chlorine	Ne	Neon
As	Arsenic	Cr	Chromium	Ni	Nickel
Ba	Barium	He	Helium	Si	Silicon
Be	Beryllium	Li	Lithium	Ti	Titanium
Br	Bromine	Mg	Magnesium	Zn	Zinc
Elements with symbols derived from Latin names					
Ag	Silver	Fe	Iron	Pb	Lead
Au	Gold	Hg	Mercury	Sb	Antimony
Cu	Copper	Na	Sodium	Sn	Tin

*Derived from Latin "kalium."

common modern representation of the table abstracts 15 members from the two long periods and prints them below the table.

The vertical columns in the table are *chemical groups.* The elements in a group chemically resemble each other more than they resemble members of adjacent groups. Groups are designated by roman numerals in two series, A and B. A few groups have such unique chemical properties that they have acquired group names. Group IA metals (excluding hydrogen) are called the alkali metals. Group IIA metals are the alkaline earth metals. On the other side of the table Group VIIA are the halogens and Group VIIIA are the noble gases.

Many of the physical and chemical properties of the elements can be deduced from their position in the periodic table. To be able to do this we need to know the reasons and laws governing its construction, and these are taken up in the next two chapters. For now it is sufficient to recognize that elements on the left side of the table are strongly metallic in character and those on the right side are nonmetallic. It is also useful to remember that what is true about one member of a group is usually true about the others in the same group.

Table 5.3

The periodic table of the elements

Groups

Periods	IA	IIA	IIIB	IVB	VB	VIB	VIIB	VIIIB			IB	IIB	IIIA	IVA	VA	VIA	VIIA	VIIIA
1	1 H																	2 He
2	3 Li	4 Be											5 B	6 C	7 N	8 O	9 F	10 Ne
3	11 Na	12 Mg											13 Al	14 Si	15 P	16 S	17 Cl	18 Ar
4	19 K	20 Ca	21 Sc	22 Ti	23 V	24 Cr	25 Mn	26 Fe	27 Co	28 Ni	29 Cu	30 Zn	31 Ga	32 Ge	33 As	34 Se	35 Br	36 Kr
5	37 Rb	38 Sr	39 Y	40 Zr	41 Nb	42 Mo	43 Tc	44 Ru	45 Rh	46 Pd	47 Ag	48 Cd	49 In	50 Sn	51 Sb	52 Te	53 I	54 Xe
6	55 Cs	56 Ba		72 Hf	73 Ta	74 W	75 Re	76 Os	77 Ir	78 Pt	79 Au	80 Hg	81 Tl	82 Pb	83 Bi	84 Po	85 At	86 Rn
7	87 Fr	88 Ra																

57 La	58 Ce	59 Pr	60 Nd	61 Pm	62 Sm	63 Eu	64 Gd	65 Tb	66 Dy	67 Ho	68 Er	69 Tm	70 Yb	71 Lu
89 Ac	90 Th	91 Pa	92 U	93 Np	94 Pu	95 Am	96 Cm	97 Bk	98 Cf	99 Es	100 Fm	101 Md	102 No	103 Lr

EXERCISE 5.3

1. Which of these elements belong to the same group?

C, N, Al, P, Cl

2. Which of these elements are in the same period?

B, H, Si, Mg, S

3. Which of the following would be expected to have similar chemical properties?

Atomic numbers 3, 9, 12, 17, 35

Formulas 5.9

The word "formula" in chemistry specifically refers to a symbolic method for expressing the relationships of elements in a compound. It does *not* refer to mixtures of things like a baby's formula or a pharmaceutical formula. Those mixtures are formulations.

The *empirical formula* gives the identity and relative number of moles of elements in a compound. The elemental composition is specified by writing the symbols of the elements next to each other with the more metallic element first. In general that means that those elements to the left and below the others in the periodic table will be written first.

EXAMPLE 5.7

1. The formula for table salt (sodium chloride) is

NaCl

Sodium is to the left of chlorine in the periodic table (more metallic) so it is written first.

2. The formula for quicklime (calcium oxide) is

CaO

Calcium is more metallic than oxygen and is written first.

The placement of hydrogen in a formula is a special case. In binary compounds (those composed of only two elements) hydrogen is customarily written after all elements in Groups IA through VA and ahead of elements in Groups VIA and VIIA.

EXAMPLE 5.8

1. methane is CH_4	BUT	4. water is H_2O
2. ammonia is NH_3		5. hydrogen sulfide is H_2S
3. phosphine is PH_3		6. hydrogen chloride is HCl

The relative number of moles of each element is indicated by a subscript numeral to the right of the symbol. Since fractional atoms do not occur in compounds the number is always a whole number.[2]

EXAMPLE 5.9

The formula for aluminum oxide is Al_2O_3. The mole ratio of oxygen to aluminum is 3:2 in the compound. We do *not* write $AlO_{1.5}$.

Many compounds consist of radicals or groups of atoms that behave as a single entity in combination with other atoms. A group occurring more than once in a formula unit is enclosed in parentheses with a subscript numeral after the right-hand parenthesis to indicate the relative number of groups.

EXAMPLE 5.10

1. The formula of aluminum hydroxide is $Al(OH)_3$. The OH group is taken 3 times, meaning there are 3 O's and 3 H's in one formula unit.
2. The formula of ammonium sulfate is $(NH_4)_2SO_4$. The NH_4 group is taken 2 times, meaning there are 2 N's and 8 H's in one formula unit.

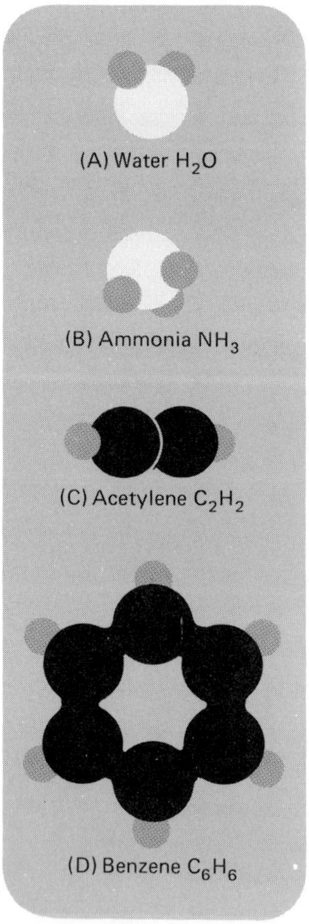

Figure 5.7 Molecular compounds.

5.10 *Molecular formulas*

Some substances such as those in Example 5.8 and Figure 5.7 exist as discrete molecules in which the atoms remain more or less permanently bound to the same partners. Other substances like sodium chloride and calcium oxide are not bound to a particular partner but instead form a lattice of charged atoms or groups of atoms (ions) held together by electrical forces (Figure 5.8). Electrical neutrality is maintained by equal numbers of positive and negative charges.

In either case the empirical formula does not necessarily indicate the size of the molecule or formula unit. The *molecular formula* describes the minimum number of atoms that behave as a discrete unit in the compound. The molecular formula cannot be obtained from knowledge of the composition alone. It is necessary also to find the molar weight by some physical means.

[2] In some crystalline compounds the mole ratio departs slightly from a ratio of small whole numbers, and this gives the crystal some interesting physical properties. Discussion of these compounds, however, is beyond the scope of this book.

EXAMPLE 5.11

1. The empirical formula of acetylene and benzene is CH. The molecular formula of acetylene is C_2H_2 and that of benzene is C_6H_6. The mole ratio of carbon to hydrogen is the same but they occur in different sized units and are bound together differently.
2. The empirical formula of mercurous chloride (calomel) is HgCl. Two mercury atoms are bound together in the compound, so the formula unit is Hg_2Cl_2.

Notice that the molecular formula is always a whole number multiple of the empirical formula.

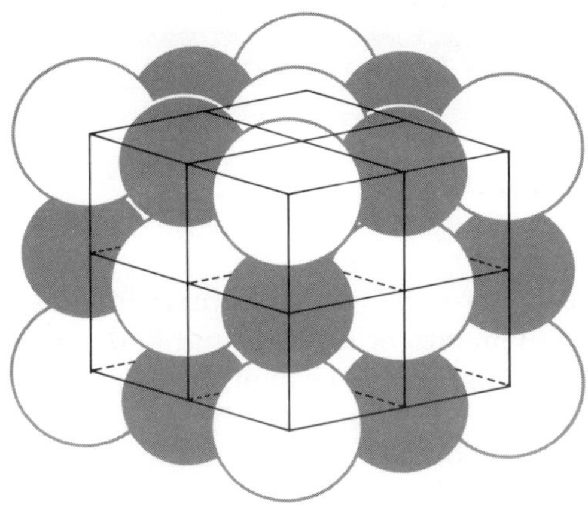

Figure 5.8 Sodium chloride.

Molar weights 5.11

When elements or compounds combine to form new compounds there is no change of mass. It follows then that one mole of a compound will weigh as much as all the moles of atoms in it. If the molecular formula is known, the molar weight can be found by summing the atomic weights of all the atoms in the compound.

EXAMPLE 5.12

1. The molecular formula of carbon dioxide is CO_2. One mole of CO_2 must weigh as much as one mole of carbon plus two moles of oxygen atoms.

$$1\ \text{mol(C)} \times 12\,\frac{\text{g}}{\text{mol(C)}} = 12\ \text{g}$$

$$2\ \text{mol(O)} \times 16\,\frac{\text{g}}{\text{mol(O)}} = 32\ \text{g}$$

$$\overline{1\ \text{mol(CO}_2)} \qquad = \overline{44\ \text{g}}$$

2. The formula of baking soda (sodium bicarbonate) is $NaHCO_3$. From a table of atomic weights we know

$$\mathrm{Na} = 23\ \mathrm{g/mol} \qquad \mathrm{C} = 12\ \mathrm{g/mol}$$
$$\mathrm{H} = 1\ \mathrm{g/mol} \qquad \mathrm{O} = 16\ \mathrm{g/mol}$$

The molar weight is the sum of the atomic weights of the *six* atoms in one formula unit of $NaHCO_3$.

$$\begin{array}{llcl}
1\ \mathrm{mol(Na)} & \times\ 23\ \mathrm{g/mol} & = & 23\ \mathrm{g} \\
1\ \mathrm{mol(H)} & \times\ \ 1\ \mathrm{g/mol} & = & 1\ \mathrm{g} \\
1\ \mathrm{mol(C)} & \times\ 12\ \mathrm{g/mol} & = & 12\ \mathrm{g} \\
\underline{3\ \mathrm{mol(O)}} & \times\ 16\ \mathrm{g/mol} & = & \underline{48\ \mathrm{g}} \\
1\ \mathrm{mol(NaHCO_3)} & & = & 84\ \mathrm{g}
\end{array}$$

The molar weight of $NaHCO_3$ is 84 g/mol.

5.12 *Percent composition*

It is often necessary to know the weight/weight relationships that exist between the elements of a compound and the compound itself. In the first place, the weights of each of the elements compared to the weight of the compound must be found by experiment in order to find the formula of the compound (Section 5.13). Second, if the formula is known, the weight of each element in a compound can be calculated from the weight of the compound and the weight/weight relationship.

For example, if we know by experiment that 3 grams of carbon can be converted chemically to 11 grams of carbon dioxide, no more, no less, then the weight fraction of carbon to carbon dioxide is $\frac{3}{11}$. This is usually expressed as percent composition by multiplying the decimal fraction by 100:

$$\frac{3}{11} = 0.273 \times 100 = 27.3\%$$

This information can be used to find the mole ratio of carbon and oxygen in the compound. The weight/weight relationship or percent composition also tells us that there are 27.3 grams of carbon in 100 grams of carbon dioxide.

If the formula is known, the weight/weight relationship can be found by dividing the weight of the element by the weight of the compound. The weight of the element is the number of moles times the atomic weight. The weight of the compound is its molar weight. The formula tells us the number of moles of each element in one mole of compound.

$$\frac{\text{weight of element}}{\text{weight of compound}} = \frac{\text{moles of element}}{\text{1 mole compound}} \times \frac{\text{atomic weight}}{\text{molar weight}} \qquad (5.3)$$

EXAMPLE 5.13

What is the (weight) percent of aluminum in aluminum oxide, Al_2O_3?

Want: $\frac{\text{weight of Al}}{\text{weight of } Al_2O_3} \times 100$

Know: **(a)** From formula, there are 2 moles Al per mole Al_2O_3
(b) The weight per mole of Al is 27 g/mol (look it up)
(c) The weight per mole of Al_2O_3 is 102 g/mol (add $2 \times 27 + 3 \times 16$)

Solution: $\% = \frac{2 \text{ mol(Al)}}{1 \text{ mol}(Al_2O_3)} \times \frac{27 \text{ g/mol(Al)}}{102 \text{ g/mol}(Al_2O_3)} \times 100 = 53\%$

Since there are six variables in equation 5.3 we could write countless variations on the same equation and bore you to death. That course of action is neither desirable nor necessary. To solve any problem in composition it is only necessary to know that

1. The *atomic weight* is the relationship between weight and moles of an element;
2. The *molar weight* is the relationship between weight and moles of a compound, and is the sum of the atomic weights of all the atoms in the compound;
3. The formula gives the mole/mole relationship of the elements in the compound.

EXAMPLE 5.14

How much iron (Fe) is in 1000 g of magnetic iron oxide, Fe_3O_4?

Want: weight of iron in grams

Have: 1000 grams of iron oxide, Fe_3O_4

Know: **(a)** From the formula, there are 3 moles Fe per mole Fe_3O_4
(b) From atomic weight table, Fe = 55.8 g/mol
O = 16.0 g/mol
(c) By addition, Fe_3O_4 = 231.4 g/mol

We want a quantity of iron; we start with a quantity of iron oxide and multiply by the relationships we know. The only relationship we know between Fe and Fe_3O_4 is the mole/mole ratio, so we must first convert weight of Fe_3O_4 to moles. We can then convert moles (Fe_3O_4) to moles (Fe) and thence to grams of Fe.

Solution:

$$1000 \text{ g} \times \frac{1 \text{ mol}(Fe_3O_4)}{231.4 \text{ g}} \times \frac{3 \text{ mol(Fe)}}{1 \text{ mol}(Fe_3O_4)} \times 55.8 \frac{\text{g}}{\text{mol(Fe)}} = 723 \text{ g}$$

5.13 *Calculation of empirical formulas*

The empirical formula of a compound only gives two pieces of information, the identity and the relative number of moles of elements in a compound. The identity is found by a process called "qualitative analysis" in the laboratory. The only other problem is to use whatever information is available to *find the number of moles* of each element in the compound, and then *compare the moles* so that their ratios are in the ratio of small whole numbers.

Information about the number of moles of each element in a sample of the compound can come in a variety of ways. If the weight of each element is known, the weight can be converted to moles using the atomic weight property.

EXAMPLE 5.15 It is found by experiment that 0.1215 gram of magnesium combines with oxygen to form 0.2015 gram of a white powder. What is the empirical formula of the product?

(a) We want to compare the moles of Mg and O.

(b) We have 0.1215 g of Mg and by difference, 0.2015 g–0.1215 g = 0.0800 g of oxygen.

(c) We know the relationship between weight and moles of an element is the atomic weight. We find

$$\text{Mg} = 24.3 \text{ g/mol} \qquad \text{O} = 16.0 \text{ g/mol}$$

$$\text{moles (Mg)} = 0.1215 \text{ g} \times \frac{1 \text{ mol(Mg)}}{24.3 \text{ g}} = 5.00 \times 10^{-3} \text{ mol(Mg)}$$

$$\text{moles (O)} = 0.0800 \text{ g} \times \frac{1 \text{ mol(O)}}{16.0 \text{ g}} = 5.00 \times 10^{-3} \text{ mol(O)}$$

(d) Compare the moles: $\frac{5.00 \times 10^{-3}\text{(Mg)}}{5.00 \times 10^{-3}\text{(O)}} = \frac{1 \text{ mol(Mg)}}{1 \text{ mol(O)}}$

The empirical formula is MgO.

EXAMPLE 5.16 A laboratory analysis of nicotine gives 74.0% carbon, 8.65% hydrogen, and 17.3% nitrogen. What is the empirical formula? Percent means parts (by weight) per hundred parts; in 100 g of nicotine there are 74.0 g of carbon, 8.65 g of hydrogen, and 17.3 g of nitrogen. Since we want to compare the moles we can compare the moles in any amount of the compound; one hundred grams is a convenient amount.

(a) Compare the moles of C, H, and N in 100 g of nicotine.

(b) Have: 74.0 g(C)
8.65 g(H)
17.3 g(N)

(c) Know: $12.0 \frac{\text{g}}{\text{mol(C)}}$; $1.0 \frac{\text{g}}{\text{mol(H)}}$; $14.0 \frac{\text{g}}{\text{mol(N)}}$

(d) Find the moles: $74.0 \text{ g} \times \frac{1 \text{ mol(C)}}{12.0 \text{ g}} = 6.16 \text{ mol(C)}$

$$8.65 \text{ g} \times \frac{1 \text{ mol(H)}}{1.0 \text{ g}} = 8.65 \text{ mol(H)}$$

$$17.3 \text{ g} \times \frac{1 \text{ mol(N)}}{14.0 \text{ g}} = 1.24 \text{ mol(N)}$$

(e) Compare the moles: On the proposition that one mole of compound must contain at least one mole of every element, we divide by the smallest number of moles:

$$\frac{6.16 \text{ mol(C)}}{1.24 \text{ mol(N)}} = \frac{5 \text{ (C)}}{1 \text{ (N)}}; \frac{8.65 \text{ mol(H)}}{1.24 \text{ mol(N)}} = \frac{7 \text{ (H)}}{1 \text{ (N)}}$$

The empirical formula is C_5H_7N.

EXAMPLE 5.17

A laboratory analysis of a compound containing only carbon and hydrogen gave 81.8% carbon and 18.2% hydrogen. What is the empirical formula?

(a) We want to compare the moles of carbon and hydrogen in 100 g of compound.

(b) We have 81.8 g of carbon and 18.2 g of hydrogen.

(c) We know the atomic weights: $12 \frac{\text{g}}{\text{mol(C)}}$; $1.0 \frac{\text{g}}{\text{mol(H)}}$

(d) Find the moles: $81.8 \text{ g} \times \frac{1 \text{ mol(C)}}{12.0 \text{ g}} = 6.82 \text{ mol(C)}$

$$18.3 \text{ g} \times \frac{1 \text{ mol(H)}}{1.0 \text{ g}} = 18.3 \text{ mol(H)}$$

(e) Compare the moles: $\frac{18.3 \text{ mol(H)}}{6.82 \text{ mol(C)}} = \frac{2.68 \text{ mol(H)}}{1.00 \text{ mol(C)}}$

We *cannot* write the formula as $CH_{2.68}$; the moles must be in the ratio of small *whole* numbers. The easiest way to find that ratio is to observe that 2.68 is close to $2\frac{2}{3}$ or $\frac{8}{3}$. The ratio of hydrogen to carbon is then seen to be $\frac{8}{3}$ mole of hydrogen to 1 mole of carbon, or 8 moles of hydrogen to 3 moles of carbon. The empirical formula is C_3H_8.

Summary 5.14

1. Compounds are formed from atoms of the elements combined in the ratio of small whole numbers.

2. The relative masses of the elements (atomic weights) can be found if the number ratio and the weight ratio of elements which combine are known:

$$\frac{\text{weight of A}}{\text{weight of B}} = \frac{\text{no. of units of A}}{\text{no. of units of B}} \times \boxed{\frac{\text{weight per unit of A}}{\text{weight per unit of B}}}$$

3. A mole is 6.02×10^{23} atoms, molecules, or formula units of a pure substance. A mole of atoms weighs as much in grams as the single atom in amu.
4. The sum of the atomic weights (in grams per mole) of all the atoms in a compound is the molar weight of the compound.
5. The formula of a compound gives the identity of the substituent elements as their symbols and the relative number of each as subscript numerals.
6. The weight fraction of the elements in a compound can be calculated if the mole ratio (from the formula) and the atomic weights of the elements are known:

$$\boxed{\frac{\text{wt of element}}{\text{wt of compound}}} = \frac{\text{no. of moles element}}{\text{1 mole compound}} \times \frac{\text{wt per mole element}}{\text{wt per mole compound}}$$

7. The relative numbers of moles of elements in a compound can be found if the weight fraction of each element and the atomic weights are known:

$$\frac{\text{weight of A}}{\text{weight of B}} = \boxed{\frac{\text{no. of moles of A}}{\text{no. of moles of B}}} \times \frac{\text{wt per mole A}}{\text{wt per mole B}}$$

EXERCISES

I. Define the following terms.

1. compound
2. element
3. mole
4. atomic weight
5. nucleus
6. proton
7. neutron
8. electron
9. atomic number
10. mass number
11. isotope
12. chemical period
13. chemical group
14. empirical formula
15. molecular formula
16. molar weight

SOLVED PROBLEMS

II. Fill in the blanks.

	Pure Substance	Weight(g)	Moles	Molar Weight(g/mol)
1.	hydrogen H_2	1.0	?	2.0
2.	potassium K	0.78	?	39.1
3.	carbon C	6.0	0.50	?
4.	magnesium Mg	3.00	0.123	?
5.	oxygen O_2	?	0.10	32.0
6.	copper Cu	?	0.25	63.5
7.	ammonia NH_3	1.7	0.10	?
8.	toluene C_7H_8	?	1.50	92.0
9.	propane C_3H_8	0.88	?	44.0
10.	butane C_4H_{10}	1000	?	58.0

III. Find the molar weight from the formula. Atomic weights are listed on the inside front cover.

		Formula
1.	hydrogen chloride	HCl
2.	nitric acid	HNO_3
3.	copper(II) sulfate	$CuSO_4$
4.	iron(III) nitrate	$Fe(NO_3)_3$
5.	calcium hydroxide	$Ca(OH)_2$

IV. Calculate the weight percent composition of

1. ammonium nitrate NH_4NO_3
2. potassium chlorate $KClO_3$
3. sodium bicarbonate $NaHCO_3$
4. iron(III) oxide Fe_2O_3
5. cellulose $(C_6H_{10}O_5)_n$
where n is a large number

V. Calculate the empirical formula

1. 27.3% C; 72.7% O
2. 56.0% N; 12.0% H; 32.0% O
3. 31.5% C; 5.3% H; 63.2% O
4. 32.6% S; 2.1% H; 65.3% O
5. 27.7% Mg; 23.6% P; 48.7% O

VI. Word Problems

1. How much copper is in 100 grams of copper oxide, CuO?
2. How much aluminum can be obtained from 1000 grams of aluminum oxide, Al_2O_3?
3. What weight of carbon is necessary to make 1 pound (454 g) of carbon dioxide, CO_2?

4. How much copper can be obtained from 2.0 grams of copper sulfate, $CuSO_4$?
5. How many grams of nitrogen is in 5.0 pounds of ammonium nitrate, NH_4NO_3?
6. How much bromine is in 500 grams of salt solution which is 2.0% sodium bromide, NaBr?
7. How much sodium fluoride, NaF, is in 20,000 gallons of drinking water if the weight fraction of fluoride to water is one part fluoride to 10^6 parts of water (1 ppm)? The density of water is 3.79 kilograms per gallon.
8. 1.500 grams of a compound containing only titanium, Ti, and chlorine react with sodium to give 0.378 gram of titanium metal. What is the empirical formula of the compound?
9. A sample of a hydrate of barium chloride was decomposed by heat to give 3.41 grams of $BaCl_2$ and 0.59 gram of water vapor. What is the mole ratio of water to barium chloride, $BaCl_2$?
10. A 0.500 gram sample of brass was chemically treated to obtain all of the copper in it as copper phosphate, $Cu_3(PO_4)_2$. The copper phosphate weighed 0.799 gram. What was the percent copper in brass?

ADDITIONAL EXERCISES

VII. Fill in the blanks.

Pure Substance	Weight(g)	Moles	Molar Weight(g/mol)
1. sodium Na	0.50	?	23.0
2. sulfur S	?	0.20	32.0
3. sodium sulfate Na_2SO_4	0.71	0.005	?
4. potassium carbonate K_2CO_3	4.14	?	138.2
5. sucrose $C_{12}H_{22}O_{11}$	?	0.25	342

VIII. Find the molar weight from the formula.

Name	Formula
1. sodium dihydrogen phosphate	NaH_2PO_4
2. sodium thiosulfate	$Na_2S_2O_3$
3. acetic acid	$C_2H_4O_2$
4. ethanol	C_2H_6O
5. benzene	C_6H_6
6. aluminum sulfate nonahydrate	$Al_2(SO_4)_3(H_2O)_9$

IX. Calculate the percent composition.

1. acetic acid $C_2H_4O_2$
2. ethanol C_2H_6O
3. ammonium sulfate $(NH_4)_2SO_4$
4. urea CH_4N_2O
5. aspirin $NaC_9H_7O_4$

X. Calculate the empirical formula.

1. 31.0% B; 69.0% O

2. 54.1% Ca; 2.7% H; 43.2% O

3. 50.0% S; 50.0% O

4. 52.9% Al; 47.1% O

5. 26.6% K; 35.4% Cr; 38.0% O

XI. Calculate the

1. weight of hydrogen in 50 grams of ammonia, NH_3.

2. weight of lead in 80 grams of lead sulfate, $PbSO_4$.

3. weight of water in 10 grams of sodium sulfate decahydrate, $Na_2SO_4(H_2O)_{10}$.

4. weight of silver in 100 grams of a solution which is 1% silver nitrate, $AgNO_3$.

5. percent by weight of mercury in 10.0 g of an ore which contains 1.50 grams of mercury sulfide, HgS.

6. weight of lead exhausted into the atmosphere when one million (10^6) gallons of gasoline containing 4.0 milliliters of tetraethyl lead, $Pb(C_2H_5)_4$, per gallon are burned. The density of tetraethyl lead is 1.66 g/ml.

SOLUTIONS

II.

1. $1.0\ g \times \frac{1\ mol(H_2)}{2.0\ g} = 0.50\ mol\ H_2$

2. $0.78\ g \times \frac{1\ mol(K)}{39.1\ g} = 0.020\ mol\ K$

3. $MW\ \frac{g}{mol} = \frac{6.0\ g}{0.50\ mol} = 12\ \frac{g}{mol(C)}$

4. $MW\ \frac{g}{mol} = \frac{3.00\ g}{0.123\ mol(Mg)} = 24.4\ \frac{g}{mol(Mg)}$

5. $weight\ g = 0.10\ mol(O_2) \times 32.0\ \frac{g}{mol(O_2)} = 3.2\ g$

6. $weight\ g = 0.25\ mol(Cu) \times 63.5\ \frac{g}{mol(Cu)} = 15.9\ g$

7. $MW\ \frac{g}{mol} = \frac{1.7\ g}{0.10\ mol(NH_3)} = 17\ \frac{g}{mol(NH_3)}$

8. $weight\ g = 1.50\ mol(C_7H_8) \times 92.0\ \frac{g}{mol(C_7H_8)} = 138\ g$

9. $mol(C_3H_8) = 0.88\ g \times \frac{1\ mol(C_3H_8)}{44.0\ g} = 0.02\ mol$

10. $mol(C_4H_{10}) = 1000\ g \times \frac{1\ mol(C_4H_{10})}{58.0\ g} = 17.2\ mol$

III. **1.**
$$\begin{array}{r@{\;=\;}r} \text{1 mol H} & \text{1.0 g} \\ \text{1 mol Cl} & \text{35.5 g} \\ \hline \text{1 mol HCl} & \text{36.5 g} \end{array}$$

2.
$$\begin{array}{r@{\;=\;}r} \text{1 mol H} & \text{1.0 g} \\ \text{1 mol N} & \text{14.0 g} \\ \text{3 mol O} & \text{48.0 g} \\ \hline \text{1 mol HNO}_3 & \text{63.0 g} \end{array}$$

3.
$$\begin{array}{r@{\;=\;}r} \text{1 mol Cu} & \text{63.5 g} \\ \text{1 mol S} & \text{32.0 g} \\ \text{4 mol O} & \text{64.0 g} \\ \hline \text{1 mol CuSO}_4 & \text{159.5 g} \end{array}$$

4. The subscript 3 after the parentheses means that the whole group (NO_3) occurs 3 times in each mole of compound. Everything inside the parentheses is multiplied by 3.

$$\begin{array}{r@{\;=\;}r} \text{1 mol Fe} & \text{55.8 g} \\ \text{3 mol N} & \text{42.0 g} \\ \text{9 mol O} & \text{144.0 g} \\ \hline \text{1 mol Fe(NO}_3)_3 & \text{241.8 g} \end{array}$$

5.
$$\begin{array}{r@{\;=\;}r} \text{1 mol Ca} & \text{40.0 g} \\ \text{2 mol O} & \text{32.0 g} \\ \text{2 mol H} & \text{2.0 g} \\ \hline \text{1 mol Ca(OH)}_2 & \text{74.0 g} \end{array}$$

IV. **1.**
$$\begin{array}{r@{\;=\;}r} \text{2 mol N} & \text{28.0 g} \\ \text{4 mol H} & \text{4.0 g} \\ \text{3 mol O} & \text{48.0 g} \\ \hline \text{1 mol NH}_4\text{NO}_3 & \text{80.0 g} \end{array}$$

$$\%N = \frac{28.0}{80.0} \times 100 = 35.0\%$$
$$\%H = \frac{4.0}{80.0} \times 100 = 5.0\%$$
$$\%O = \frac{48.0}{80.0} \times 100 = 60.0\%$$

2.
$$\begin{array}{r@{\;=\;}r} \text{1 mol K} & \text{39.1 g} \\ \text{1 mol Cl} & \text{35.5 g} \\ \text{3 mol O} & \text{48.0 g} \\ \hline \text{1 mol KClO}_3 & \text{122.6 g} \end{array}$$

$$\%K = \frac{39.1}{122.6} \times 100 = 31.9\%$$
$$\%Cl = \frac{35.5}{122.6} \times 100 = 29.0\%$$
$$\%O = \frac{48.0}{122.6} \times 100 = 39.1\%$$

3.
$$\begin{array}{r@{\;=\;}r} \text{1 mol Na} & \text{23.0 g} \\ \text{1 mol H} & \text{1.0 g} \\ \text{1 mol C} & \text{12.0 g} \\ \text{3 mol O} & \text{48.0 g} \\ \hline \text{1 mol NaHCO}_3 & \text{84.0 g} \end{array}$$

$$\%Na = \frac{23.0}{84.0} \times 100 = 27.4\%$$
$$\%H = \frac{1.0}{84.0} \times 100 = 1.2\%$$
$$\%C = \frac{12.0}{84.0} \times 100 = 14.3\%$$
$$\%O = \text{remainder} = 57.1\%$$

4.
$$\begin{array}{r@{\;=\;}r} \text{2 mol Fe} & \text{111.6 g} \\ \text{3 mol O} & \text{48.0 g} \\ \hline \text{1 mol Fe}_2\text{O}_3 & \text{159.6 g} \end{array}$$

$$\%Fe = \frac{111.6}{159.6} \times 100 = 69.9\%$$
$$\%O = \text{remainder} = 30.1\%$$

5. The value of *n* does not affect the composition.

$C_6H_{10}O_5 = 162$ g/mol

$$\%C = \frac{6 \times 12}{162} \times 100 = 44.4\%$$
$$\%H = \frac{10 \times 1}{162} \times 100 = 6.2\%$$
$$\%O = \text{remainder} = 49.4\%$$

V. **1.** Compare the moles of C and O in 100 g of compound.

Have: 27.3 g carbon; 72.7 g oxygen

Know: C = 12 g/mol; O = 16 g/mol

$$\text{mol C} = 27.3\text{ g} \times \frac{1\text{ mol C}}{12.0\text{ g}} = 2.28\text{ mol(C)}$$

$$\text{mol O} = 72.7\text{ g} \times \frac{1\text{ mol O}}{16.0\text{ g}} = 4.54\text{ mol (O)}$$

Compare the moles: $\frac{4.54\text{ mol O}}{2.28\text{ mol C}} = \frac{2\text{ mol O}}{1\text{ mol C}}$

The formula is CO_2.

2. **Have:** 56 g of N; 12.0 g of H; 32.0 g of O

Know: N = 14 g/mol; H = 1 g/mol; O = 16 g/mol

$$\text{mol N} = 56\text{ g} \times \frac{1\text{ mol N}}{14.0\text{ g}} = 4.0\text{ mol N}$$

$$\text{mol H} = 12\text{ g} \times \frac{1\text{ mol H}}{1.0\text{ g}} = 12.0\text{ mol H}$$

$$\text{mol O} = 32\text{ g} \times \frac{1\text{ mol O}}{16.0\text{ g}} = 2.0\text{ mol O}$$

Compare the moles: $\frac{4.00\text{ mol N}}{2.00\text{ mol O}} = \frac{2\text{ N}}{1\text{ O}};\ \frac{12.0\text{ H}}{2.0\text{ O}} = \frac{6\text{ H}}{1\text{ O}}$

The formula is N_2H_6O.

3. $\text{mol C} = 31.5\text{ g} \times \frac{1\text{ mol C}}{12.0\text{ g}} = 2.63\text{ mol C}$ $\qquad \frac{3.95\text{ O}}{2.63\text{ C}} = \frac{1.5\text{ O}}{1\text{ C}} = \frac{3\text{ O}}{2\text{ C}}$

$\text{mol H} = 5.3\text{ g} \times \frac{1\text{ mol H}}{1.0\text{ g}} = 5.3\text{ mol H}$

$\text{mol O} = 63.2\text{ g} \times \frac{1\text{ mol O}}{16.0\text{ g}} = 3.95\text{ mol O}$ $\qquad \frac{5.3\text{ H}}{2.63\text{ C}} = \frac{2\text{ H}}{1\text{ C}}$

The formula is $C_2H_4O_3$.

4. $32.6\text{ g} \times \frac{1\text{ mol S}}{32.0\text{ g}} = 1.02\text{ mol S}$ $\qquad \frac{4.08\text{ mol O}}{1.02\text{ mol S}} = \frac{4\text{ O}}{1\text{ S}}$

$2.1\text{ g} \times \frac{1\text{ mol H}}{1.0\text{ g}} = 2.1\text{ mol H}$ $\qquad \frac{2.1\text{ mol H}}{1.02\text{ mol S}} = \frac{2\text{ H}}{1\text{ S}}$

$65.3\text{ g} \times \frac{1\text{ mol O}}{16.0\text{ g}} = 4.08\text{ mol O}$

The formula is H_2SO_4

5. $27.7\text{ g} \times \frac{1\text{ mol Mg}}{24.3\text{ g}} = 1.14\text{ mol Mg}$ $\qquad \frac{1.14\text{ Mg}}{0.76\text{ P}} = \frac{1.50\text{ Mg}}{1\text{ P}} = \frac{3\text{ Mg}}{2\text{ P}}$

$23.6\text{ g} \times \frac{1\text{ mol P}}{31.0\text{ g}} = 0.76\text{ mol P}$ $\qquad \frac{3.04\text{ O}}{0.76\text{ P}} = \frac{4\text{ O}}{1\text{ P}}$

$48.7\text{ g} \times \frac{1\text{ mol O}}{16.0\text{ g}} = 3.04\text{ mol O}$

The formula is $Mg_3P_2O_8$.

VI. In this and subsequent solution sets we will adopt a convention that the symbol or formula will stand for *one mole* of the substance so identified.

1. Want: weight of copper

Have: 100 g of copper oxide, CuO

Know: **(a)** there is one mole of Cu per mole of CuO: $\frac{\text{Cu}}{\text{CuO}}$

(b) atomic wt of copper: $63.5\ \frac{\text{g}}{\text{Cu}}$

(c) molar wt of copper oxide: $79.5\ \frac{\text{g}}{\text{CuO}}$

Solution: $100\text{ g} \times \frac{1\text{ CuO}}{79.5\text{ g}} \times \frac{1\text{ Cu}}{\text{CuO}} \times \frac{63.5\text{ g}}{\text{Cu}} = 79.9\text{ g of copper}$

2. Want: weight of aluminum

Have: 1000 g of aluminum oxide, Al_2O_3

Know: $\frac{2\text{ Al}}{\text{Al}_2\text{O}_3}$; $27\ \frac{\text{g}}{\text{Al}}$; $102\ \frac{\text{g}}{\text{Al}_2\text{O}_3}$

Solution: $1000\text{ g} \times \frac{\text{Al}_2\text{O}_3}{102\text{ g}} \times \frac{2\text{ Al}}{\text{Al}_2\text{O}_3} \times \frac{27\text{ g}}{\text{Al}} = 529\text{ g of aluminum}$

3. Want: weight of carbon

Have: 454 g of carbon dioxide, CO_2

Know: $\frac{1\text{ C}}{\text{CO}_2}$; $12\ \frac{\text{g}}{\text{C}}$; $44\ \frac{\text{g}}{\text{CO}_2}$

Solution: $454\text{ g} \times \frac{\text{CO}_2}{44\text{ g}} \times \frac{1\text{ C}}{\text{CO}_2} \times \frac{12\text{ g}}{\text{C}} = 124\text{ g of carbon}$

4. Want: grams of copper

Have: 2.0 g of copper sulfate, $CuSO_4$

Know: $\frac{1\text{ Cu}}{\text{CuSO}_4}$; $63.5\ \frac{\text{g}}{\text{Cu}}$; $159.5\ \frac{\text{g}}{\text{CuSO}_4}$

Solution: $2.0\text{ g} \times \frac{\text{CuSO}_4}{159.5\text{ g}} \times \frac{1\text{ Cu}}{\text{CuSO}_4} \times \frac{63.5\text{ g}}{\text{Cu}} = 0.80\text{ g of copper}$

5. Want: grams of nitrogen

Have: 5.0 lb of ammonium nitrate, NH_4NO_3

Know: $\frac{2\text{ N}}{\text{NH}_4\text{NO}_3}$; $14\ \frac{\text{g}}{\text{N}}$; $80\ \frac{\text{g}}{\text{NH}_4\text{NO}_3}$; $454\ \frac{\text{g}}{\text{lb}}$

Solution: $5.0\text{ lb} \times \frac{454\text{ g}}{1\text{ lb}} \times \frac{\text{NH}_4\text{NO}_3}{80\text{ g}} \times \frac{2\text{ N}}{\text{NH}_4\text{NO}_3} \times \frac{14\text{ g}}{\text{N}}$

$= 795\text{ g of nitrogen}$

6. Want: grams of bromine (as bromide, Br^-)

Have: 500 grams of salt solution

Know: $\frac{2\text{ g(of NaBr)}}{100\text{ g of sol'n}}$; $79.9\ \frac{\text{g}}{\text{Br}}$; $102.9\ \frac{\text{g}}{\text{NaBr}}$; $\frac{1\text{ Br}}{\text{NaBr}}$

Solution: $500\text{ g} \times \frac{2.0\text{ g}}{100\text{ g}} \times \frac{1\text{ NaBr}}{102.9\text{ g}} \times \frac{1\text{ Br}}{\text{NaBr}} \times \frac{79.9\text{ g}}{\text{Br}}$

$= 7.8$ g of bromine

7. **Want:** grams of sodium fluoride, NaF

Have: 20,000 gallons of drinking water

Know: $\frac{1\text{ g (of F)}}{10^6\text{ g of water}}$; $3.79 \times 10^3 \frac{\text{g of water}}{\text{gallon}}$; $19 \frac{\text{g}}{\text{F}}$; $42 \frac{\text{g}}{\text{NaF}}$; $\frac{1\text{ F}}{\text{NaF}}$

Solution: $20{,}000\text{ gal} \times \frac{3.79 \times 10^3\text{ g}}{\text{gallon}} \times \frac{1\text{ g of F}}{10^6\text{ g}} \times \frac{1\text{ F}}{19\text{ g}} \times \frac{\text{NaF}}{\text{F}}$

volume — to weight — to wt of F — to mol of F — to mol of NaF

$\times \frac{42\text{ g}}{\text{NaF}} = 168$ g of sodium fluoride

to wt of NaF

8. **Want:** to compare the moles of Ti and Cl

Have: 0.378 g of Ti; by difference $1.500\text{ g} - .378\text{ g} = 1.122$ g Cl

Know: $47.9 \frac{\text{g}}{\text{Ti}}$; $35.5 \frac{\text{g}}{\text{Cl}}$

$0.378\text{ g} \times \frac{\text{Ti}}{47.9\text{ g}} = 0.0079\text{ Ti}$; $1.122\text{ g} \times \frac{\text{Cl}}{35.5\text{ g}}$

$= 0.0316$ Cl

Solution: $\frac{0.0316\text{ Cl}}{0.0079\text{ Ti}} = \frac{4.0\text{ Cl}}{1\text{ Ti}}$; $TiCl_4$

9. **Want:** to compare the moles of water and $BaCl_2$

Have: 3.41 g of $BaCl_2$; 0.59 g of water

Know: $18 \frac{\text{g}}{H_2O}$; $208 \frac{\text{g}}{BaCl_2}$ $\quad 3.41\text{ g} \times \frac{BaCl_2}{208\text{ g}} = 0.0164\ BaCl_2$

Solution: $\frac{0.0328\ H_2O}{0.0164\ BaCl_2} = \frac{2\ H_2O}{1\ BaCl_2}$ $\quad 0.59\text{ g} \times \frac{H_2O}{18\text{ g}} = 0.0328\ H_2O$

10. **Want:** % Cu in brass, i.e., $\frac{\text{wt of copper}}{\text{wt of brass}} \times 100$

Have: wt of brass = 0.500 g; 0.799 g of copper phosphate obtained from that sample of brass

Know: $63.5 \frac{\text{g}}{\text{Cu}}$; $380.5 \frac{\text{g}}{Cu_3(PO_4)_2}$; $\frac{3\text{ Cu}}{Cu_3(PO_4)_2}$

Solution: $\frac{0.799\text{ g}}{0.500\text{ g}} \times \frac{Cu_3(PO_4)_2}{380.5\text{ g}} \times \frac{3\text{ Cu}}{Cu_3(PO_4)_2} \times \frac{63.5\text{ g}}{\text{Cu}} \times 100 = 80\%$

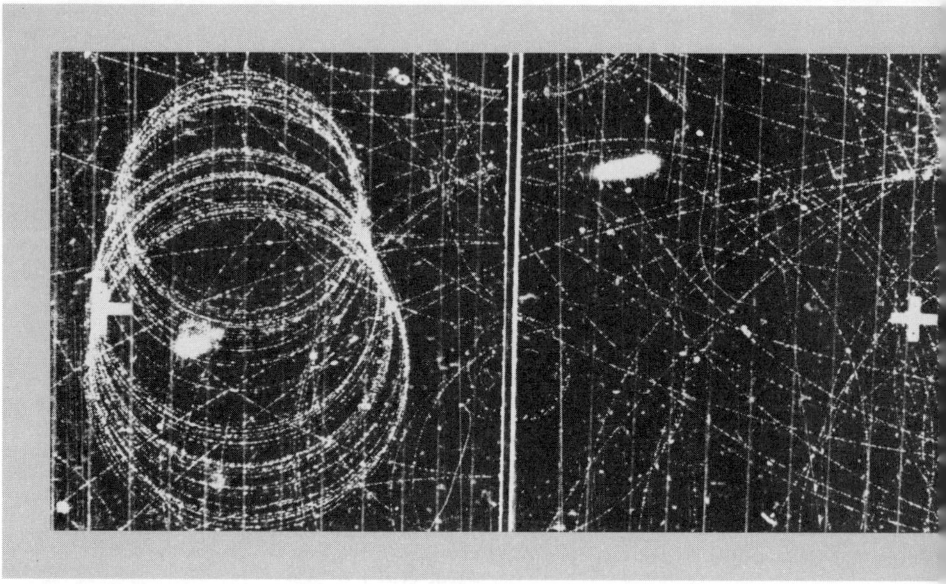

The explanation for the ways that atoms combine lies in the structures of the atoms.

After reading this chapter you should be able to

1. Name the experiments that made it necessary to change the laws of motion as applied to microscopic particles;
2. Describe the experiment that led to the concept of a nuclear atom;
3. Describe the Uncertainty Principle;
4. Define an orbital;
5. List the allowed orbitals for each energy level;
6. State the Pauli Exclusion Principle;
7. Prescribe the order of filling orbitals in a many-electron atom;
8. Write the electron configuration of an atom;
9. Designate the chemically significant (valence) electrons of an atom;
10. Relate the outer electron configuration to the chemical groups.

atomic structure 6

The first eight sections of this chapter describe in a qualitative way some of the experiments and concepts in physics that have led to our modern picture of the atom. If this introduction makes you feel that the submicroscopic regions are different from anything you have ever known, it is sufficient. A detailed description of the experiments would require a lengthy excursion into physics and mathematics, and that is not the purpose of this course. Nevertheless, we can appreciate the beauty and simplicity of the results and apply the concepts to chemistry.

Prelude to a revolution 6.1

The nineteenth century was a most satisfying time for scientists. All the great riddles of the universe appeared to have been solved, and it seemed that scientists could spend the twentieth century filling in minor details and enjoying the fruits of their wisdom.

The mathematics of Sir Isaac Newton's laws of motion had been refined to the point where the French mathematician Marquis de Laplace could exultantly declare that it was possible in principle to predict all future events if the position and momentum of every particle in the universe could be known at one instant.

The concept of the atom proposed by Dalton and refined by Avogadro and Cannizzaro was a triumph. All the results of chemistry could be satisfactorily explained by invoking the existence of atoms of various kinds, which combined in small whole number ratios. Dmitri Mendeleev discovered that atoms differed from each other in a systematic way. He found that the chemical properties of the elements repeated in a cyclic way if the elements were arranged in increasing order of atomic weights. The chemical similarities of groups of elements were so striking that he was able to predict the existence of undiscovered elements. This not only encouraged many people to look for the undiscovered elements, but also stimulated speculation about the inner structure of atoms. What are the parts of an atom, and in what order are they arranged? Somehow or other, atoms must be built up from smaller parts in a definite order, and the structure that accounts for the chemical properties must be repeated every so often.

Organic chemistry was born in the nineteenth century also. The compounds isolated from living creatures were thought to be peculiar to life, synthesized by some "living principle," but Friedrich Wöhler showed that organic compounds could be made in a test tube from indisputably inorganic substances. The evidence of organic chemistry showed that molecules have definite shapes and are not just clumps of atoms. Somehow, a model of an atom must explain the shapes of the compounds formed from them as well.

Electricity was related to the atom by Sir Humphry Davy and Michael Faraday, but the "atom" of electricity was not discovered until the latter part of the nineteenth century by George Johnstone Stoney. He called it the "electron." Later, J. J. Thomson proved that electrons were contained within atoms and were negatively charged. To explain why atoms were neutral, he proposed that electrons were embedded in a mass of positively charged matter, like rice grains in a meatball.

Radioactivity was discovered near the end of the century and within a few years Pierre and Marie Curie isolated radium, a powerful source of radioactive particles.

Newton had originally construed light to be particle-like. The discovery of diffraction or interference of light in which two rays of light apparently cancel each other out in some areas and reinforce each other in others made it necessary to treat light as a wave train (Figure 6.1). It was impossible to conceive of two particles eliminating each other. Then James Clerk Maxwell and Heinrich Hertz showed that light and radio waves were electromagnetic waves that travel through space carrying energy and are caused by an accelerated electric charge. Wiggle an electric charge, somehow, and a pulse of radiation is sent out with the same frequency as the wiggle. The *frequency* is the number of wave disturbances received per second.

So that is how things stood at the turn of the century. Matter and electricity (electrons) come in lumps. Light and other electromagnetic radiation come in waves. Anything else was unthinkable.

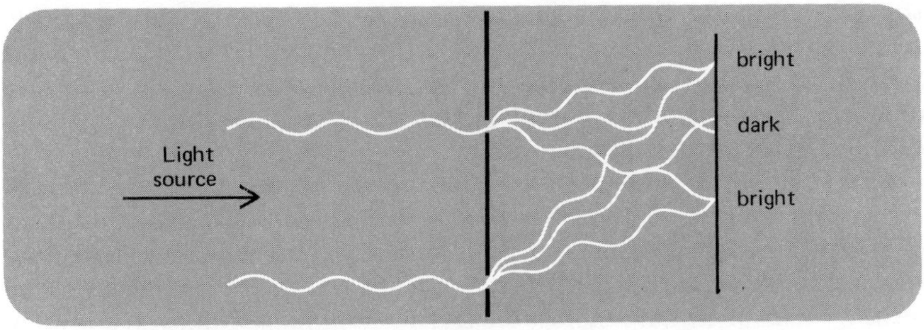

Figure 6.1 Diffraction of light. The difference in path length from the slits is a whole number of wavelengths in the bright regions and a $\frac{1}{2}$ wavelength in the dark regions. When two crests arrive together they reinforce each other. When a trough arrives with a crest, they cancel each other.

The ultraviolet catastrophe 6.2

The year 1900 brought the first of a series of experiments that cast doubt on the structure of classical physics. Max Planck was investigating the relationship of temperature to the radiation that a hot object emits. Everyone knows a hot body radiates. You can feel it as heat or see it as light if it is hot enough.

The radiation emitted by a hot body is spread over a fairly wide range of frequencies, but most of the energy comes in a small range of predominant frequencies (Figure 6.2). An object that absorbs and emits all frequencies is called a "black body." At low temperatures

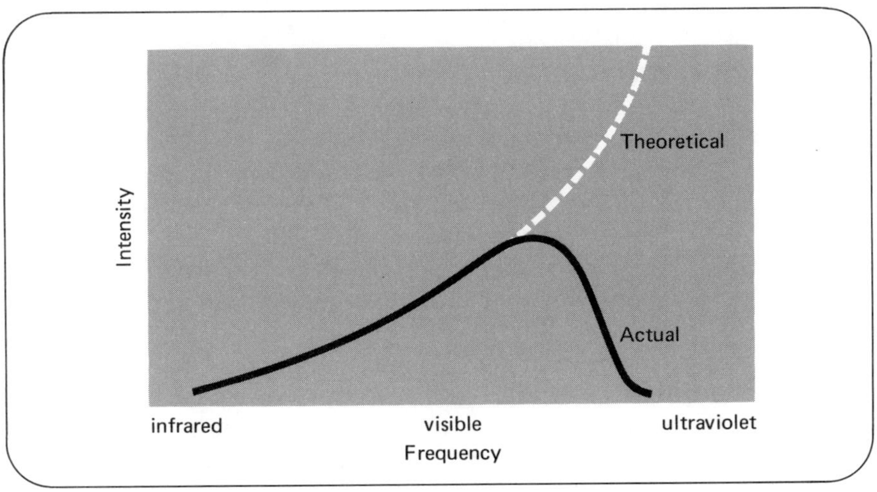

Figure 6.2 The ultraviolet catastrophe. The Rayleigh-Jeans theory predicted that every black body would emit an infinite amount of ultraviolet, X rays, and even higher energy radiation!

the predominant frequencies are in the radio range or invisible infrared (heat). At about 600–800°C the object glows red. At about 2000°C the radiation contains all the colors of the visible spectrum and looks white hot. Still higher temperatures look blue-white and contain much ultraviolet radiation.

Maxwell had shown that radiation is the result of wiggling a charge; presumably atoms were wiggling their electrons as a result of their thermal motions and that sent out the radiation. Lord Rayleigh and Sir James Jeans attempted to explain the radiation curve on the basis of kinetic molecular theory applied to Maxwell's laws of electricity and radiation. Their reasoning was consistent with everything that was known, but the results were terrible. Their theory predicted that every black body radiated an infinite amount of ultraviolet, X rays, and even higher energy radiation! The evidence that the theory was wrong horrified physicists so much that they called it the "ultraviolet catastrophe."

Planck formulated an equation that matched the actual frequency/intensity curve. To explain why the equation worked in the face of Maxwell's laws, he had to postulate that an atom could not change its energy smoothly but only in discrete jumps.[1] The change in energy is accompanied by the emission or absorption of a packet (quantum) of radiation. The size of the energy change is directly proportional to the frequency (ν) of the radiation.

$$E = h\nu \tag{6.1}$$

The proportionality constant h is called Planck's constant and is equal to 6.6×10^{-34} joule-seconds.

6.3 *The photoelectric effect*

So maybe light comes in lumps, too. This idea was given additional experimental validity by Albert Einstein. When light strikes the surface of a metal, electrons are emitted. This effect, used in "electric eye" machines, is called the "photoelectric effect." Einstein's apparatus (Figure 6.3) measured the number of electrons emitted and the kinetic energy of the electrons.

He found that the number of electrons emitted per second depends on the intensity of the light, and the energy of the electrons depends on the frequency of the light. Below a certain frequency *no* electrons are emitted, no matter how long the metal is irradiated or

[1] For a marvelous explanation of the development of Planck's hypothesis, as well as other developments in early quantum theory, read *Thirty Years That Shook Physics: The Story of Quantum Theory* by George Gamow (Doubleday-Anchor, 1966).

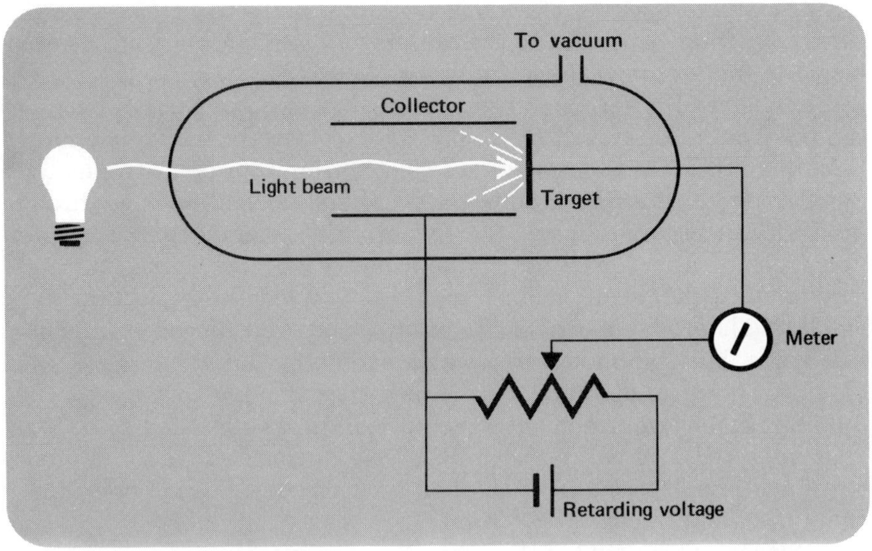

Figure 6.3 Photoelectric apparatus. When light of the proper frequency shines on the target current will flow in the circuit. The retarding voltage stops the current until light of a higher frequency is used.

how intense the light. This cannot be explained by any wave theory. It can only be explained by assuming the existence of packets or quanta of radiation, each of which can knock loose one electron if its energy is high enough. The energy of the light quanta was found to be exactly that given by Planck's equation (6.1).

The Rutherford atom 6.4

The discovery that matter interacts with light only in discrete energy packets (quanta) whose energy content is proportional to their frequencies stimulated much speculation about the structures of the atoms themselves. Dalton's idea of indestructible spheres had to be modified to account for the evidence that atoms contain positively and negatively charged particles. J. J. Thomson proposed the theory that atoms consisted of jelly-like blobs of positive matter in which negative electrons were imbedded like grapes in a Jello fruit salad. Collisions with light or other atoms would cause the electrons to vibrate around their equilibrium positions, and thus absorb or emit light at certain frequencies. The model was plausible but it failed to predict either the frequencies observed or the chemical properties of the elements.

One way to examine a complicated mechanism is to smash it to pieces and examine the parts. After radioactivity was discovered,

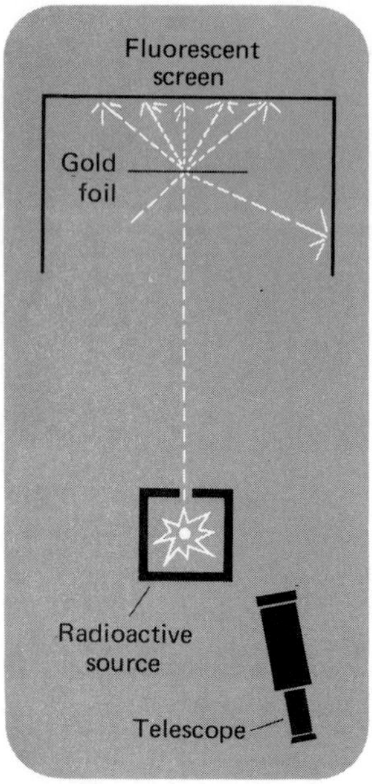

Figure 6.4 Rutherford's alpha particle scattering experiment. A radioactive source sends alpha particles through a gold foil. A pinpoint of light is emitted when the particles hit the surrounding fluorescent screen. The position and number of flashes are observed with a telescope.

physicists had a source of atomic-sized bullets to shoot at atoms. By observing what happened to the bullets and to the atomic pieces, they hoped to find out how the atom was constructed.

Ernest Rutherford had discovered that the radioactive particles coming from radium (called "alpha particles") were positively charged helium atoms. Alpha particles are ejected from radium at tremendous speeds and behave like atomic-sized bullets when they pass through matter. Rutherford's apparatus (Figure 6.4) was designed to count and measure the deflection of alpha particles after they had passed through a thin foil of gold atoms. He expected the deflection to be small, if Thomson's model of the atom was correct, because Jello does not deflect bullets to any appreciable extent.

Most of the alpha particles did indeed go straight through the gold foil. But a few were deflected at quite large angles, and some were even reflected backward (Figure 6.5)! As Rutherford said, "It was as though we had fired a 15-inch shell at a piece of tissue paper and it came back and hit us!"

Rutherford concluded that the atom must consist of a very small, very dense, positively charged nucleus surrounded by orbiting electrons. Otherwise there would not be enough mass and charge density to reflect an alpha particle. This model of the atom was called "the nuclear atom" or the "planetary atom" because it was analogous to the planets orbiting the sun.

Rutherford knew that the atom had to have a small dense nucleus, but he could not explain a severe difficulty predicted by Maxwell's laws of electricity. The negative electrons are attracted to the nucleus by the coulombic force, and the electrons must resist that force by a force generated by their motion. If the electrons move in an orbit around the nucleus, they would be accelerated. A change in direction of motion is an acceleration as much as a change in speed. But Hertz and Maxwell showed that an accelerated electric charge will emit energy as electromagnetic radiation. If the electron radiated energy it would lose speed until it spiraled into the nucleus. Since atoms are present this event obviously does not happen.

6.5 *The Bohr atom*

Niels Bohr, working in Rutherford's laboratory, proposed a way out of the dilemma and made a valuable connection at the same time. To show the connection, we must digress for a brief explanation.

We already know that matter will absorb and emit radiation, especially when it is hot. The radiation spectrum from solid or liquid matter is more or less continuous; that is, it has a wide range of frequencies in it. Sunlight has a wide range of frequencies in it, and when they are spread apart by a prism we see a rainbow of colors (Figure 6.6).

Matter in the *atomic* state is more particular. Free atoms absorb or emit radiation only as particular frequencies. Atoms excited in a flame will give off characteristic colors, and when the colors are spread apart by a prism or a grating, a spectrum like that in Figure 6.7 is obtained.

People who study such things are called spectroscopists, and they found that the systems of lines were characteristic of each element. In fact, for the simpler elements such as hydrogen the frequencies of the lines followed a fairly simple mathematical law.

Niels Bohr's contribution was to propose a model of the atom that was consistent with Rutherford's nuclear atom, but that would not collapse and would link Planck's constant with the observed spectral lines of hydrogen. He simply proposed that any orbit of the electron with an angular momentum[2] that was a whole number times Planck's constant was stable and did not radiate. Absorption or emission of energy by radiation had to occur in quanta and corresponded to a

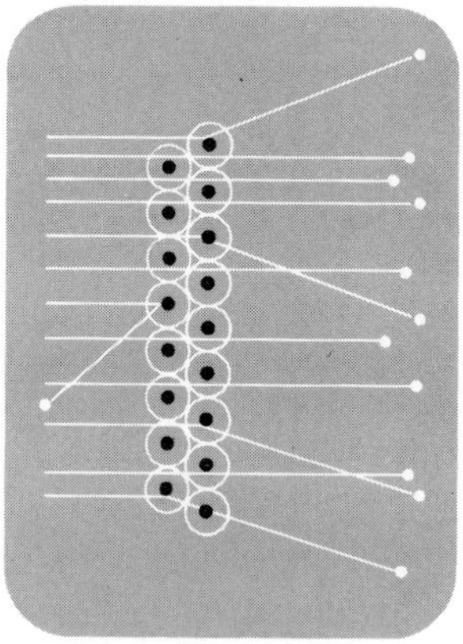

Figure 6.5 Scattering of alpha particles by gold atoms. The results were interpreted by Rutherford to mean that atoms must have small, dense nuclei.

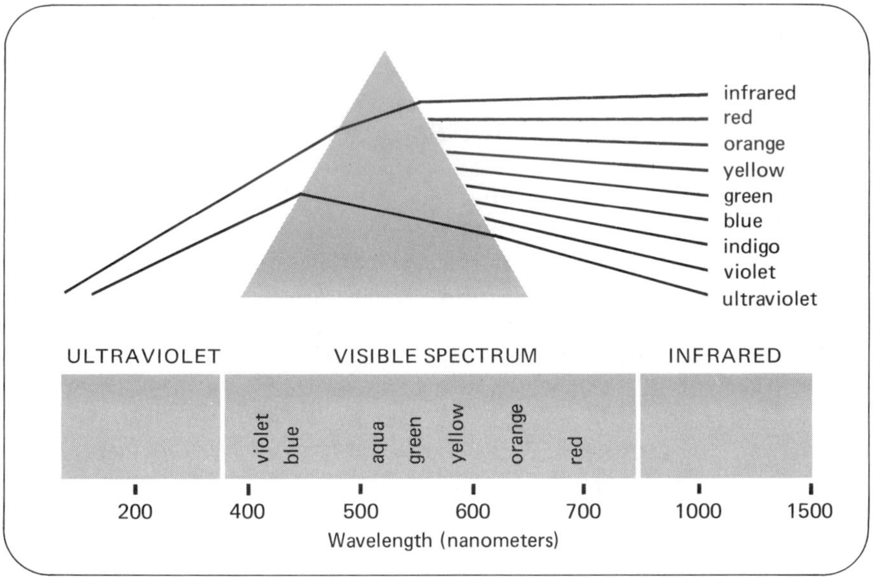

Figure 6.6 Dispersion of sunlight by a prism. The short wavelengths are *refracted* (bent) more than the longer wavelengths by the transparent prism. This separates the white light into distinct bands of color.

[2] Angular momentum is an inertial property of objects that are rotating or revolving, just as linear momentum is an inertial property of objects that are moving in a straight line. An object's linear momentum changes only if a *force* acts on it; the angular momentum changes only if a *torque* (twisting force) acts upon it. The spin that a figure skater achieves by pulling her arms close to her body is a result of the conservation of angular momentum.

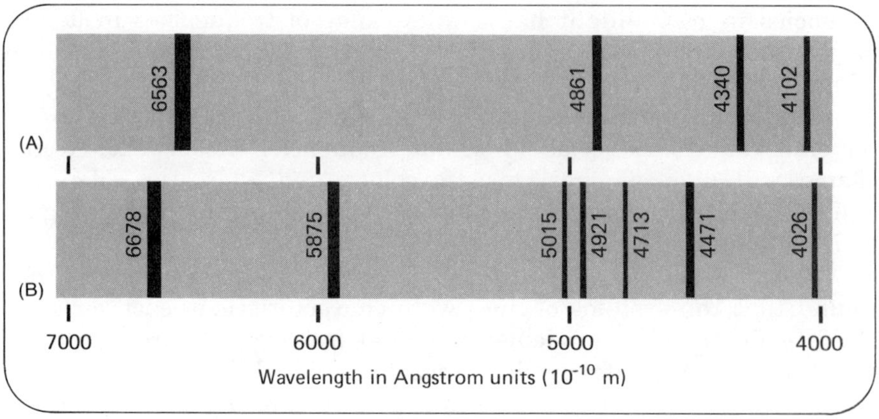

Figure 6.7 Emission spectra. (A) Hydrogen and (B) helium.

jump from one orbit to another (Figure 6.8). He didn't say *why* a quantized orbit was stable, but his proposal was very successful in explaining the spectrum of hydrogen. The next ten years were spent in trying to fit the theory to the other elements without much success.

6.6 *De Broglie and the matter wave*

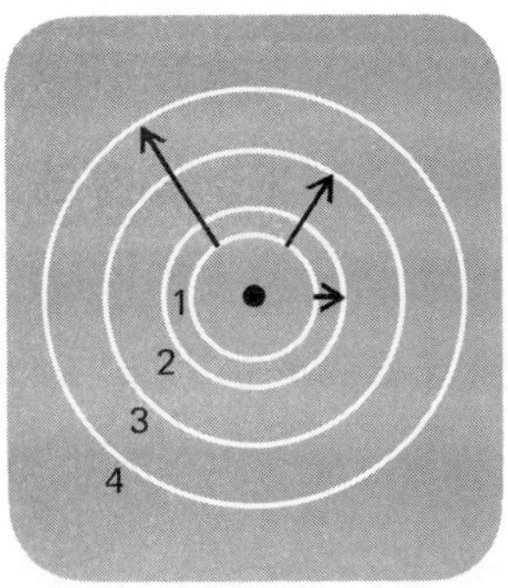

Figure 6.8 The Bohr atom. The only stable orbits are those having a whole number of units of angular momentum. The only energy quanta that the atom can absorb or emit correspond to a jump from one orbit to another.

By 1924 physicists had learned to live with the strange dual character of light. They found that when they looked at single events, such as collisions of light quanta (now called *photons*) with matter, light behaved like particles. And when they looked at group properties, such as the pathways followed by rays or beams of light, light behaved like waves. So be it. That is the way that light behaves.

A young physicist named Louis de Broglie wondered if the same dual character did not also apply to things that were definitely particles, such as electrons. Nature was notorious for being impartial and symmetrical, and it did not seem reasonable that light could be so complex and electrons so simple. Besides, except for the hydrogen atom, the simple concept of electrons whirling around an atom in Bohr orbits was not getting anywhere.

De Broglie's calculations indicated that very small particles, such as atoms and electrons, might exhibit wave properties like diffraction under the right conditions. His proposal was soon tested by experiment and found to be true. *Electrons behaved just like light,* inasmuch as an experiment that looked at *events* made out an electron to be a particle, and experiments that looked at the *pathways* that an electron might take made it necessary to treat them as waves. Except for a difference in scale, electrons behave just as light does in Figure 6.1.

The uncertainty principle 6.7

According to Newtonian mechanics, if an object's position and momentum are known at one instant, its future position can be predicted as accurately as desired. The idea of the orbits in Bohr's model of the atom is Newtonian. It implies that, once the orbit is found, one can know where the electron *is now* and where it *will be later*.

As a matter of fact, it is impossible to know exactly the position and momentum of a particle at one instant. The property of momentum (mv) includes its velocity. To measure the velocity one must measure the position at two places at two instants of time and divide the distance interval by the time interval.

To know the position of a particle we have to "see" it, that is to interact with the particle in some way, perhaps with a photon of light. But *light carries energy and momentum* and the interaction will change the momentum of the electron.

Werner Heisenberg first stated the Uncertainty Principle. Formally it says that the product of the uncertainty in the position (Δs) and the momentum (Δmv) of a particle must be equal or greater than Planck's constant h.

$$\Delta s \times \Delta mv \geq h$$

This statement is at the foundation of the wave properties of microscopic matter. It means that we cannot make laws or predictions based on classical mechanics but instead must use wave mechanics.

A nucleus is of the order of a few femtometers across. If an electron fell into the nucleus, its position would be known precisely. But wave mechanics and the Uncertainty Principle say that it must then have a very large momentum—entirely too large to stay in the nucleus!

Rearranging the Heisenberg equation, we see that the uncertainty in the momentum must be greater than Planck's constant divided by the uncertainty in position.

$$\Delta mv \geq \frac{h}{\Delta s}$$

As Δs becomes very small, the uncertainty in momentum must become very large. Therein lies the explanation for the stable nuclear atom. The electron *cannot* fall into the nucleus, because its energy would have to be too high to exist in such a small volume.

6.8 *Quantum theory**

When we attempt to explain a phenomenon, we call upon various laws of experience which, taken together, show the event to be a logical consequence of what we know already. The idea of atoms and molecules as tiny masses in motion is not hard to take, because we are familiar with baseballs and bullets, and it is not unreasonable to extrapolate the behavior of baseballs down to a very small size.

In a similar fashion we have learned about waves. A wave can be described generally as a disturbance that travels through space with a characteristic velocity. Water waves, sound waves, vibrations in structures, and electromagnetic waves all fit that description.

Waves can pass through each other and their form and velocity remain unchanged. Waves can be transmitted, reflected, refracted, and diffracted. In our experience none of these properties can be remotely connected with the behavior of a baseball or a bullet. Now science is faced with explaining how something can be *both* a wave *and* a particle.

De Broglie thought that light and electrons were both particles, but guided in their motions by a sort of "pilot wave":

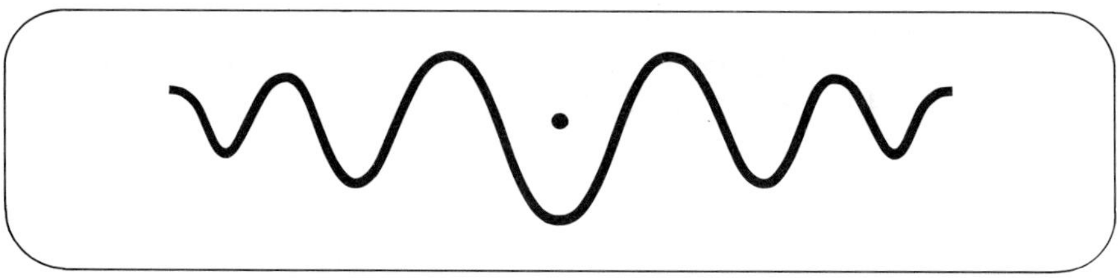

Others thought that the particle-like nature was the result of many waves superimposed on each other. Both concepts ran into difficulties.

The modern interpretation was suggested by Max Born. He proposed to treat electrons and photons as lumps whose behavior does not follow classical laws, but rather is controlled by probability laws of a form resembling waves as far as the mathematical form of them is concerned. So there is no wave, but there is a *wave equation* that describes the forces acting on the electron in its motions around the nucleus.

The solutions of the wave equation are a series of complex mathematical functions, called wave functions or psi functions. The physical interpretation of these wave functions is that the square of these functions gives the *probability per unit volume* of finding the electron in a given small region of space. Evaluation of the probability for small increments of volume at various distances from the nucleus gives us a *probability density function.*

* Optional topic.

A plot of the probability density for an electron in three dimensions outlines a region of space in which we expect to find the electron. This region of space is commonly called an *orbital.* The size and shape of the orbitals are determined by the wave function.

Atomic orbitals 6.9

If we insist on the requirement that the electron stays in the neighborhood of the nucleus for all time, neither getting lost nor crashing into it, the wave equation tells us that the electron can only have certain fixed energy values or steps. A disturbance from the outside, like a collision with another atom or a photon, can only cause the electron to change energy in jumps, or quanta.

Associated with each energy level are a number of *orbitals.* An orbital is a region of space where there is a high probability of finding an electron. The situation is something like what occurs when a guitar string is plucked. The string can vibrate a number of ways simultaneously depending only on the length and weight of the string and the tension (Figure 6.9), but there is always a whole number of half-waves between the nut and the bridge.

The vibrations of a guitar string are described by a wave equa-

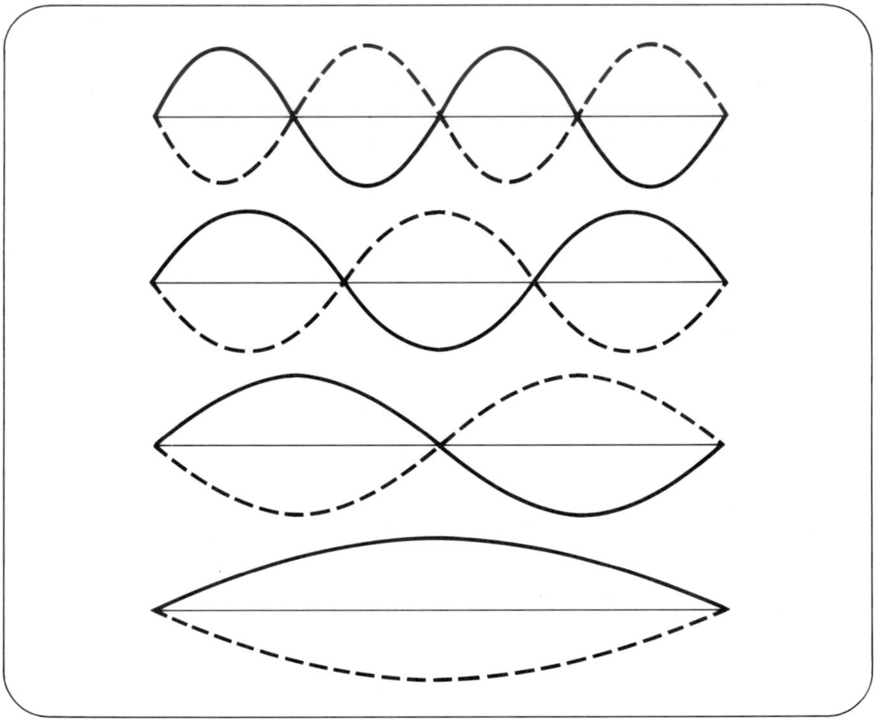

Figure 6.9 Possible modes of vibration of a string.

tion. In the same way the probability of an electron's location depends on a wave equation, so it is not surprising that there are similarities between them.

The solutions to the wave equation, the wave functions, each have a characteristic set of whole numbers which determine the size and shape of the orbital. These numbers are called *quantum numbers.* The energy of the electron depends to a large extent on the principal quantum number, which may take any whole number value (1, 2, 3, 4, . . .).

The lowest energy level of an electron around the nucleus has a principal quantum number of 1. It is the energy level in which the electron has the smallest average distance to the nucleus, and there is only one orbital in that level. There is an equal probability of finding the electron in any direction, but we don't expect to find it very far away or very close to the nucleus. So if we colored the probability of the electron's location, we would have a sphere with weak coloring far out and close in and a maximum somewhere between. This kind of orbital is called an *s* orbital.

A curve showing the probability of finding the electron at a specified distance from the nucleus is a *probability density* curve, and it is obtained from the square of the wave function. The curve is symmetrical about the nucleus in all directions for an *s* orbital. The 1*s* orbital of hydrogen has a maximum probability at 0.53 angstrom (Å) (Figure 6.10). An angstrom unit is 10^{-10} meter.

The second energy level of an atom may have two kinds of orbitals. One of the orbitals has the same spherical shape as the 1*s* orbital. The other kind has a probability density that depends on direction. It is

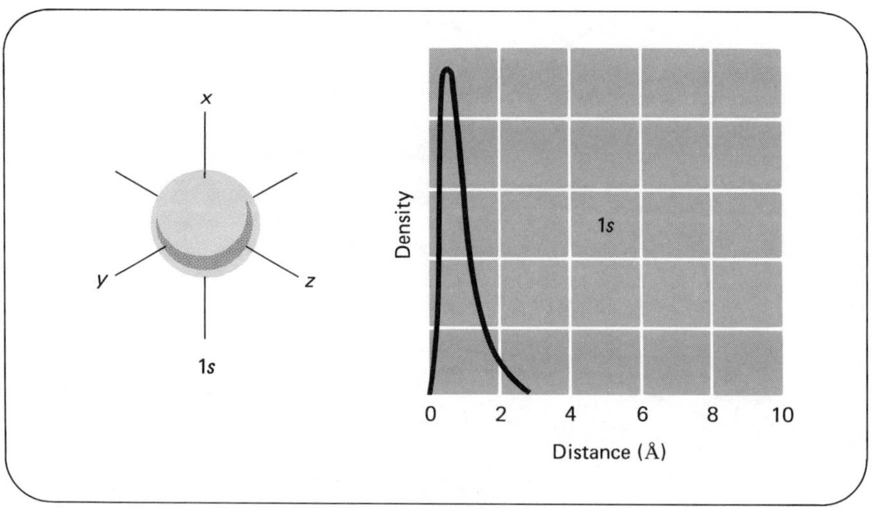

Figure 6.10 The 1*s* orbital and its density vs. distance in angstrom units from the center of the atom.

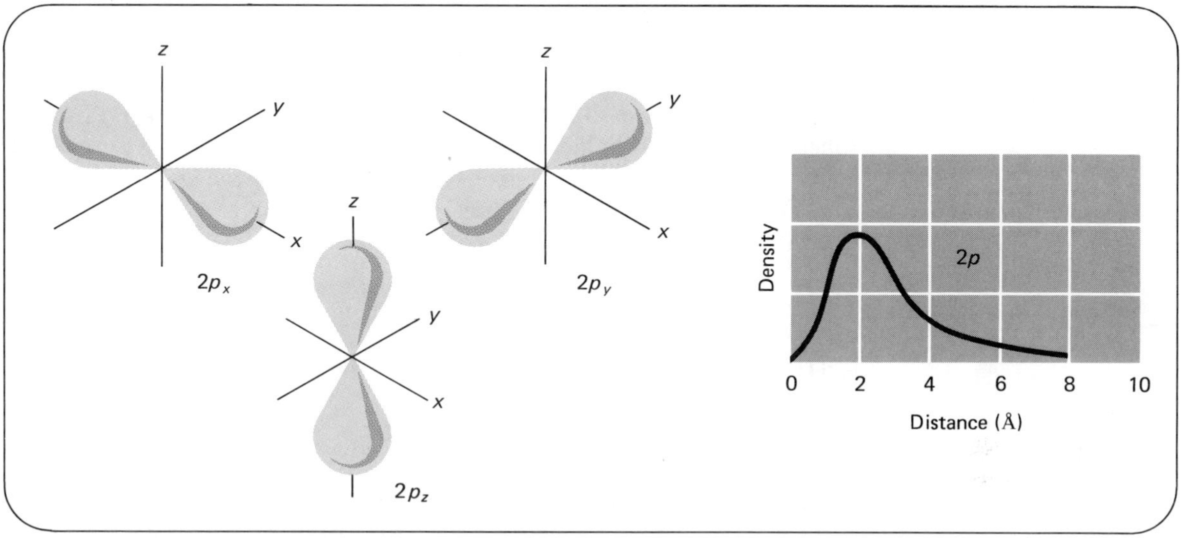

Figure 6.11 The sizes and shapes of the three *2p* orbitals.

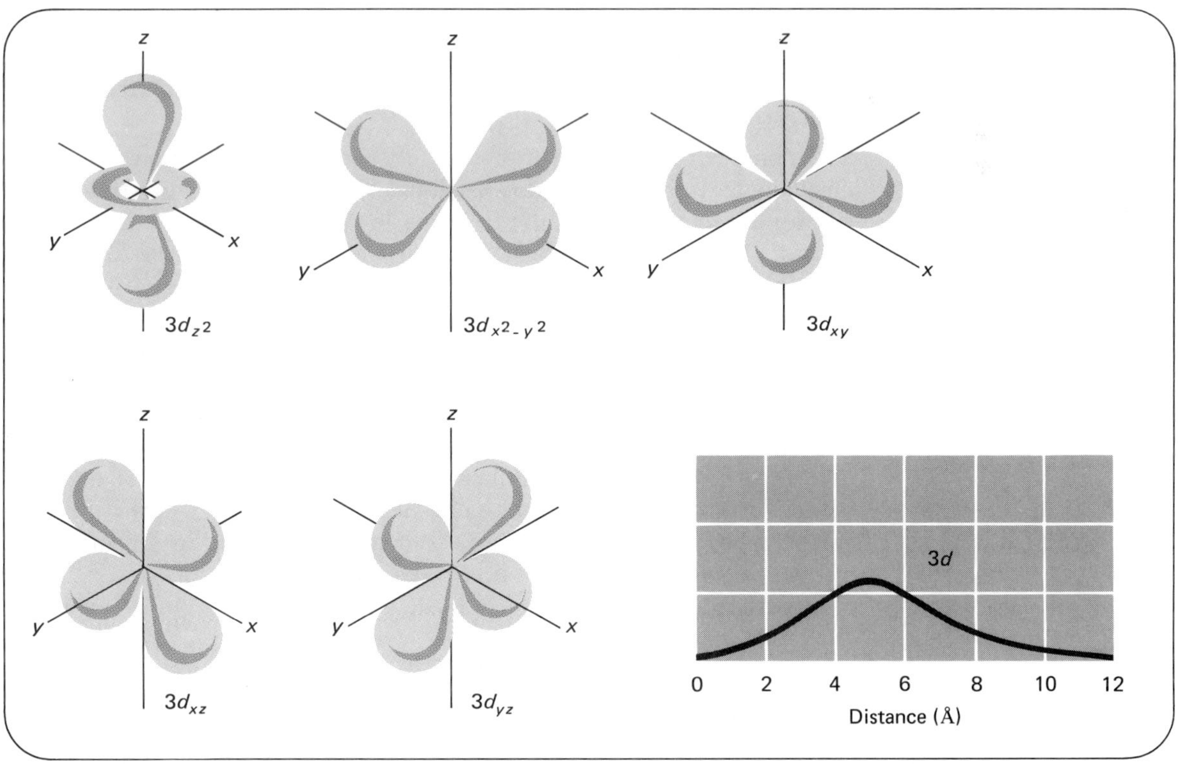

Figure 6.12 The sizes and shapes of the five 3*d* orbitals.

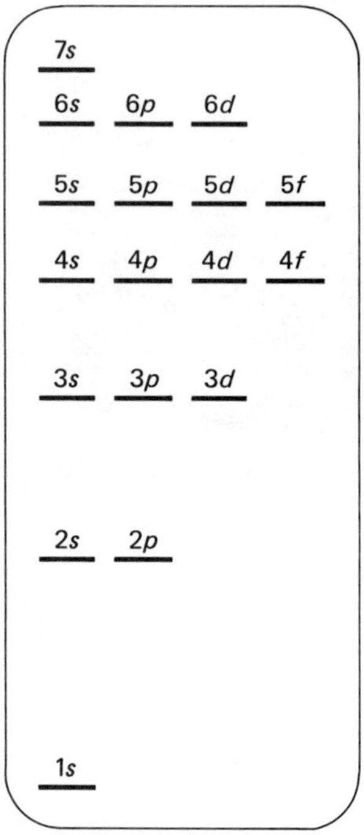

Figure 6.13 Orbital diagram of hydrogen.

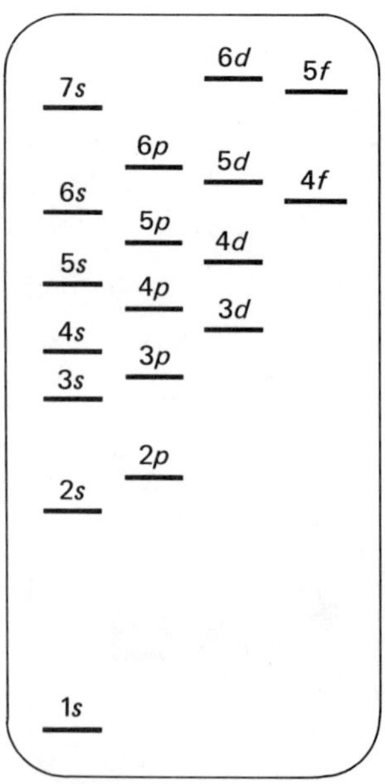

Figure 6.14 Orbital diagram of multielectron atoms.

dumbbell shaped, and there are three of them at right angles to each other, called *p* orbitals (Figure 6.11).

The third energy level in an atom may have three different kinds of orbitals. Two kinds are *s* and *p* orbitals, but the third kind has four lobes, like a four-bladed propeller or two dumbbells lying across each other (Figure 6.12). There are five of them, called *d* orbitals.

The fourth energy level has, in addition to one *s*, three *p*, and five *d* orbitals, seven orbitals of a complicated shape called *f* orbitals.

6.10 *Energy differences and the Aufbau principle*

Orbitals can be compared to rooms in a hotel built on a hillside. Orbitals are available spaces for electrons, and the cheap rooms (lowest energy) are on the bottom. In the hydrogen atom all the orbitals on a particular energy level have the same energy (Figure 6.13).

The single electron of the hydrogen atom normally occupies the 1*s* orbital on the ground floor (ground state), so to speak, but if energy quanta are received from the surroundings the electron may be "excited" to the empty upper levels. When the electron falls back to the ground state, the radiation spectrum characteristic of hydrogen is emitted.

The hydrogen atom has only one electron, so the only force the electron feels is that of the nucleus. The other elements have more than one electron and the electrons interact with each other to some extent. This changes the energy of the *p, d,* and *f* orbitals slightly with respect to the *s* orbitals. The energy diagram for multielectron atoms is that of Figure 6.14.

If it were not for a certain property of electrons, all of them in a many-electron atom would occupy the ground floor and there would be no science of chemistry. That property is called "spin." It comes in two varieties (like people) and *no two electrons with the same spin can occupy the same orbital.* This is a statement of the Pauli exclusion principle, and it appears to be one of the fundamental laws of nature. With respect to atoms, it means that an orbital can hold a maximum of two electrons, and they must have opposite spins.

Now that the structure is built we can put the guests in it. The "house rules" are these:

1. In a neutral atom the number of electrons is equal to the atomic number;
2. Electrons will be placed in the lowest available energy state (orbital);
3. Each orbital can accommodate two electrons of opposite spin.

These three principles constitute the basis for the "aufbau" principle. Aufbau means "building up" in German. To show the electron configuration of a neutral atom, we put electrons into the orbitals with the lowest energy, no more than two electrons to an orbital, until we have added the same number as the atomic number.

The order of filling orbitals in a multielectron atom can be remembered most easily by this device. Upon an orbital diagram like that in Figure 6.13, diagonals are drawn from the bottom up and to the left. The order of filling the orbitals is given by following the diagonals from the bottom (1*s*), connecting the head of one diagonal to the tail of the next (Figure 6.15).

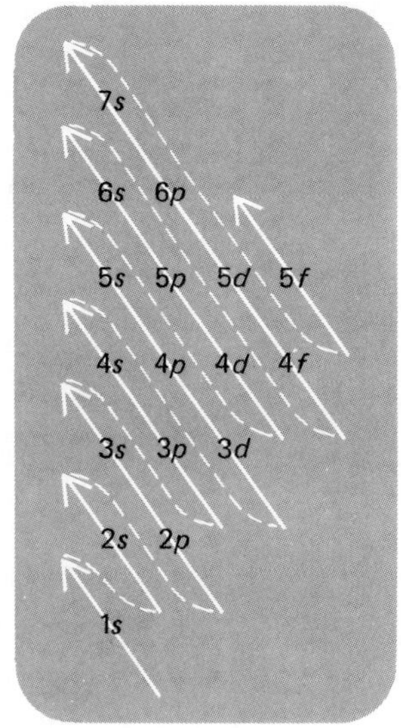

Figure 6.15 Order of filling orbitals.

Electron configuration of an atom 6.11

The complete list of orbitals which contain electrons in an atom is called its "electron configuration." In an isolated atom there is no difference in energy between the three *p* orbitals in each level. Likewise all *d* orbitals with the same quantum number are alike. They are said to be "degenerate." These orbitals are filled with one electron each, until each has one electron. Additional electrons are then paired with those present, until each orbital has two electrons with opposite spins.

The notation for electron configurations lumps all of the electrons with the same energy together and shows the number as a superscript on the orbital name.

EXAMPLE 6.1

1. The atomic number of hydrogen is 1. The neutral hydrogen atom will hold one electron, and it will normally be in a 1*s* orbital. This is the orbital with the lowest energy. The electron configuration for hydrogen is $1s^1$.
2. The atomic number of helium is 2. Helium holds two electrons, and both can be in a 1*s* orbital. The electron configuration for He is $1s^2$.
3. The atomic number of lithium is 3. Lithium can hold three electrons, but only two can go into the 1*s* orbital. The next one must take the next higher energy orbital, 2*s*. The electron configuration of Li is $1s^2\ 2s^1$.
4. The atomic number of nitrogen is 7. A neutral nitrogen atom must have seven electrons; two in the 1*s* orbital, two in the 2*s* orbital, and three in the 2*p* orbitals. The electron configuration is written: $1s^2\ 2s^2\ 2p^3$.
5. The atomic number of iron is 26. A neutral iron atom must hold 26 electrons. Notice that the 3*d* orbitals are slightly higher in energy than the 4*s* orbitals. This means that after the 3*p* orbitals are filled, the next orbital to be filled is the 4*s*. After the 4*s* is filled, the 3*d* electrons are added. The electron configuration for iron is written:

$$_{26}\text{Fe} = 1s^2\ 2s^2\ 2p^6\ 3s^2\ 3p^6\ 4s^2\ 3d^6$$

The sum of the superscripts equals 26.

Table 6.1

The order of filling atomic orbitals in the form of the periodic table

$1s^1$																	$1s^2$
$2s^1$	$2s^2$											$2p^1$	$2p^2$	$2p^3$	$2p^4$	$2p^5$	$2p^6$
$3s^1$	$3s^2$											$3p^1$	$3p^2$	$3p^3$	$3p^4$	$3p^5$	$3p^6$
$4s^1$	$4s^2$	$3d^1$	$3d^2$	$3d^3$	$3d^4$	$3d^5$	$3d^6$	$3d^7$	$3d^8$	$3d^9$	$3d^{10}$	$4p^1$	$4p^2$	$4p^3$	$4p^4$	$4p^5$	$4p^6$
$5s^1$	$5s^2$	$4d^1$	$4d^2$	$4d^3$	$4d^4$	$4d^5$	$4d^6$	$4d^7$	$4d^8$	$4d^9$	$4d^{10}$	$5p^1$	$5p^2$	$5p^3$	$5p^4$	$5p^5$	$5p^6$
$6s^1$	$6s^2$		$5d^2$	$5d^3$	$5d^4$	$5d^5$	$5d^6$	$5d^7$	$5d^8$	$5d^9$	$5d^{10}$	$6p^1$	$6p^2$	$6p^3$	$6p^4$	$6p^5$	$6p^6$
$7s^1$	$7s^2$																
		$5d^1$	$4f^1$	$4f^2$	$4f^3$	$4f^4$	$4f^5$	$4f^6$	$4f^7$	$4f^8$	$4f^9$	$4f^{10}$	$4f^{11}$	$4f^{12}$	$4f^{13}$	$4f^{14}$	
		$6d^1$	$5f^1$	$5f^2$	$5f^3$	$5f^4$	$5f^5$	$5f^6$	. . .	. . .							

Electron configuration and the periodic table 6.12

If we put the orbital notation for the last (highest energy) electron to go into an element into the spot occupied by the element in the periodic table, it immediately becomes apparent that the electron configuration and the periodic table are related (Table 6.1). In fact we can say that the order of filling atomic orbitals stipulated by wave mechanics accounts for the form of the periodic table.

The reason for similarities in chemical behavior within groups (vertical columns) of elements becomes apparent, too. The electron configurations within a group differ only in the principal quantum numbers. That is, all the elements in a group have a similar appearance with respect to the outer occupied orbitals. All the alkali metals (Li, Na, K, etc.) have an ns^1 configuration. All the halogens (F, Cl, Br, I) have an ns^2np^5 configuration, and so on.

The fact that elements within groups are found to have similar chemical properties and the fact that they have similar outer electron configurations lead us to suspect that the outer electrons are responsible for the chemical properties of the elements.

The electrons involved in the chemical transformations of an element are called "valence electrons." They are usually the *s* and *p* electrons with the highest quantum number in the atom. If the atom has *d* orbitals, these may sometimes be involved, especially if there are no *p* electrons outside of them. Electrons in *f* orbitals do not contribute to chemical activity.

A period seems to end with a filled set of *p* orbitals. All of the noble gases except helium have an ns^2np^6 configuration and they represent minimal chemical activity. In every case, the next electron must occupy an *s* orbital with the next higher quantum number. Evidently a full set of occupied *p* orbitals is an exceptionally stable configuration.

The fact that electrons in a completed period do not contribute to chemical activity has given rise to a terminology that is convenient, if unfortunate in its implications. The electrons in a given period are said to be in a "shell." The first period is named the *K* shell, the second is named the *L* shell, and so on. An element with filled *s* and *p* orbitals is said to have a "completed shell," and the outer *s* and *p* (and sometimes *d*) electrons belong to the "valence shell."

This terminology is all right as long as it does not lead to visualizing the atom as an onion. Shells really refer to principal energy levels and not to a layer of electrons.

The order in which the orbitals are filled accounts for the difference between the number of elements in a period and the number of electrons with a given quantum number. Table 6.2 summarizes the orbital types in each period.

Table 6.2

Orbital types in each period

Period	Orbitals				Members	Shell
1	1*s*				2	*K*
2	2*s*			2*p*	8	*L*
3	3*s*			3*p*	8	*M*
4	4*s*		3*d*	4*p*	18	*N*
5	5*s*		4*d*	5*p*	18	*O*
6	6*s*	4*f*	5*d*	6*p*	32	*P*
7	7*s*	5*f*	6*d*	—	incomplete	

6.13 *Summary*

1. The beginning of modern physics occurred when Planck explained black-body radiation in terms of the quantum effect, and Einstein demonstrated the quantum behavior of light.

2. The existence of the atomic nucleus was discovered by Rutherford on the evidence of alpha particle scattering.

3. The Bohr model of the atom was the first to quantize the motions of the electron, and thereby show the relationship between atomic spectra and the atom.

4. The motions of the electron cannot be described exactly, as in the Bohr atom, but must be described by a wave equation.

5. The solutions of the wave equation that correspond to conditions placed on a real atom are complex mathematical functions called wave functions. Each allowed wave function is characterized by a series of whole numbers, called quantum numbers.

6. (optional) The wave functions that are the solutions to the wave equation for hydrogen are called *orbitals* and are characterized by their quantum numbers (1*s*, 2*s*, 2*p*, etc.).

7. (optional) The square of the wave function, evaluated over the space around a nucleus, gives the region of space in which the electron is likely to be found. This region of space is also called an orbital.

8. Orbitals are filled in an atom by placing electrons in the orbital with the lowest energy, no more than two electrons to an orbital, until the number of electrons equals the atomic number of the element.

9. The order of filling orbitals stipulated by wave mechanics accounts for the form of the periodic table of the elements. A period ends with a noble gas possessing filled *s* and *p* orbitals.

10. Elements in the same chemical group have the same outer configuration of electrons. The electrons that determine the chemical properties of an element (the valence electrons) are therefore the electrons that come after the last completed period (shell).

EXERCISES

I. Theory Review

1. What three facts forced scientists to modify the classical mechanical and classical electrodynamic theory so that they were compatible?

2. Why was the Rutherford model of the atom impossible according to classical theory?

3. What was fundamentally wrong with Niels Bohr's model of an atom? What was right about it?

4. What do we mean by the wave character of an electron?

5. What force exists between an electron and a nucleus?

6. Why doesn't the electron fall into the nucleus?

7. What is an orbital?

8. Under what circumstances could an electron change orbitals?

9. What is the "aufbau principle"?

10. Why don't all of the electrons occupy the lowest orbital?

PROBLEMS

II. **1.** Write the electron configurations for

Be	C	O	Ne
Na	P	Cl	Zn
Ca	Ti	Mn	Br

2. Write the "outer" electron configuration for all the elements in

(a) Group IA
(b) Group IIIA
(c) Group IIIB
(d) Group VA
(e) Group VIIA

SUGGESTED READING

Brackenridge and Rosenberg, *The Principles of Physics and Chemistry* (McGraw-Hill, Inc., 1970), Chapters Twenty-three and Twenty-four.

Feynman, Leighton, and Sands, *The Feynman Lectures on Physics* (Addison-Wesley, Inc., 1963), Chapter 37.

George Gamow, "The Principle of Uncertainty," *Scientific American* (January, 1958), p. 51.

George Gamow, "The Exclusion Principle," *Scientific American* (July, 1959), p. 74.

George Gamow, *Thirty Years That Shook Physics: The Story of Quantum Theory* (Doubleday-Anchor, 1966 (paperback)).

R. Hochstrasser, *Behavior of Electrons in Atoms* (W. A. Benjamin, Inc., 1965 (paperback)).

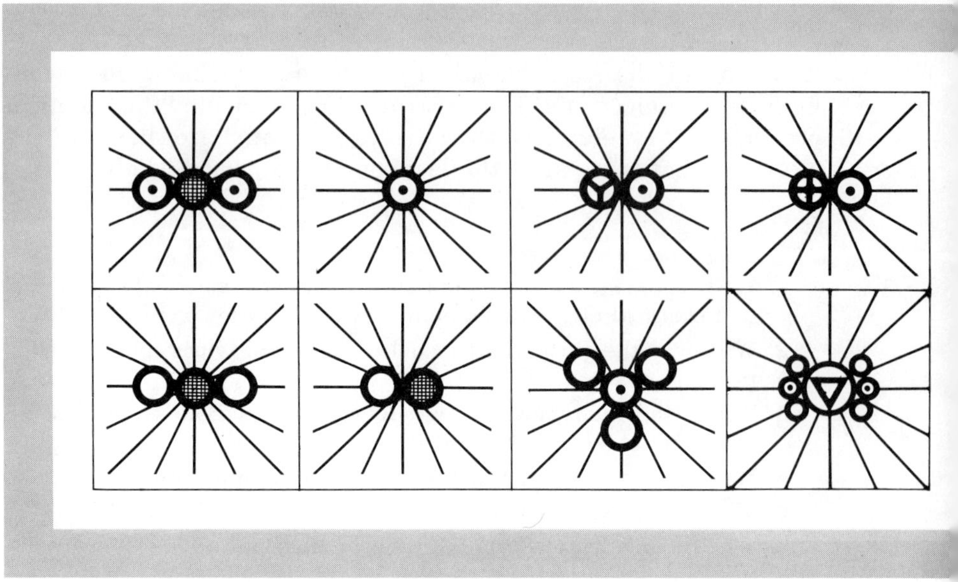

We use our knowledge of atomic structure to explain the ways that atoms can combine.

After reading this chapter you should be able to

1. Define the concept of valence;
2. Write electron dot structures of atoms from electron configurations;
3. Write Lewis diagrams of molecules and ions;
4. Write formulas of ionic compounds by balancing charges;
5. Describe the concept of electronegativity;
6. Classify compounds as principally ionic, covalent, or polar covalent on the basis of electronegativity differences;
7. Describe molecular shapes;
8. Compute oxidation numbers of atoms.

models of compounds

7

Four year olds playing with a toy car are not really concerned that the toy is not a faithful representation of the real thing. As long as the toy has wheels, they can supply the noise and imagination will take care of the rest. No matter how crude the model is, they begin to develop an intuition about momentum and friction as well as collisions and upsets.

Twelve year olds will want a more sophisticated model. They know that real cars need engines and steering wheels and the like. In the process of assembling the model they will probably learn in a limited way how the real one works.

Aeronautical engineers designing a new airplane use a still different kind of model. They put their experiences about aerodynamics and materials and engines into a computer. The relationships between these things are represented by voltages, capacitors, switches, and mathematical equations. In a sense, the model exists only in the engineers' imaginations. The computer tells them how the hypothetical airplane would perform if it were built to their specifications and if all their assumptions were correct.

In each case someone is using a model to predict or represent the behavior of a real object that is not accessible to direct investigation. The degree of complication or refinement in the model depends on two things: (1) the model must be based upon a collection of real experiences or observations, and (2) it needs only the parts necessary to explain or predict the behavior being studied. The engineer responsible for the power-to-weight ratio of the plane does not need a complete model of the galley, only its weight.

The chemist's model of the atom and of the compounds formed from atoms began as a nondescriptive clump which explained the weight and number relationships only. The need to explain the interactions of matter with light resulted in a model that exists only as mathematical equations, because no macroscopic model can act in the way atoms and electrons must behave.

If we are satisfied for the present to explain (1) the combining capacity of atoms, (2) the existence of compounds as ions and molecules, and (3) the shapes of molecules, we can construct a reasonably simple model consistent with much of the known chemistry.

7.1 *Valence*

Each element has a potential or capacity to combine with other atoms. This characteristic combining capacity is the same for all members of a chemical group. It is one of the salient properties that caused Mendeleev to construct a table of groups in the first place. It means that all the chlorides of Group III metals have the general formula $MeCl_3$; all the oxides of Group I have the formula M_2O; all the hydrides of Group V have the formula XH_3, and so on. At one time it was useful to define a *unit combining capacity* for each group in terms of their capacity to bind hydrogen and oxygen. The name given this power to bind was the *valence* of the element.

Table 7.1

Formulas of the hydrides and oxides of period 2 elements

Group	**I**	**II**	**III**	**IV**	**V**	**VI**	**VII**
Hydrides	LiH	BeH_2	BH_3	CH_4	NH_3	H_2O	HF
Oxides	Li_2O	BeO	B_2O_3	CO_2	N_2O_3	–	OF_2

If hydrogen is given a valence of one, the valence of any element is equal to the number of hydrogens it holds in a binary compound. Since oxygen holds two hydrogens, oxygen has a valence of two. The valence of any element combined with oxygen is two times the mole ratio of oxygen atoms to the element in the formula. By these definitions, we would assign a valence of 2 to sulfur in H_2S, but 4 to sulfur in SO_2.

This concept of valence based upon hydrogen as a reference did not really explain anything and can be very confusing. Some elements have fractional valences in some compounds, which is embarrassing, and some have so many different valences that the idea loses all value.

Valence and the "rule of eight" 7.2

At a time when the Bohr atom was the only model that chemists had, G. N. Lewis made a proposal which accounted for the observed valences of the elements including the "fractional" valences and other embarrassments. He took his clue from the noble gases. Observing that helium has two electrons and that neon and argon have eight electrons in their outer shells, he proposed that each element would *gain, lose,* or *share* enough electrons in a chemical reaction until it had a noble gas configuration around it.

The *Lewis diagram* (also called the electron dot formula) is a means of representing the outer electrons of an atom. The electrons in the outer *s* and *p* orbitals are represented by dots around the symbol for the element. No distinction is made between electrons in *s* and *p* orbitals. Table 7.2 gives the Lewis diagrams of some common elements (valence electrons are underlined).

Table 7.2

Lewis diagrams of some common elements

Element	Electron Configuration	Lewis Diagram
hydrogen	$\underline{1s^1}$	H·
lithium	$1s^2\underline{2s^1}$	Li·
beryllium	$1s^2\underline{2s^2}$	·Be·
boron	$1s^2\underline{2s^22p^1}$	·Ḃ·
carbon	$1s^2\underline{2s^22p^2}$	·Ċ̣·
nitrogen	$1s^2\underline{2s^22p^3}$	·Ṇ̇:
oxygen	$1s^2\underline{2s^22p^4}$	:Ọ̇:
fluorine	$1s^2\underline{2s^22p^5}$	:Ḟ̤:
sodium	$1s^22s^22p^6\underline{3s^1}$	Na·
chlorine	$1s^22s^22p^6\underline{3s^23p^5}$	:Ċ̤l:

Ionic bonding 7.3

Electrons carry a unit negative charge. If an atom has lost one or more electrons, there will be an excess corresponding positive charge remaining on the atom, and we call the charged atom a *cation*. Likewise, an atom that has gained electrons will have an excess of negative charge, and the charged atom is called an *anion.*

The metallic elements on the left side of the periodic table react by transferring electrons from their valence shell to an acceptor atom or atoms. In doing so, they become a positively charged ion with the electron configuration of the *previous* noble gas. Sodium loses its 3*s* electron and becomes Na^+, with the electron configuration of neon. Calcium loses its 4*s* electrons and becomes Ca^{2+}, with the configuration of argon.

When the acceptor atom is a nonmetal the atom acquires a negative charge and the electron configuration of the *next* noble gas. Chlorine accepts one electron and becomes Cl^-, with the configuration of argon. Oxygen accepts two electrons and becomes O^{2-} with the configuration of neon.

Figure 7.1 Sodium chloride (Na^+Cl^-), an ionic crystal. The dark spheres represent sodium ions, Na^+, and the light spheres chloride ions, Cl^-

Anions and cations attract each other with a coulombic force that extends in all directions. This means that the ions will be bonded to a number of oppositely charged ions around them but will not have a shared bond with any particular one. The compound consists of endless rows of alternating positive and negative ions (Figure 7.1).

The electrical fcrces holding the ions together are strong and extend beyond the immediate neighbors of each ion. Ionic compounds are usually hard, brittle, crystalline substances with a high melting point.

The relative number of each kind of ion in a compound depends on the charge on the ions. If the charges are equal, there will be the same number of each kind. If the charge on the anion is two times that of the cation, there must be twice as many cations as anions, so that the overall charge is balanced.

EXAMPLE 7.1

1. What is the number ratio of Mg^{2+} to Br^- in a compound?
   There must be the same number of positive and negative charges in the compound; that is, the algebraic sum of the charges must be zero.
   $$(2+) + 2(-) = 0$$
   There are two Br^- to one Mg^{2+}: $MgBr_2$
2. What is the number ratio of Al^{3+} to S^{2-} in a compound?
   $$2(3+) + 3(2-) = 0$$
   There are two Al^{3+} to three S^{2-}: Al_2S_3

Lewis diagrams of ionic compounds simply show the ions and their charges, thus:

$$:\ddot{\underset{\cdot\cdot}{Br}}:^{-}\ Mg^{2+}\ :\ddot{\underset{\cdot\cdot}{Br}}:^{-}$$

No bond is shown between the positive and negative ions.

Covalent bonds 7.4

Hydrogen, oxygen, nitrogen, and the halogens all exist as diatomic molecules in the elemental state. Since the atoms are identical, there is no reason for one to be positive and the other negative. Nevertheless, they are strongly bonded. The attraction seems to be satisfied with just one partner and does not go on indefinitely like the ionic compounds.

Nonmetals have little tendency to lose electrons for reasons we will discuss presently. They do have a tendency to fill their *p* orbitals. When nonmetals combine with each other, a bond is formed by merging or combining orbitals in such a way that one pair of electrons occupies a mutual orbital (called a molecular orbital). We cannot go into the details of the method in this book, but we can show what happens as two atoms approach each other.

Imagine two hydrogen atoms, each with one electron in its 1*s* orbital (Figure 7.2A). As the atoms approach each other, the orbitals distort as each electron begins to "feel" the presence of the other nucleus (B). Finally, the two orbitals have merged (C) into a new orbital around both nuclei. The shape suggests that the two electrons are most likely to be found between the two nuclei where they exert an attractive force on both of them.

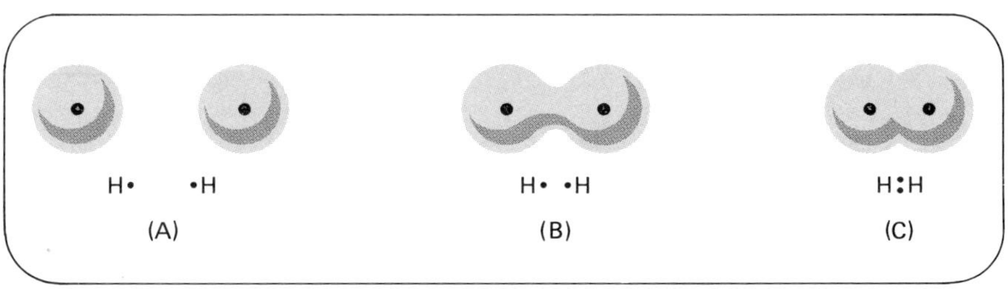

Figure 7.2 The combination of *s* orbitals of hydrogen to form a bonding molecular orbital.

We can also imagine how an *s* orbital and a *p* orbital might combine (Figures 7.3 and 7.4). Chlorine atoms have two filled *p* orbitals and one half-filled (or half-empty, if you prefer). The *s* orbital of a hydrogen atom can combine with or overlap the unfilled *p* orbital of

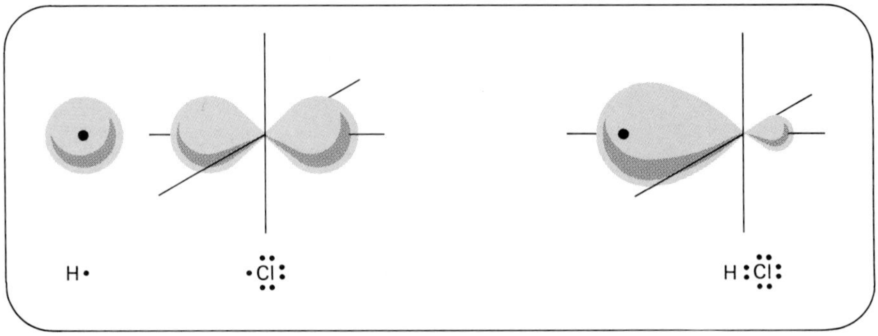

Figure 7.3 The combination of the *s* orbital of hydrogen and one *p* orbital of chlorine form a molecular orbital. The other *p* orbitals of chlorine are omitted for clarity.

chlorine, thereby giving hydrogen and chlorine a complete molecular orbital.

Hydrogen has a single orbital which can hold two electrons. The only possibilities for reaction are

- H· loses one electron to another atom and becomes H^+;
- H· gains one electron from a metal and becomes $H:^-$;
- H· contributes one electron to a bonding orbital (a shared pair) with another atom.

This accounts for hydrogen never being bonded to more than one other atom; it is always monovalent.

Carbon has four electrons that can be used to form bonding orbitals. A noble gas configuration could be formed by losing four or gaining four electrons, but a charge of 4+ or 4− on a single atom is much more than the little thing can bear. So it contributes its four electrons to four shared pairs with other atoms. Carbon avoids a high charge concentration that way but manages to complete its outer shell.

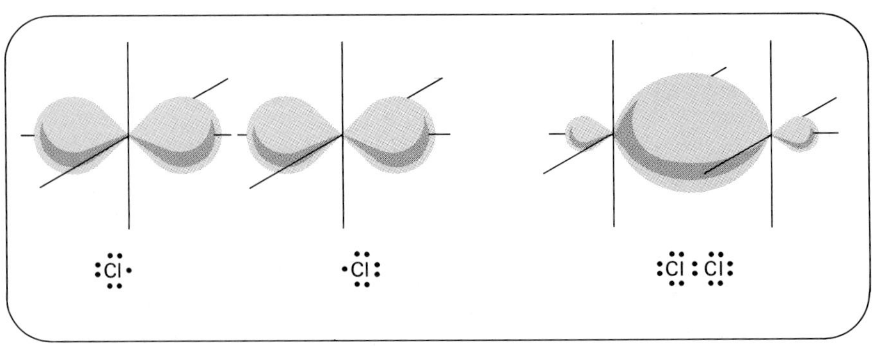

Figure 7.4 The *p* orbitals may also overlap to form molecular orbitals. Here chlorine uses its unfilled *p* orbitals to form a bond with another chlorine atom.

$$\cdot\dot{\underset{\cdot}{\text{C}}}\cdot + 4\text{H}\cdot \longrightarrow \text{H}:\overset{\text{H}}{\underset{\text{H}}{\ddot{\underset{\cdot\cdot}{\text{C}}}}}:\text{H}$$

Carbon is tetravalent in almost all of its compounds.

Nitrogen has five electrons in its outer shell. Three of the electrons are unpaired and can be used in bonding orbitals by pairing them with electrons from other atoms.

$$\cdot\ddot{\underset{\cdot}{\text{N}}}\cdot + 3\text{H}\cdot \longrightarrow \text{H}:\underset{\text{H}}{\ddot{\underset{\cdot\cdot}{\text{N}}}}:\text{H}$$

Nitrogen is trivalent in this compound.

By the same process we would predict that oxygen would be divalent and fluorine would be monovalent. They can complete their outer orbitals by gaining or sharing electrons.

An atom like sulfur has two half-filled *p* orbitals. It can use them to bond two atoms to sulfur. Because *p* orbitals are perpendicular to each other the molecule will not be linear like

$$\text{H}:\ddot{\underset{\cdot\cdot}{\text{S}}}:\text{H}$$

but is bent like

$$\text{H}:\underset{\text{H}}{\ddot{\underset{\cdot\cdot}{\text{S}}}}:$$

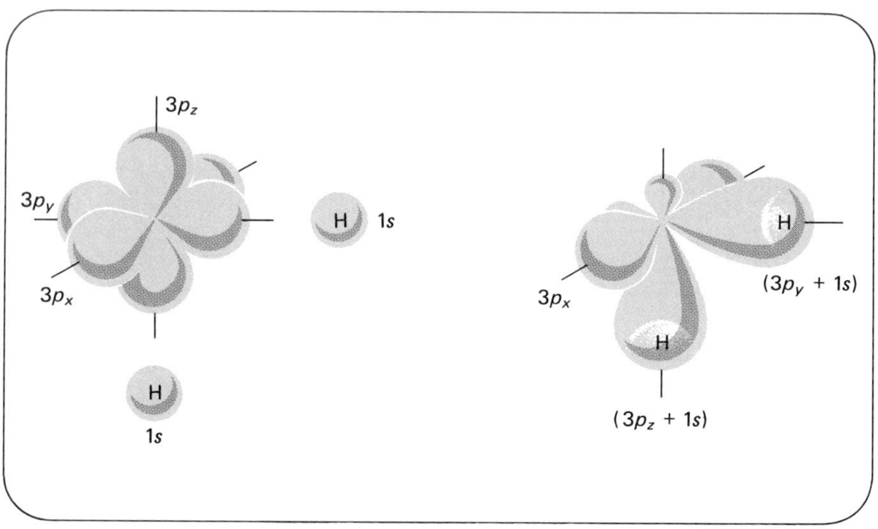

Figure 7.5 Bonding of two hydrogens to sulfur. The $3p_y$ and $3p_z$ orbitals of sulfur have one electron each. A bond is formed when the hydrogen 1*s* orbital overlaps one lobe of the *p* orbital. The other lobe becomes smaller, reflecting the smaller likelihood of finding the electrons on the side away from the hydrogen. For clarity, the 3*s* orbital and the inner electrons of sulfur are not shown.

The formation of bonding orbitals from two *p* orbitals is shown in Figure 7.5.

Water is also a bent molecule. The hydrogens are at an angle of about 104 degrees to each other in water.

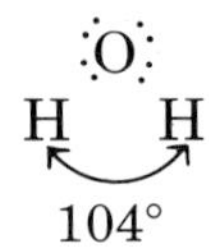

The bond formed by the overlapping of atomic orbitals is called a covalent bond. It implies that there is a sharing of valence electrons, but the sharing is not necessarily equal. In the water molecule and the HCl molecule, oxygen and chlorine have the greater share of the electrons in the bond. The center of negative charge will be closer to the oxygen or chlorine, leaving the positive hydrogen nucleus relatively exposed. This results in the bond having a somewhat positive end and a negative end, so it is said to be a *polar covalent bond.* The degree of polarity in the bonds of a molecule greatly influences the physical and chemical properties of the substance.

EXERCISE 7.1

(a) Write electron dot formulas for

CCl_4 CHF_3 N_2H_4 OF_2

(b) What structure can be drawn for a compound containing four H, one C, and one O?

7.5 *Multiple bonds*

Carbon, nitrogen, oxygen, and sulfur are sometimes bonded to fewer atoms than their normal valence indicates. An unused valence is a very reactive thing, so these atoms satisfy their valence requirements by forming double or triple bonds with each other.

EXAMPLE 7.2

1. A molecule of carbon dioxide consists of one atom of carbon and two atoms of oxygen. The Lewis diagram of carbon dioxide can be formed by making bonding pairs between all of the unpaired electrons of carbon and oxygen:

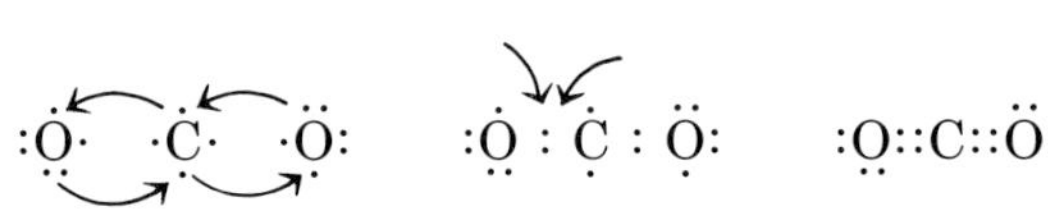

Carbon and oxygen each satisfy their valence by forming two double bonds.

2. A molecule of hydrogen cyanide has one atom each of hydrogen, carbon, and nitrogen. Carbon and nitrogen satisfy their unused valences and achieve an octet by forming a triple bond.

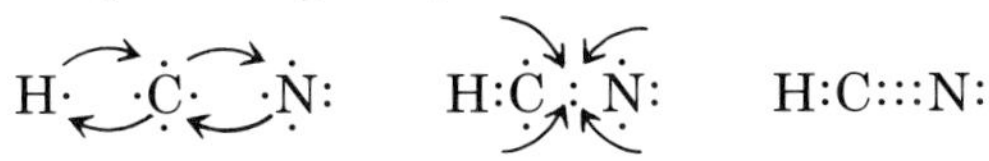

3. A molecule of formaldehyde has two hydrogens, one carbon, and one oxygen. Carbon has three atoms bound to it, but has four electrons to share. Oxygen has one atom bound to it but has two electrons to share. The extra valences are used by forming a double bond.

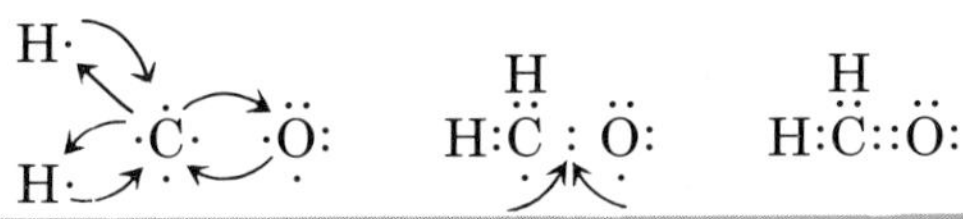

We now have a revised definition of valence. The *valence* of an atom is the number of electrons it has *gained* or *lost*, or has *contributed to shared bonds* in a compound.

Coordinate covalent bonds 7.6

No more than two electrons may occupy a molecular orbital and these must have opposite spins according to the exclusion principle. However, there is no requirement that each atom supply one electron. Each atom must supply one *orbital*; the electrons needed to occupy the orbital may come from one atom. Bonds like this are called *coordinate covalent* bonds, but they are not chemically different from ordinary covalent bonds. We distinguish between them only if it is necessary to keep track of the valences of the atoms.

EXAMPLE 7.3

The "donated" electron pair is indicated by an arrow from the donor atom:

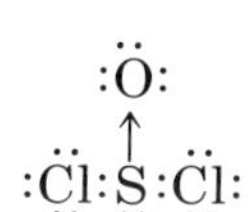

thionyl chloride. S is tetravalent because it has contributed four electrons to bonding.

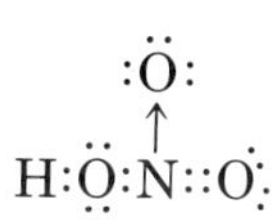

nitric acid. N is pentavalent because it has contributed all five of its electrons to bonds.

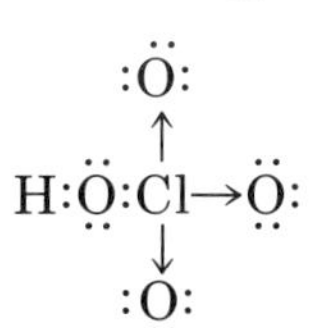

perchloric acid. Cl is heptavalent because it has involved all seven of its electrons in bonds.

Multiple bonds form in preference to coordinate covalent bonds if, in doing so, more electrons are not put in an atom's valence shell than it can hold.

EXAMPLE 7.4

1. We could show the formation of carbon dioxide this way:

But now carbon only has four electrons in its outer shell, along with two empty orbitals. We are not forced to write coordinate covalent bonds here; we can put in double bonds.

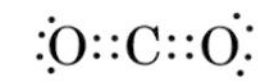

2. We cannot write

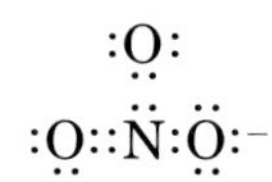

for nitrate ion, because nitrogen can only put eight electrons in the valence shell. We are forced to write one of the nitrogen-oxygen bonds as a coordinate covalent bond.

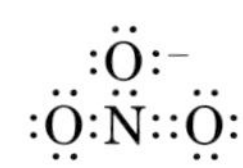

3. We cannot write

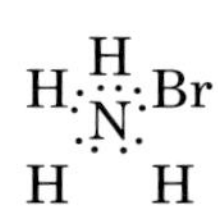

for ammonium bromide for the same reason. The bond between nitrogen and bromine must be an ionic bond.

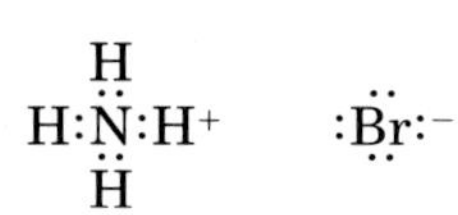

7.7 *Limitations of the Lewis diagrams*

The "rule of eight" was popular because it explained the observed valences of the elements in the first and second periods. Third period elements often fit the rule, too. Much of organic chemistry is concerned with only these elements, so Lewis diagrams were useful for depicting organic molecules. Lewis diagrams have contributed much to the understanding of acids and bases (see Chapter 12) and to the understanding of reaction sites in molecules. The Lewis theory is far from being a complete theory, however. (Remember it was proposed before orbitals were invented). Here are some facts that are contrary to Lewis theory or are left unexplained by it.

1. Many molecules exist in which there are not eight electrons around each atom. Some have fewer than eight:

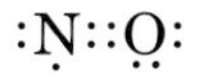
nitric oxide

boron trifluoride

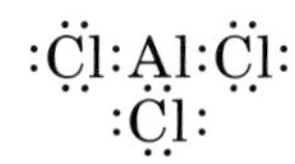
aluminum chloride

Some elements in the third period and beyond have more than eight:

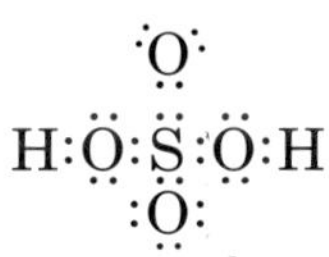
sulfuric acid

phosphorus pentachloride

2. Lewis theory offered no explanation for the fact that some elements combine with ions and neutral molecules with no apparent regard for any rules of valence:

$$Cu(NH_3)_4^{2+} \qquad Fe(CN)_6^{3-} \qquad Co(H_2O)_6^{2+}$$

3. The true shapes of the molecules are not shown by Lewis diagrams.

Electronegativity 7.8

How can we predict whether a bond will be mostly ionic, polar covalent, or pure covalent? One of the requirements for the formation of molecular orbitals is that the two atomic orbitals not be too different in energy. An approximate idea of the relative energy of an atomic orbital can be obtained by measuring the work necessary to pull off the highest energy electron in the atom. This work is called the *ionization potential.* The value of the first ionization potential of the elements in the first four periods is shown in Figure 7.6. Notice that the very active metals all have very low ionization potentials, and the noble gases have the highest. The active nonmetals also have high ionization potentials.

Free electrons are quickly captured by atoms. The energy released

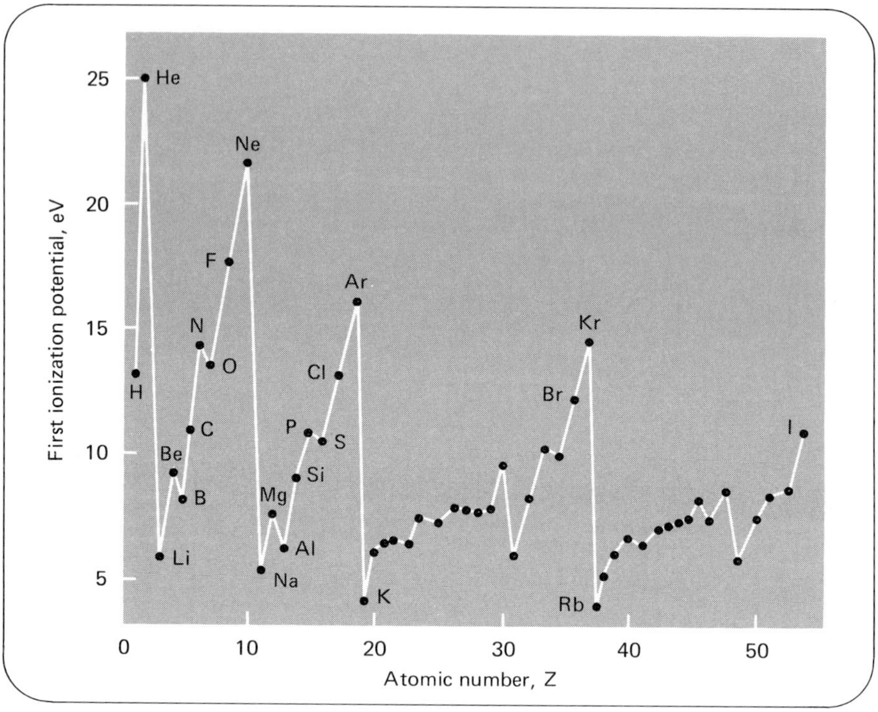

Figure 7.6 Plot of the first ionization potentials of the elements of the first four periods. The work needed to pull off the first electron is given in electron volts, an atomic-sized energy unit.

when an electron falls into an atomic orbital is called the *electron affinity* of the atom. The electron affinity of nonmetals is higher than that of metals.

These two quantities have been combined by Linus Pauling and others into a semiquantitative expression of the tendency of an atom to attract electrons in a bond. This tendency is called the *electronegativity* of the atom.

Pauling's values for the electronegativities of elements in the first three periods are given in Table 7.3. These are comparative numbers based on F = 4.0 (no dimensions).

Table 7.3
Electronegativity values

H						
2.1						
Li	Be	B	C	N	O	F
1.0	1.5	2.0	2.5	3.0	3.5	4.0
Na	Mg	Al	Si	P	S	Cl
0.9	1.2	1.5	1.9	2.1	2.5	3.0

The difference between the electronegativity numbers of two elements is an indication of the amount of polarity in a single bond between the two isolated atoms. If the difference is large (1.0 unit or more) it means that there are large differences in the ionization potentials and electron affinities. The bond will be mostly ionic, as if the electrons spent most of the time around the more negative atom.

As the difference between electronegativity numbers becomes smaller the bond exhibits more covalent character. The atoms become bonded to each other and are not separated by thermal motions or electrical forces. The degree of polarity in a bond can be roughly estimated from the size of the electronegativity difference. The polarity of a bond does depend on the rest of the molecule and is strongly affected by a polar environment like water, so the numerical value of the electronegativity difference should not be taken too literally.

Molecular shapes and structural formulas 7.9

Molecules are, of course, three dimensional. The shape, size, and polarity of a molecule or ion determines its behavior to a large extent. Visual representations unfortunately are two dimensional so we must use conventions and lots of imagination to visualize the molecular model from its representation. A representation that shows the bonds between individual atoms and the space relationships between them is a *structural formula.*

A simple and effective rule for predicting the shape of a covalent molecule is based on the fact that electrons with the same spin state repel each other strongly. This means that electron pairs will occupy orbitals as far from each other as possible. The four pairs of electrons in an ordinary valence shell will be directed toward the four corners of a tetrahedron (Figure 7.7).

An atom with six electrons in its valence shell would take the form of a triangle; elements like phosphorus or sulfur, which can hold as many as five or six singly bonded atoms, take the most symmetrical arrangement possible. Most compounds of interest to us at this stage with the exception of hydrogen have four pairs of electrons around each atom.

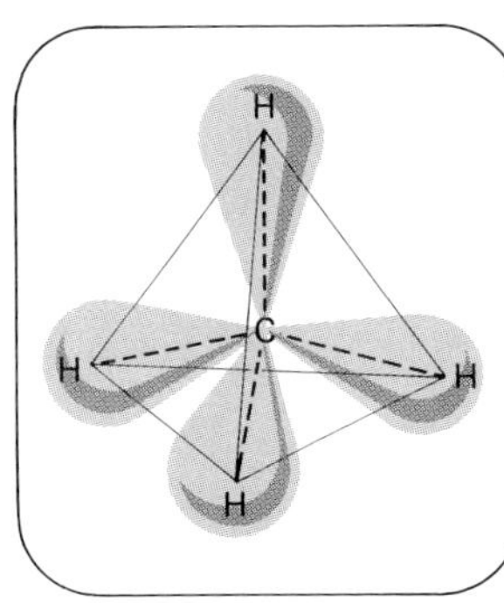

Figure 7.7 Methane, CH_4

Let us represent a pair of shared electrons (a bond) by a line projecting from the center of the atom toward the corner of a tetrahedron as in Figure 7.7. Nonbonding electrons are best represented by dots to remind us that they do take up space and affect the direction of the bonds. They occupy the other corners of the tetrahedron.

If we imagine that single bonds are formed by sharing corners of tetrahedra (Figure 7.8A), double bonds are shared edges of tetrahedra

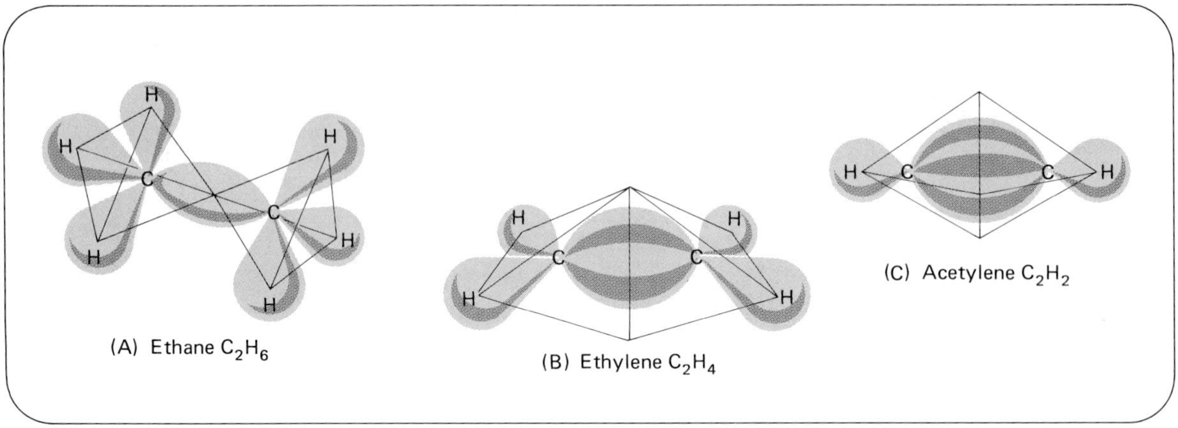

Figure 7.8 (A) Ethane C_2H_6; **(B)** ethylene C_2H_4; **(C)** acetylene C_2H_2.

(Figure 7.8B), and triple bonds are shared faces of tetrahedra (Figure 7.8C), then we have a molecular model which closely resembles the observed geometry of real molecules.

Table 7.4 compares the three formula conventions of a number of molecules and ions.

7.10 *Oxidation numbers*

The combining capacity of an atom in a compound is its valence. The valence is only a counting number and does not specify what the atom can combine *with.* A chloride ion cannot combine with a bromide ion, even if both are monovalent. The valence is therefore given a positive or negative sign, depending on the number of electrons donated or accepted by the atom in its particular bonding condition. This number now represents the combining capacity of the atom for atoms of opposite sign. It is called the *oxidation number* of the atom.

The oxidation number of a monatomic ion is the charge on the ion:

Ion	*Oxidation Number*
H^+	+1
Na^+	+1
Ca^{2+}	+2
S^{2-}	−2
Br^-	−1

The oxidation number of covalently bound atoms is calculated by assuming that the more electropositive atom *donates* electrons to a bond and the more electronegative atom *accepts* them. Each donated electron counts as one *plus*; each accepted electron as one *minus*.

Table 7.4
Structural formulas of some simple compounds based on Lewis diagrams

Compound	Molecular Formula	Lewis Diagram	Structural Formula
1. water	H_2O	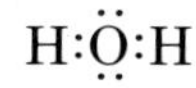	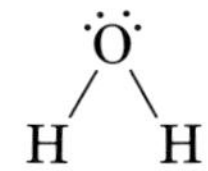
2. methane	CH_4	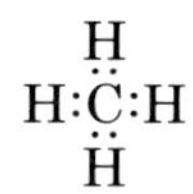	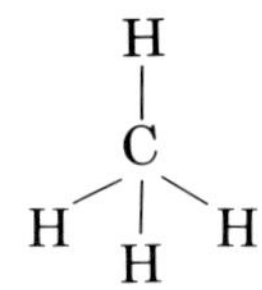
3. ammonia	NH_3	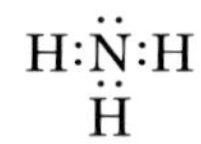	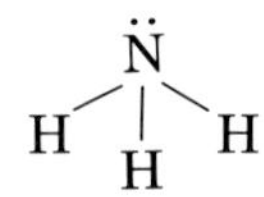
4. methanol	CH_4O	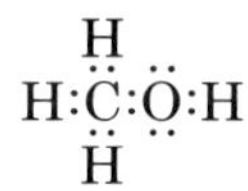	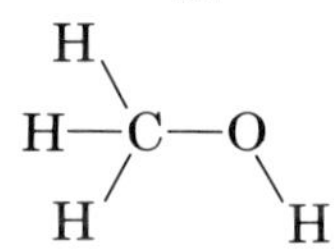
5. formaldehyde (methanal)	CH_2O	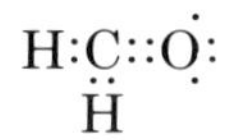	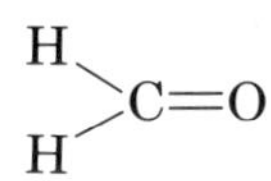
6. hydrogen cyanide	HCN	H:C:::N:	H—C≡N
7. sulfate ion[a]	SO_4^{2-}	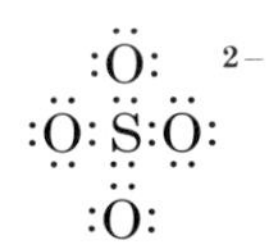	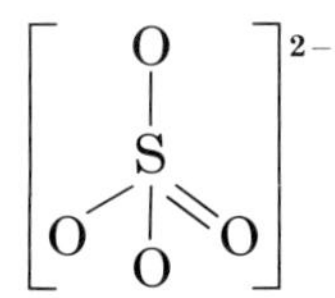
8. nitrate ion[a]	NO_3^-	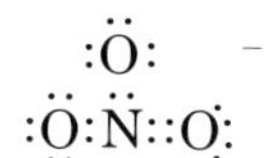	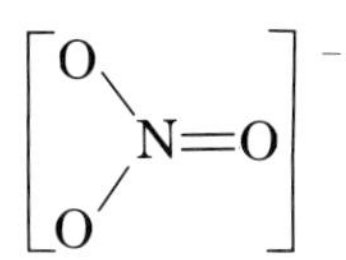
9. ammonium ion	NH_4^+	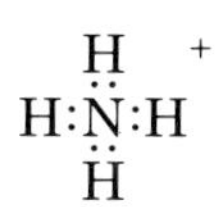	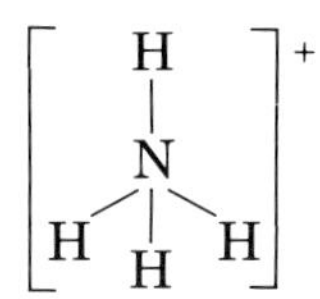
10. bicarbonate ion	HCO_3^-	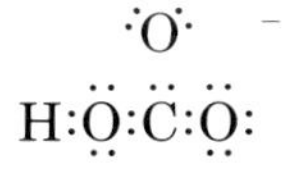	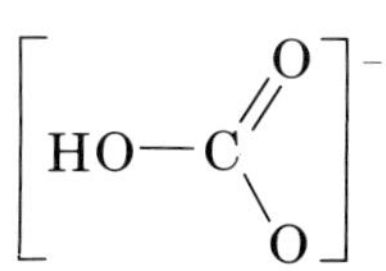

[a] Note: the S—O bonds in sulfate and the N—O bonds in nitrate are all identical, but this cannot be shown with this model.

The algebraic sum of the pluses and minuses is the oxidation number of the atom.

Oxidation numbers should refer to the relative positive or negative valence of an atom in a compound. The charges on polyatomic ions should be called ion charges.

Since the ion charge and oxidation number of monatomic ions are identical, sometimes people call the charge on a polyatomic ion the oxidation number. We purists will not invite them to our clambake, will we?

Ion charges are used to compute combining capacity of ions so that the formulas we write have neutral charges.

Oxidation numbers are mostly useful for determining whether an atom in a compound is relatively rich or poor in electrons. It is also used as a bookkeeping device for electron transfers in oxidation-reduction reactions. *The oxidation number is not the charge on the atom in covalent compounds!*

EXAMPLE 7.6

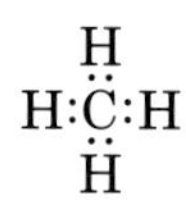

methane

Carbon is more negative than hydrogen; each hydrogen "donates" one electron, carbon "accepts" four electrons: C = −4; H = +1

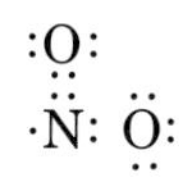

nitrogen dioxide

Oxygen is more negative than nitrogen. Nitrogen "donates" four electrons to the bonds; each oxygen "accepts" two: N = +4; O = −2

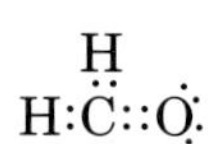

formaldehyde

Oxygen is more negative than carbon; carbon is more negative than hydrogen. Carbon "accepts" two from hydrogen and "donates" two to oxygen:
H = +1; C = 0; O = −2

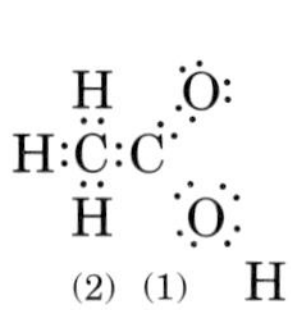

acetic acid

A bond shared between identical atoms is a trade-off; it is assumed that the electrons are shared equally. However, number 2 carbon has accepted three electrons from hydrogen; number 1 carbon has donated three electrons to the two oxygens:
$C_1 = +3$, $C_2 = -3$, H = +1, O = −2

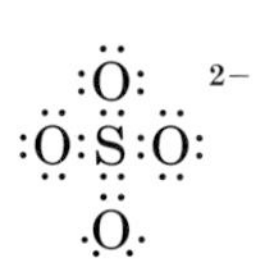

sulfate ion

Oxygen is more electronegative than sulfur; sulfur has "donated" six electrons to bonding. The ion has two extra electrons from whatever cations make up the compound.
S = +6; O = −2

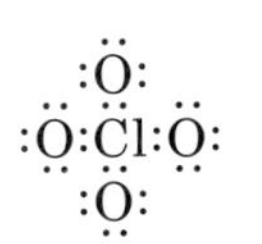

perchlorate ion

Oxygen is more negative than chlorine; the latter has donated seven electrons to the bonds.
Cl = +7; O = −2

hydrogen peroxide

An O—O bond does not count. Each O has accepted one electron from hydrogen.
O = −1; H = +1

Note that the sum of the oxidation numbers of all the atoms in a compound is zero. The sum of the oxidation numbers of the atoms in a polyatomic ion is the charge on the ion.

Summary 7.11

1. Atoms combine with each other by gaining, losing, or sharing electrons from their outermost *s*, *p* and (sometimes) *d* orbitals.
2. The valence of an atom is the number of electrons it has lost, gained, or contributed to shared bonds.
3. Ionic compounds are held together by the electrostatic attraction of oppositely charged particles (ions). The charge on a positive ion is equal to the number of electrons lost. The charge on a negative ion is equal to the number of electrons gained.
4. Atoms are bonded covalently when they share a mutual orbital containing a pair of electrons. Covalent bonds are possible when atoms do not differ too much in electronegativity.
5. Electron dot (Lewis) structures identify which atoms are bonded and the approximate electron distribution in a molecule or polyatomic ion.
6. Structural formulas also indicate which atoms are bonded and may be drawn to indicate the shape of the molecule.
7. The oxidation number of an atom is the algebraic (signed) number equal to the sum of the number of electrons the atom has donated to more electronegative atoms *minus* the number of electrons accepted from more electropositive atoms.

EXERCISES

I. Draw the electron dot diagram of these atoms:

A. He Be C N F Na Al S Ar Ca

B. H B O Ne Mg Si P K Cl Zn

II. Draw the electron dot diagram of these ions:

A. Li^+ Mg^{2+} F^- O^{2-}

B. Na^+ Al^{3+} S^{2-} Cl^-

III. Draw the electron dot diagram of these molecules and ionic compounds:

A. LiBr H_2S PCl_3 $AlCl_3$ H_2CO_3
Na_2SO_4 C_2H_6 $B(OH)_3$ HOCl $(NH_4)H_2PO_4$

B. KI SiH_4 $Al(OH)_3$ $HClO_4$ H_2CO_2
$CHCl_3$ C_2H_5OH HNO_3 $KMnO_4$ $(NH_4)_2SO_4$

IV.A. What is the valence of nitrogen in these compounds?

1. (a) NH_3 **(b)** N_2H_4 **(c)** NH_2OH **(d)** N_2 **(e)** NO
(f) NO_2 **(g)** HNO_3 **(h)** $(CH_3)_3N$ **(i)** $(CH_3)_3N \rightarrow O$

2. What is the oxidation number of nitrogen in those compounds?

B. What is the valence of carbon in these compounds?

(a) CH_4 **(b)** C_2H_6 **(c)** CH_3OH **(d)** C (diamond) **(e)** CO
(f) CO_2 **(g)** H_2CO_3 **(h)** CCl_4 **(i)** $CHCl_3$ **(j)** CH_2Cl_2
(k) CH_2O

2. What is the oxidation number of carbon in those compounds?

V.A. Draw structural formulas for all compounds in IVA.
B. Draw structural formulas for all compounds in IVB.

VI.A. Draw the possible structures of compounds formed from these collections of atoms:

1. 1B, 3Cl
2. 1N, 3Cl
3. 2S, 2Cl
4. 1C, 1N, 1H
5. 1C, 1N, 5H
6. 2C, 2H
7. 2C, 4H
8. 2C, 6H
9. 1O, 2F
10. 1Cl, 3F, 1C

B. Draw the possible structures of compounds formed from these collections of atoms:

1. 1P, 5Cl
2. 1Si, 4Cl
3. 2P, 5O
4. 1P, 3H
5. 1C, 2S
6. 1C, 1S, 4H
7. 1C, 1O, 2Cl
8. 1P, 1O, 3Cl
9. 1P, 4O, 3H
10. 1S, 3O, 2H

VII.A. Draw the possible Lewis structures of these polyatomic ions:

1. ClO_3^-	**4.** $Cr_2O_7^{2-}$
2. $S_2O_3^{2-}$	**5.** SCN^-
3. $C_2O_4^{2-}$	**6.** CrO_4^{2-}

B.

1. ClO_4^-	**4.** $P_2O_7^{4-}$
2. NO_2^-	**5.** OCN^-
3. HCO_2^-	**6.** MnO_4^-

SOLUTIONS

I.A. 

II.A. $Li^+ = He$, $Mg^{+2} = Ne$, $F^{-1} = Ne$, $O^{-2} = Ne$

III.A. 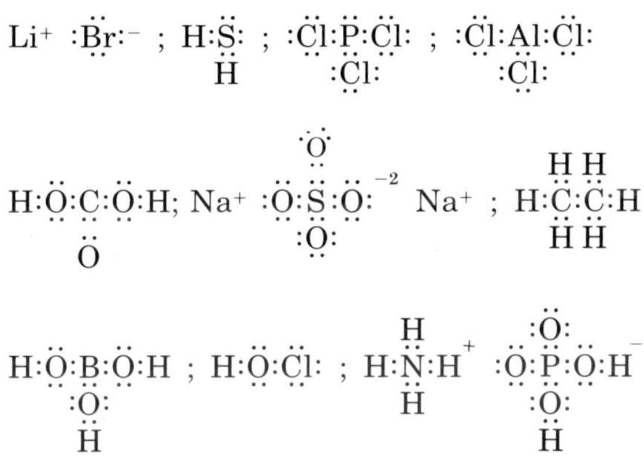

IV.A. **1.** **(a)** 3 **(b)** 3 **(c)** 3 **(d)** 3 **(e)** 2
(f) 4 **(g)** 5 **(h)** 3 **(i)** 5

2. **(a)** −3 **(b)** −2 **(c)** −1 **(d)** zero **(e)** +2
(f) +4 **(g)** +5 **(h)** −3 **(i)** −1

V.A. 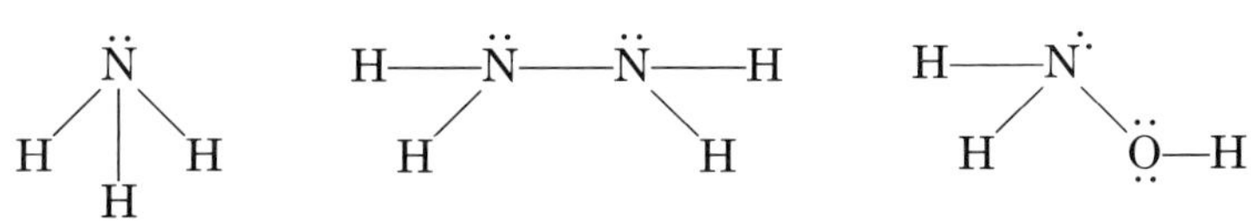

(a) **(b)** **(c)**

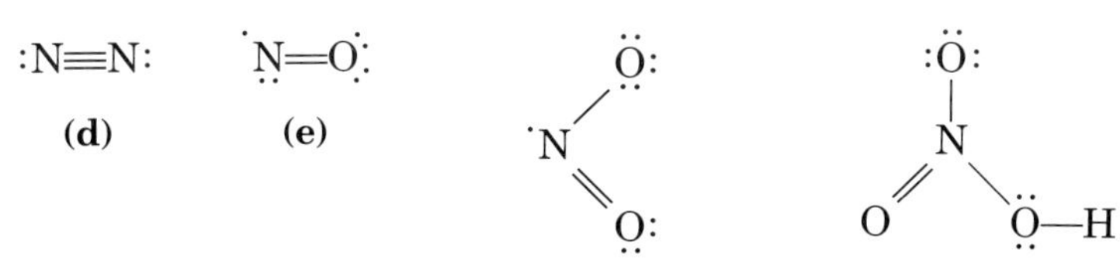

(d) **(e)** **(f)** **(g)**

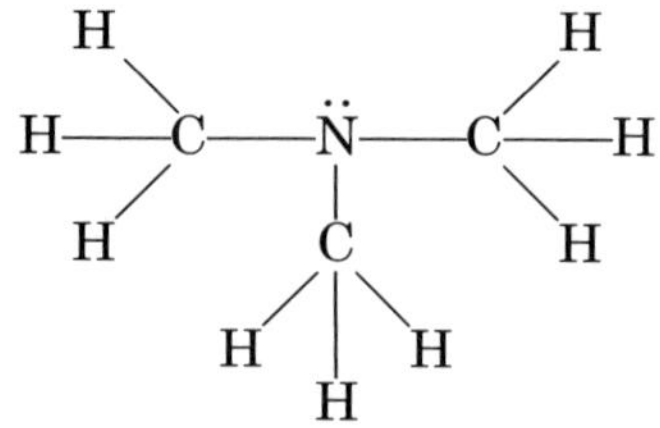

(h)

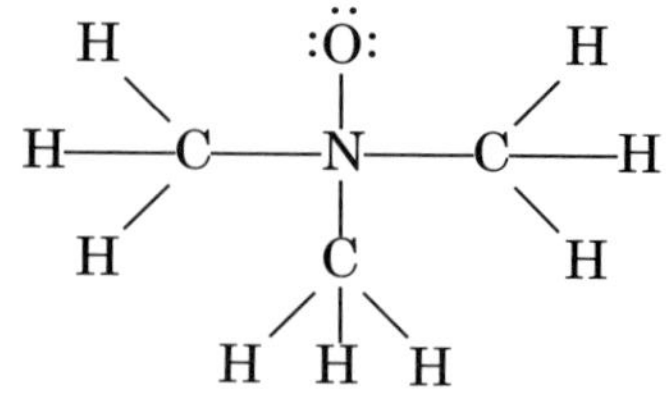

(i)

VI.A.

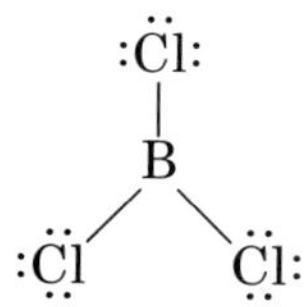

(1)

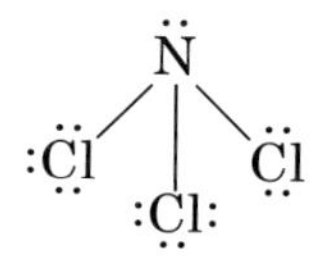

(2)

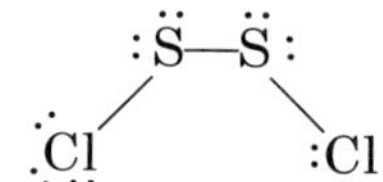

(3)

H—C≡N

(4)

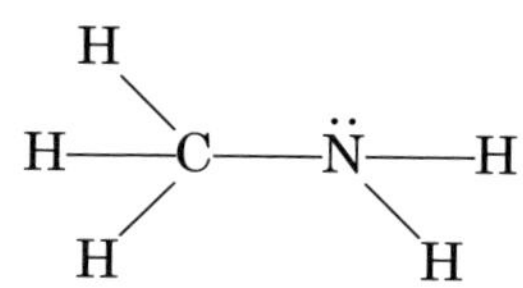

(5)

H—C≡C—H

(6)

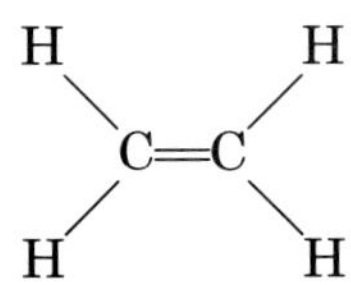

(7)

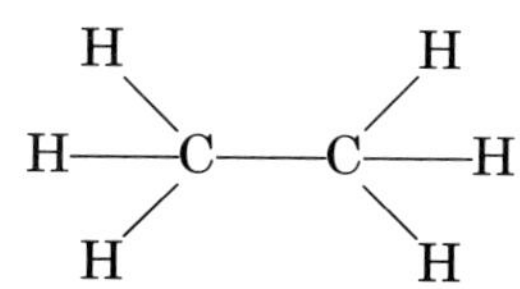

(8)

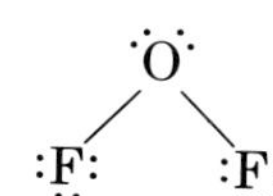

(9)

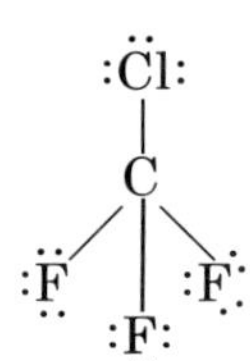

(10)

VII.A.

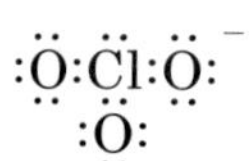

(1)

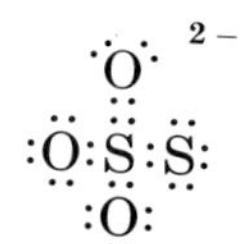

(2)

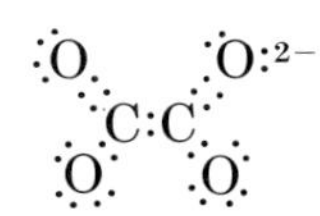

(3)

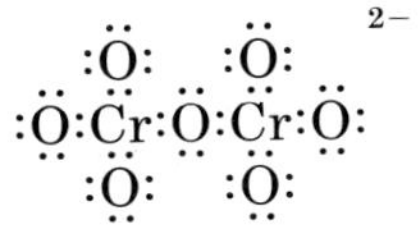

(4)

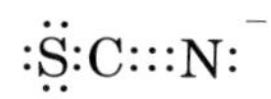

(5)

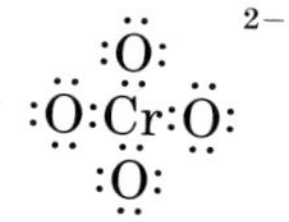

(6)

A systematic method of naming compounds ensures that each substance will have a unique name—even those that have not yet been made!

After reading this chapter you should know

1. The lowest (-ide) oxidation numbers of the nonmetals;
2. The classical and Stock system for naming binary compounds;
3. The chemical groups with invariant oxidation numbers;
4. The structures, names, and ion charges of the common polyatomic ions;
5. The names of some common acids and acid salts;
6. How to write formulas from names;
7. How to calculate oxidation numbers from rules and reference atoms;
8. How to write formulas and names of hydrates.

chemical nomenclature

8

The name of a pure substance should convey complete information about its composition to anyone concerned with using it. This ideal situation contrasts with the practices of the proprietary trades where every effort is made to conceal the composition with meaningless trade names. Presumably a consumer with a little knowledge of chemistry could make his own judgment about the use and efficacy of products if their composition were known. This idea seems to horrify ad agencies.

Chemists, being more interested in exchanging information than concealing it, have devised a system of nomenclature that attaches a unique name to every possible compound. Since some 400,000 inorganic compounds and upward from 2,500,000 organic compounds have been made or isolated, you can be sure that they are not named in the same haphazard way that people are.

The nomenclature system was recommended by a committee of the International Union of Pure and Applied Chemistry (IUPAC). The rules as given here are by no means comprehensive, but they are the foundation of inorganic chemical nomenclature and will enable you to recognize thousands of inorganic compounds.

Trivial or common names 8.1

Certain substances have been known for so long and are so familiar that it is not worth the effort to convert their names to conform to the system. Except for those given in Table 8.1, the use of trivial names is discouraged.

Table 8.1

Commonly accepted trivial names

Name	Formula	Name	Formula
water	H_2O	phosphine	PH_3
ammonia	NH_3	silane	SiH_4
methane	CH_4	ethylene	C_2H_4
ethane	C_2H_6	acetylene	C_2H_2
benzene	C_6H_6	acetic acid	$HC_2H_3O_2$
formic acid	HCO_2H		

8.2 *Names of binary compounds*

Given the formula of a compound it may be presumed that the most positive element is written first. The name must indicate

1. the identity of the elements,
2. that the elements are combined with each other, and
3. the relative number of each element, especially if two or more compounds of these same elements exist.

Two commonly accepted nomenclatures communicate this information. Both systems indicate compound formation by adding a suffix *-ide* to the root name of the more negative (second) element. The -ide ending of the root name of an element indicates that it is in its lowest (most negative) oxidation state (Table 8.2).

Table 8.2

Roots and oxidation states of the more negative elements

Element	Root	Negative Element in Compound	Oxidation Number
hydrogen	hydr-	hydride	−1
carbon	carb-	carbide	−4
nitrogen	nitr-	nitride	−3
phosphorus	phosph-	phosphide	−3
oxygen	ox-	oxide	−2
sulfur	sulf-	sulfide	−2
fluorine	fluor-	fluoride	−1
chlorine	chlor-	chloride	−1
bromine	brom-	bromide	−1
iodine	iod-	iodide	−1

A few other ions also carry an -ide ending as part of their names (Table 8.3).

Table 8.3
Some polyatomic ions ending in -ide

Name	Ion
hydroxide	OH^-
amide	NH_2^-
cyanide	CN^-
acetylide	C_2^{2-}
peroxide	O_2^{2-}
disulfide	S_2^{2-}

The two systems are identical with respect to the identity and combining capacity of the positive elements whose oxidation numbers are always the same. These are the Group IA metals, Group IIA metals, Group IIIA metals, and Group IIB metals. The identity is given simply by the name of the element, and the oxidation number is assumed to be known.

EXAMPLE 8.1

Formula	Name
NaH	sodium hydride
K_2O	potassium oxide
$AlCl_3$	aluminum chloride
CaF_2	calcium fluoride
ZnS	zinc sulfide
Al_2O_3	aluminum oxide

The same convention follows with those polyatomic ions ending in -ide:

$Mg(CN)_2$	magnesium cyanide
$Ca(OH)_2$	calcium hydroxide
$NaNH_2$	sodium amide

The other elements, both metallic and nonmetallic, can have a number of different combining states, and the systems differ in the method of imparting that information. The classical system will be outlined first, because its use is widespread and it should be recognized.

Classical system of nomenclature 8.3

A. Metals combined with nonmetals

If a metal commonly exists in two oxidation states the higher is indicated by an *-ic* ending to the root name of the metal, and the lower by an *-ous* ending. If there are more than two oxidation states of a metal, some other method must be used. To add to the difficulty some of the roots are taken from the Latin name of the element.

EXAMPLE 8.2

Formula	Name	Oxidation State of Metal
$FeCl_2$	ferrous chloride	+2
$FeCl_3$	ferric chloride	+3
Cu_2O	cuprous oxide	+1
CuO	cupric oxide	+2
Hg_2Cl_2	mercurous chloride	+1
$HgCl_2$	mercuric chloride	+2
CrO	chromous oxide	+2
Cr_2O_3	chromic oxide	+3

B. Binary compounds of nonmetals

Nonmetals can usually combine covalently in several different combinations. This makes the -ous/-ic method inoperative. One alternative is to specify the number of each element with a Greek prefix (so now you have to learn to count in Greek!) (Table 8.4).

Table 8.4

Greek number prefixes

Number	Prefix
one	mono-
two	di-
three	tri-
four	tetra-
five	penta-
six	hexa-
seven	hepta-
eight	octa-
nine	ennea- (or nona-)
ten	deca-

EXAMPLE 8.3

Formula	Name
As_2S_3	diarsenic trisulfide
SiO_2	silicon dioxide
CO	carbon monoxide
SO_2	sulfur dioxide
P_4O_{10}	tetraphosphorus decoxide
N_2O_3	dinitrogen trioxide
CS_2	carbon disulfide
CCl_4	carbon tetrachloride

8.4 *The IUPAC (Stock) system*

Although the use of the prefix notation is still recognized by the IUPAC, it is strongly recommended that the Stock system be used exclusively. The system is simple and requires little prior knowledge.

In the Stock system the more metallic element is named first (without a suffix) followed by a roman numeral in parentheses to indicate its oxidation state. The more negative element or anion is then named.

EXAMPLE 8.4

Formula	Stock Name (compare Example 8.2)
Cu_2O	copper(I) oxide
CuO	copper(II) oxide
PbO	lead(II) oxide
Pb_2O_3	lead(II, IV) oxide (lead does not occur in the +3 state)
Pb_3O_4	lead(II, II, IV) oxide
PbO_2	lead(IV) oxide
FeO	iron(II) oxide
Fe_2O_3	iron(III) oxide
Fe_3O_4	iron(II, III) oxide
CrO	chromium(II) oxide
Cr_2O_3	chromium(III) oxide
CrO_3	chromium(VI) oxide

If an element occurs in two oxidation states in the same compound, both states are specified. If two or more compounds exist with the same two oxidation states, then the relative number of each state must be specified.

The system for compounds of nonmetals is exactly the same. The designation of an oxidation number by the Stock system says nothing about the degree of ionic character in a bond (cf. Chapter 7) but only provides a unique name from the formula and vice versa.[1]

EXAMPLE 8.5

Formula	Old Name	Stock Name
N_2O	nitrous oxide	nitrogen(I) oxide
NO	nitric oxide	nitrogen(II) oxide
NO_2	nitrogen dioxide	nitrogen(IV) oxide
NCl_3	nitrogen trichloride	nitrogen(III) chloride
SCl_2	sulfur dichloride	sulfur(II) chloride
PCl_3	phosphorus trichloride	phosphorus(III) chloride
ICl	iodine chloride	iodine(I) chloride

[1] In fact the accepted order of writing the elements does not always write the more negative element last. The halogens (except fluorine) behave as the more positive element in their compounds with nitrogen but are conventionally written after nitrogen. Hydrogen is listed after nitrogen in inorganic compounds, also.

8.5 *Polyatomic anions*

As noted previously, the -ide ending means that the ultimate element in a binary compound is in its lowest oxidation state. In many compounds the negative part (anion) is a covalently bound group of atoms with an over all negative charge. The nomenclature of these anions does demand that you *memorize* the charge and structure of the anion.

The relative oxidation state of the central atom in polyatomic anions is given by a suffix to the root name of the central atom. The most common higher oxidation state is given the suffix *-ate*. The next lower oxidation state is given the suffix *-ite*.

EXAMPLE 8.6

Formula	Oxidation Number of Central Atom	Name
NO_3^-	+5	nitrate
NO_2^-	+3	nitrite
PO_4^{3-}	+5	phosphate
PO_3^{3-}	+3	phosphite
SO_4^{2-}	+6	sulfate
SO_3^{2-}	+4	sulfite
ClO_3^-	+5	chlorate
ClO_2^-	+3	chlorite

The -ate or -ite suffix to a root means that the element named by the root is an anion surrounded by and bonded to a number of oxygens. These structures are called *oxyanions.*

The attachment of atoms or groups *other than* oxygen is indicated by a prefix formed from the root name of the attached atom or group followed by *-o-*.

EXAMPLE 8.7

Formula of Anion	Name	Attached Atom or Group	Prefix
BF_4^-	fluoborate ion	F^-	fluo-
$S_2O_3^{2-}$	thiosulfate ion	S	thio-
$Al(OH)_4^-$	tetrahydroxo-aluminate ion	OH^-	hydroxo-
$Fe(CN)_6^{3-}$	hexacyano-ferrate(III) ion	CN^-	cyano-
$PtCl_6^{2-}$	hexachloro-platinate(IV) ion	Cl^-	chloro-

Higher or lower oxidation states of the central atom are indicated by a prefix to the root name. The prefix *per-* means higher than, all, or complete—unless it means peroxide. This ambiguity has not been resolved but most be gotten from the context.[2] The prefix *hypo-* means under, lower than, or less than.

EXAMPLE 8.8

ClO_4^- perchlorate means higher than chlorate ClO_3^-.
ClO^- hypochlorite means lower than chlorite ClO_2^-.

Table 8.5 gives the formulas, names, and ion charges of a number of common anions. These should be memorized.

Table 8.5

Common polyatomic anions

Charge −	Charge 2−	Charge 3−
OH^- hydroxide	CO_3^{2-} carbonate	PO_4^{3-} phosphate
CN^- cyanide	$C_2O_4^{2-}$ oxalate	PO_3^{3-} phosphite
ClO^- hypochlorite	SO_4^{2-} sulfate	AsO_4^{3-} arsenate
ClO_3^- chlorate	SO_3^{2-} sulfite	
ClO_4^- perchlorate	$S_2O_3^{2-}$ thiosulfate	
NO_3^- nitrate	CrO_4^{2-} chromate	
NO_2^- nitrite	$Cr_2O_7^{2-}$ dichromate	
SCN^- thiocyanate		
MnO_4^- permanganate		
$C_2H_3O_2^-$ acetate		

Acids and acid salts 8.6

When the negative charge on an anion is neutralized by a positive hydrogen ion (H^+) the polarity of the resulting covalent bond may be highly polar or relatively nonpolar. Those compounds with hydrogen-to-nonmetal bonds that are more polar than those in water are called *acids* (see Chapter 12). Acids may be named like the salts, with hydrogen substituted for the metal, but it is more common to name their aqueous solutions as acids. The only rule to be followed is that acids of the -ate salts end in *-ic acid,* those of the -ite salts end in *-ous acid,* and those of the -ide salts are *hydro*(root)*ic acids*. The formulas of the acids have the acidic H's written first (Table 8.6).

[2] A solution to this undesirable situation has been proposed (R. C. Brasted, *J. Chem. Ed. 35*:136–9 (1958)) but has not yet been adopted.

Table 8.6

Some common acids

Mono-H Acids	Di-H Acids	Tri-H Acids
HF hydrofluoric	H_2CO_3 carbonic	H_3PO_4 phosphoric
HCl hydrochloric	$H_2C_2O_4$ oxalic	H_3BO_3 boric
HBr hydrobromic	H_2SO_4 sulfuric	
HI hydriodic	H_2CrO_4 chromic	
HCN hydrocyanic	$H_2C_8H_4O_4$ phthalic	
$HClO_4$ perchloric	H_2S hydrosulfuric	
HNO_3 nitric		
$HC_2H_3O_2$ acetic		

The di- and trivalent anions may be neutralized by two or more different cations. Naming these is no problem; the cations are simply listed in order. Alkali metals are first, followed by the other metals, ammonium ion, and hydrogen. The compound $K(NH_4)HPO_4$, for example, would be named potassium ammonium hydrogen phosphate.

A special case is observed with monohydrogen-substituted salts of divalent anions. These are often called acid salts or bi-salts.

EXAMPLE 8.9

	$NaHCO_3$	sodium hydrogen carbonate, or, sodium acid carbonate, or, sodium bicarbonate
	$NaHSO_4$	sodium hydrogen sulfate, or, sodium acid sulfate, or, sodium bisulfate
	$KHC_8H_4O_4$	potassium hydrogen phthalate, or, potassium acid phthalate, or, potassium biphthalate
BUT		
	Na_2HPO_4	disodium hydrogen phosphate, or, sodium monohydrogen phosphate
NEVER		sodium biphosphate

8.7 *Calculation of oxidation numbers*

The oxidation numbers of central elements in a formula may be calculated from a few rules and reference atoms.

Rule 1. The oxidation number of uncombined elements is zero.

Rule 2. The algebraic sum of the oxidation numbers is the charge on the ion (zero for neutral compounds).

Rule 3. The oxidation number of the most negative element in the compound is its lowest oxidation state.

Reference States of Atoms: fluoride is −1 always. Oxygen is −2, unless combined with itself (peroxides) or with fluorine. The alkali metals and ammonium (Group IA) are always +1. The alkaline earth metals (Group IIA) are always +2, as are the Group IIB metals zinc and cadmium (mercury is sometimes +1). Aluminum is always +3.

EXAMPLE 8.10

1. What is the oxidation number of sulfur in $NaHSO_4$?

   The most negative element is oxygen = −2 $4 \times -2 = -8$
   Sodium and hydrogen are each +1 $2 \times +1 = +2$
   -6

   The charge on the compound is zero; the sum of the oxidation number of sulfur and −6 must equal zero. Therefore the oxidation number of sulfur is +6.

2. What is the oxidation number of manganese in MnO_4^-?

   Manganese must have an oxidation number which when added to the four oxygens' −8 equals −1, the charge on the ion.

   $$x + (-8) = -1$$
   $$x = +7$$

   Manganese = +7.

3. What is the oxidation number of iron in Fe_3O_4?

   The oxidation numbers of *three* iron atoms must equal +8 (neutralizing $4 \times -2 = -8$). The only way to do this is to have two +3 atoms and one +2 atom (4 and 1 are unknown oxidation states for iron). The name of the compound is therefore iron(II, III) oxide.

Formulas from names 8.8

Writing the symbols in a formula from the name is no problem, provided of course that you have memorized the symbols. Problems develop in assigning the proper number of atoms and groups to the formula.

Rule 1. The charges on the atoms and groups must add algebraically to zero.

Rule 2. There can be no fractional atoms or groups (polyatomic ions).

EXAMPLE 8.11

1. What is the formula for aluminum sulfide?

   The -ide ending indicates that it is a binary compound and that sulfur is in its lowest oxidation state (−2). Aluminum is one of those invariant atoms that must be known to be +3. There must be enough +3's and enough −2's to cancel each other. The common multiple of 3 and 2 is 6:

$$\begin{aligned} 2 \times +3 &= +6 \\ 3 \times -2 &= \underline{-6} \\ & \quad\; 0 \end{aligned}$$

   Therefore, the formula is Al_2S_3

2. What is the formula for iron(III) nitrate?

   The oxidation number of iron is given (+3); it must be remembered that the charge on nitrate is −1. It requires three nitrates to neutralize one iron; that fact is indicated in the formula by putting parentheses around the nitrate ion and labeling the final parenthesis with a subscript 3:

$$Fe(NO_3)_3$$

   The parentheses go around only the group that is repeated.

3. What is the formula for barium cyanide?

   Barium is one of the Group IIA metals; the charge on the ion is therefore +2. Cyanide (along with hydroxide, amide, acetylide, etc.) is one of those -ide exceptions and is known to be CN^-. It therefore requires two cyanides per barium and that fact requires parentheses and a subscript around cyanide:

$$Ba(CN)_2$$

   Note that it is *not* $BaCN_2$. That formula is barium cyanamide and it is a completely different compound.

4. What is the formula for iron(II) ammonium sulfate?

   The oxidation number for iron is given (+2). The ammonium ion must be recognized as NH_4^+, a positively charged polyatomic ion. Sulfate is SO_4^{2-}. Writing the symbols in order, we have

$$Fe(NH_4)_2(SO_4)_2$$

8.9 *Hydrates*

Certain compounds crystallize from water with a definite mole ratio of water bound into the crystal in a variety of ways. This *water of hydration* sometimes gives special forms or colors to the crystal so it is really part of a pure compound. The formulas of hydrates depart somewhat from the format; it is customary to put the number of moles of water associated with one mole of the compound after the formula of the compound, separated by a dot.

EXAMPLE 8.12

$$Na_2SO_4 \cdot 10H_2O = Na_2SO_4(H_2O)_{10}$$

The number of moles of water is given either by a Greek prefix or by English numerals.

$Na_2SO_4 \cdot 10H_2O$, sodium sulfate decahydrate; or,
sodium sulfate ten water-of-hydration

The former is preferred.

Summary of nomenclature rules 8.10

I. Names from formulas

A. Binary Compounds

1. The more electropositive element is named first.

a. It is identified by the given name of the element if it only has one common oxidation state.

b. Classical system. If the element occurs in two common oxidation states, the oxidation state of the element is indicated by a suffix to the root name of the element: -ic for the higher and -ous for the lower oxidation state.

c. Stock system. If the element occurs in two or more oxidation states, the oxidation state is given by a roman numeral after the given name of the element. If two or more oxidation states of the element occur in the same compound both oxidation states are specified. If necessary to avoid ambiguity the relative number of each state is given.

d. Binary compounds of nonmetals. The relative number of each element may be given by a Greek number prefix (mono, di, tri, . . .). Alternatively the Stock system may be used.

2. The more electronegative element is named last.

a. It is identified by the stem name of the element plus an -ide ending.

b. The second element in a binary compound is always in its lowest oxidation state except when identical atoms are bonded to each other.

B. Polyatomic Ions

1. Oxyanions. The stem name of an element with an -ite or an

-ate ending indicates that the element occurs in an anion surrounded by a number of covalently bound oxygens. The -ite ending signifies the lower common oxidation state of the central element, and the -ate ending signifies the next higher oxidation state.

2. Other Polyatomic Anions. Atoms and groups which are substituted for oxygen in -ite and -ate anions are specified by a prefix. The number is given by a Greek prefix and the identity by the stem name of the atom or group followed by -o-.

C. Acids

1. Acids derived from -ide salts are named hydro(root)ic acids.
2. Acids derived from -ate salts are (root)ic acids.
3. Acids derived from -ite salts are (root)ous acids.

II. Calculation of oxidation numbers

A. The sum of the oxidation numbers of the elements in a compound is zero.

B. The sum of the oxidation numbers of the elements in an ion is the charge on the ion.

1. The oxidation number of a monatomic ion is the charge on the ion.

C. The most electronegative element in a compound or a polyatomic ion is in its lowest oxidation state (when not bonded to an identical atom).

III. Formulas from names

A. From information about identity, oxidation states, and ion charges, the formula is written so that the sum of the ion charges and oxidation numbers is zero.

EXERCISES

I.A. Name the following compounds.

1. $AlBr_3$	**6.** CdS
2. Al_4C_3	**7.** Mg_3N_2
3. Al_2O_3	**8.** KCl
4. $BaBr_2$	**9.** Na_2S
5. BeO	**10.** ZnO

B. Write formulas for the following.

1. aluminum nitride
2. cadmium chloride
3. calcium oxide
4. calcium acetylide
5. magnesium fluoride
6. potassium cyanide
7. barium iodide
8. strontium bromide
9. sodium sulfide
10. zinc chloride

II.A. Write names for the following using the -ous/-ic suffix:

1. $CuBr$
2. $CuCl_2$
3. $FeBr_2$
4. Fe_2O_3
5. $MnCl_2$
6. HgO

B. Write formulas for the following:

1. cobaltous chloride
2. cuprous oxide
3. cupric oxide
4. ferric chloride
5. plumbous sulfide
6. stannous chloride

III.A. Write names for the following using prefix notation:

1. BN
2. BrF_3
3. CBr_4
4. CS_2
5. Cl_2O_7
6. ICl_3
7. PCl_5
8. SCl_2
9. SF_6
10. SO_2

B. Write formulas for the following:

1. diiodine pentoxide
2. nitrogen trichloride
3. dinitrogen oxide
4. silicon dioxide
5. phosphorus trichloride
6. disulfur dichloride
7. silicon tetrafluoride
8. sulfur trioxide
9. dinitrogen pentoxide
10. chlorine trifluoride

IV.A. Write names for the compounds in IIIA using IUPAC (Stock) notation.

B. Write formulas for the following:

1. ammonia
2. methane
3. hydrogen peroxide
4. nitrogen(IV) oxide
5. nitrogen(V) oxide
6. sulfur(VI) oxide
7. nitrogen(II) oxide
8. chlorine(VII) oxide
9. iodine(III) chloride
10. phosphorus(V) oxychloride

V.A. Write names for the following using Stock notation where necessary.

1. $Al(C_2H_3O_2)_3$
2. $Al(OH)_3$
3. NH_4Br
4. $(NH_4)_2Cr_2O_7$
5. $(NH_4)_2CO_3$
6. $NH_4H_2PO_4$
7. NH_4BF_4
8. $SbCl_3$
9. $SbCl_5$

10. $BaCrO_4$
11. $Ca(OH)_2$
12. $Cr(NO_3)_3$
13. Cr_2O_3
14. $Cr(OH)_2$
15. $CuSO_4$
16. $AuCl_3$
17. $Fe(OH)_3$
18. Li_2CO_3
19. $LiAlH_4$
20. $MgSO_4$

B. Write formulas for the following:

1. aluminum nitrate
2. ammonium acetate
3. ammonium perchlorate
4. antimony(III) sulfide
5. barium amide
6. barium sulfate
7. boron oxide
8. calcium carbonate
9. calcium hypochlorite
10. chromium(IV) oxide
11. chromium(VI) oxide
12. cobalt(II) acetate
13. copper(I) carbonate
14. gold(I) chloride
15. iron(III) phosphate
16. lead(II) hydroxide
17. lead(II) oxalate
18. lithium carbonate
19. lithium perchlorate
20. magnesium carbonate
21. manganese(IV) oxide
22. potassium permanganate
23. mercury(I) chloride
24. nickel(II) carbonate
25. potassium hydroxide
26. potassium nitrate
27. potassium iodate
28. silver nitrate
29. sodium hypochlorite
30. sodium bicarbonate

VI.A. Write names for the following acids:

1. HBr
2. HCl
3. HCN
4. HNO_3
5. H_2SO_4
6. H_3PO_4
7. $HClO_4$
8. H_3AsO_4

B. Write formulas for the following acids:

1. hydrofluoric acid
2. hydriodic acid
3. hydrosulfuric acid
4. boric acid
5. hypochlorous acid
6. periodic acid

VII.A. Write names for the following hydrates:

1. $Al(BrO_3)_3 \cdot 9H_2O$
2. $AlCl_3 \cdot 6H_2O$
3. $BaCl_2 \cdot 2H_2O$
4. $CaSO_4 \cdot 2H_2O$
5. $Na_2B_4O_7 \cdot 10H_2O$
6. $Fe(C_2H_3O_2)_2 \cdot 4H_2O$

B. Write formulas for the following:

1. aluminum arsenate octahydrate
2. iron(II) sulfate heptahydrate
3. magnesium oxalate dihydrate
4. sodium thiosulfate pentahydrate
5. sodium permanganate trihydrate
6. sodium dihydrogen phosphate dihydrate

SOLUTIONS

I.A. **1.** aluminum bromide
3. aluminum oxide
5. beryllium oxide
7. magnesium nitride
9. sodium sulfide

B. **1.** AlN
3. CaO
5. MgF_2
7. BaI_2
9. Na_2S

II.A. **1.** cuprous bromide
3. ferrous bromide
5. manganous chloride

B. **1.** $CoCl_2$
3. CuO
5. PbS

III.A. **1.** boron nitride
3. carbon tetrabromide
5. dichlorine heptoxide
7. phosphorus pentachloride
9. sulfur hexafluoride

B. **1.** I_2O_5
3. N_2O
5. PCl_3
7. SiF_4
9. N_2O_5

IV.A. **1.** boron nitride
3. carbon(IV) bromide
5. chlorine(VII) oxide
7. phosphorus(V) chloride
9. sulfur(VI) fluoride

B. **1.** NH_3
3. H_2O_2
5. N_2O_5
7. NO
9. ICl_3

V.A. **1.** aluminum acetate
3. ammonium bromide
5. ammonium carbonate
7. ammonium fluoborate
9. antimony(V) chloride
11. calcium hydroxide
13. chromium(III) oxide
15. copper(II) sulfate
17. iron(III) hydroxide
19. lithium aluminum hydride

B. **1.** $Al(NO_3)_3$
3. NH_4ClO_4
5. $Ba(NH_2)_2$
7. B_2O_3
9. $Ca(ClO)_2$
11. CrO_3
13. Cu_2CO_3
15. $FePO_4$
17. PbC_2O_4
19. $LiClO_4$
21. MnO_2
23. Hg_2Cl_2
25. KOH
27. KIO_3
29. NaClO

VI.A. **1.** hydrobromic acid
3. hydrocyanic acid
5. sulfuric acid
7. perchloric acid

B. **1.** HF
3. H_2S
5. HOCl

VII.A. **1.** aluminum bromate ennea- (or nona-) hydrate
3. barium chloride dihydrate
5. sodium tetraborate decahydrate (borax)

B. **1.** $AlAsO_4 \cdot 8H_2O$
3. $MgC_2O_4 \cdot 2H_2O$
5. $NaMnO_4 \cdot 3H_2O$

Elements and compounds can interact with each other to break old bonds and form new bonds. The "how" and "why" of these interactions is the main concern of chemistry.

After reading this chapter you should be able to

1. Write and balance a chemical equation, given the reactants and products;
2. Describe the main regions of chemical reactivity in the periodic table;
3. Predict the products of direct combination of metals with nonmetals, nonmetals with nonmetals, and water with oxides of metals and nonmetals;
4. Predict the products of displacement of metals with water, acids, and bases;
5. Predict the products of combination and displacement of ions with metals, neutral substances, and other ions;
6. Write a net ionic equation;
7. Recognize decomposition, rearrangement, and polymerization reactions.

9 chemical reactions and equations

A person unfamiliar with chemistry asking the question, "What is it?" about some consumable item really means, "What does it do?" The ecology minded, or just plain careful, person should also ask

- "How much is needed to do the job?"
- "What are the possible dangers in its use?"
- "What is the final disposition of it?"

A chemist asking the same question wants to know the composition of the item, preferably in terms of the structural formulas of the components. A *structural formula* is a three-dimensional representation of a molecule that shows which atoms are bonded to each other and what their space relationships are. From the size, shape, and distribution of charge in the molecule the chemist can make a pretty good guess about its physical and chemical interactions.

A chemist can do this in spite of there being millions of compounds because there are only a limited number of ways that they can react. Furthermore, elements and compounds fall into natural groups that undergo similar reactions, so that knowing the group to which a substance belongs enables the chemist to predict much of its chemical behavior.

In this chapter you will learn

(a) how a chemist writes a chemical reaction;

(b) some of the ways atoms and molecules undergo chemical change; and

(c) how you can begin to group atoms and molecules into some kind of organization.

9.1 *How to write a chemical reaction: chemical equations*

A chemical equation is a sentence, written in symbols, that tells what has been experimentally found to happen when two or more substances (reactants) change into different substances (products). An equation supplies the same information about a reaction that a molecular formula does about a compound:

1. A molecular formula gives us the identity and number of elements in a compound;
2. An equation gives us the identity and number of compounds in a reaction.

Chemical reactions are events, or happenings, and the equation represents that event. The "before" condition is on the left side, and the "after" condition is on the right side. The number and identity of the *atoms* are not changed by the transformation. They have simply changed partners.

For example, we find by experiment that butane combines with oxygen to form carbon dioxide and water (Figure 9.1). Let us write the equation for the reaction.

1. What are the molecular formulas of reactants and products?

butane $= C_4H_{10}$	carbon dioxide $= CO_2$
oxygen $= O_2$	water $= H_2O$

2. Write the skeleton equation:

$$C_4H_{10} + O_2 \longrightarrow CO_2 + H_2O$$

3. One mole of C_4H_{10} contains 4 moles of C and 10 moles of H; we must have the same number on the right side. This requirement is satisfied by writing a *coefficient* 4 in front of CO_2 and a coefficient 5 in front of H_2O.

$$4 \text{ mol C} \times \frac{1 \text{ mol } CO_2}{1 \text{ mol C}} = 4 \text{ mol } CO_2$$

$$10 \text{ mol H} \times \frac{1 \text{ mol } H_2O}{2 \text{ mol H}} = 5 \text{ mol } H_2O$$

Note that we *do not change the subscripts* on the molecular formulas; that would be equivalent to changing the identity of the products!

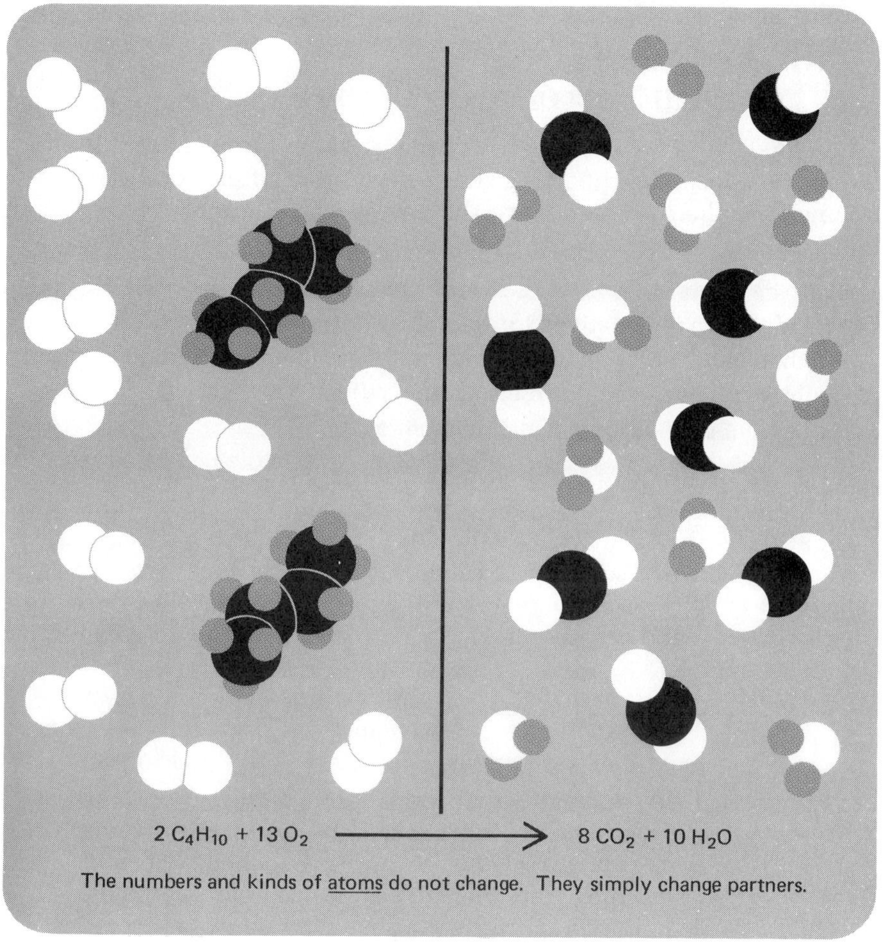

Figure 9.1

4. Four moles of CO_2 and five moles of H_2O require 13 moles of O atoms or $\frac{13}{2}$ moles of O_2 molecules

$$4 \text{ mol } CO_2 \times \frac{2 \text{ mol O}}{1 \text{ mol } CO_2} = 8 \text{ mol O}$$

$$5 \text{ mol } H_2O \times \frac{1 \text{ mol O}}{1 \text{ mol } H_2O} = 5 \text{ mol O}$$

$$\text{Total mol O} = \overline{13}$$

We can supply that by putting the coefficient $\frac{13}{2}$ in front of O_2.

$$C_4H_{10} + \frac{13}{2}\, O_2 \longrightarrow 4\ CO_2 + 5\ H_2O$$

It is preferable to avoid fractional coefficients so we multiply *everything* by 2 and clear the fraction.

5. $$2\ C_4H_{10} + 13\ O_2 \longrightarrow 8\ CO_2 + 10\ H_2O$$

The equation is balanced. There are the same numbers and kinds of atoms on each side (reflecting the "before" and "after" condition of the atoms) and the coefficients are the lowest whole numbers that we can write.

EXERCISE 9.1

Write and balance equations for the following:

1. methane plus chlorine give carbon tetrachloride plus hydrogen chloride.
2. ammonia plus oxygen give nitrogen(II) oxide plus water.
3. aluminum plus oxygen give aluminum oxide.

Equations are written to convey information. The *minimum* information is the identity and the number of species that react. Additional symbols are sometimes used to impart qualitative information about the conditions of the reaction, the physical states of the reactants and products, and whether the reaction goes sensibly to completion.

Symbol	Meaning
$\longrightarrow$	This reaction goes to completion from left to right. At least one reactant is completely converted to product(s).
$\rightleftarrows$	This reaction does not go to completion. Appreciable quantities of reactants *and* products are present when a steady state (equilibrium) is reached. The reaction is reversible, and the final state may be approached from either side of the equation. One arrow may be drawn longer to show which species predominate at equilibrium.
$\xrightarrow[\Delta]{}$	(over or under the arrow) This reaction needs to be heated to occur.
$\xrightarrow{\text{cat.}}$	(over the arrow) This reaction needs a catalyst. The identity of the catalyst may be printed in small letters over the arrow. A catalyst is a substance that affects the rate of a reaction without undergoing chemical change itself.
(c) or (s) (l) or (liq) (g) (aq)	The physical state of the reactants and products may be indicated by lowercase letters in parentheses after the formula, if this information is necessary. (c) or (s) means the substances is in the crystalline or solid state; (l) means a liquid; (g) means it is a gas; and (aq) means the substance is in aqueous solution. The latter usually means the substance is ionized in solution as well (see Section 9.8).

An insoluble substance is sometimes indicated by an underline or a vertical arrow:

$$BaSO_4(c) = \underline{BaSO_4} = BaSO_4 \downarrow$$

An insoluble substance formed from a mixture of two solutions is a *precipitate.*

A gaseous product may be similarly indicated by an overline or a vertical arrow.

$$CO_2(g) = \overline{CO_2} = CO_2 \uparrow$$

EXAMPLE 9.1

Iron(III) sulfate reacts with sodium hydroxide in solution to form insoluble iron(III) hydroxide and a solution of sodium sulfate. Write the equation for the reaction.

1. What are the formulas of the reactants and products?
   Reactants: $Fe_2(SO_4)_3$ and NaOH
   Products: $Fe(OH)_3$ and Na_2SO_4
2. Write the skeleton reaction:

$$Fe_2(SO_4)_3(aq) + NaOH\ (aq) \longrightarrow \underline{Fe(OH)_3} + Na_2SO_4(aq)$$

3. One mole of iron(III) sulfate contains 2 moles of Fe^{3+} and 3 moles of SO_4^{2-}. We must therefore have 2 moles of $Fe(OH)_3$ and 3 moles of Na_2SO_4 to account for the number of iron(III) and sulfate ions.

$$Fe_2(SO_4)_3(aq) + NaOH\ (aq) \longrightarrow 2\ \underline{Fe(OH)_3} + 3\ Na_2SO_4(aq)$$

4. To make 2 moles of $Fe(OH)_3$ and 3 moles of Na_2SO_4, it is necessary to have 6 moles of Na^+ ions and 6 moles of OH^- ions available.

$$2\ \text{mol}\ Fe(OH)_3 \times \frac{3\ \text{mol}\ OH^-}{\text{mol}\ Fe(OH)_3} = 6\ \text{mol}\ OH^-$$

$$3\ \text{mol}\ Na_2SO_4 \times \frac{2\ \text{mol}\ Na^+}{\text{mol}\ Na_2SO_4} = 6\ \text{mol}\ Na^+$$

5. These ions may be supplied if 6 moles of NaOH are present, so we put a coefficient 6 in front of NaOH.

$$Fe_2(SO_4)_3(aq) + 6\ NaOH\ (aq) \longrightarrow 2\ \underline{Fe(OH)_3} + 3\ Na_2SO_4(aq)$$

Observe that the OH^- group and the SO_4^{2-} group are not changed in this reaction. Since the groups are unchanged we may count them as groups and not individual atoms. It makes the arithmetic much simpler.

The periodic table of the elements 9.2

The idea of family resemblances between groups of elements was conceived in the last century by Johann Döbereiner. A little later,

Mendeleev made a more successful and detailed grouping based upon chemical similarities, which could be used to predict the existence of undiscovered elements. As knowledge about atomic structure unfolded, a still more detailed grouping, the form we use today, was made. Now we know that the form of the periodic table is determined by the outer electron configuration of the elements.

Table 9.1 reproduces the periodic table with the symbols of the elements set off in characteristic regions of activity.

9.3 *The elements: metals, nonmetals, and semimetals*

Imagine that all the elements are arrayed before you in their natural uncombined state. The trends in their chemical and physical properties will be most evident if they are arrayed in the form of the periodic table.

Metals

Your first impression might be that most of them are metallic. They have the properties of being shiny, opaque, electrical conductors. One is a liquid (Hg, mercury), but the rest have melting points above room temperature. The light metals on the left side of the table tend to be soft and are easily cut or bent. Some of the heavy metals (Cu, copper; Pb, lead; Au, gold) are fairly soft, but most of them are quite hard. All of them have the property of being ductile (capable of being drawn into a wire) or malleable (bent with a hammer) at some temperature below their melting point. The elements in regions A, B, C, and D in our array are metals.

Nonmetals

The elements in regions F and G are not electrical conductors. Most of them are gases. We recognize hydrogen (H_2), nitrogen (N_2), oxygen (O_2), fluorine (F_2), and chlorine (Cl_2) as diatomic molecular gases. Bromine (Br_2) is a reddish brown liquid. Sulfur (S), phosphorus (P), and iodine (I_2) are solids with moderately low melting and boiling points.

Boron (B) and carbon (C) have very high melting temperatures, but are not metallic. Carbon as graphite will conduct electricity, but it cannot be melted, drawn, or pounded into anything but powder. Carbon as diamond will conduct a girl to the altar, but it won't conduct electricity.

Semimetals

The elements in region E are chameleons. They behave like both metals and nonmetals depending on conditions. All of them exist in two forms (called "allotropes"): a metallic electrically conducting form and a nonmetallic nonconducting form. The metallic forms are not good conductors though, and that property makes some of them useful for making transistors. A *transistor* transmits electricity in one direction and resists its flow in the opposite direction. Germanium (Ge)

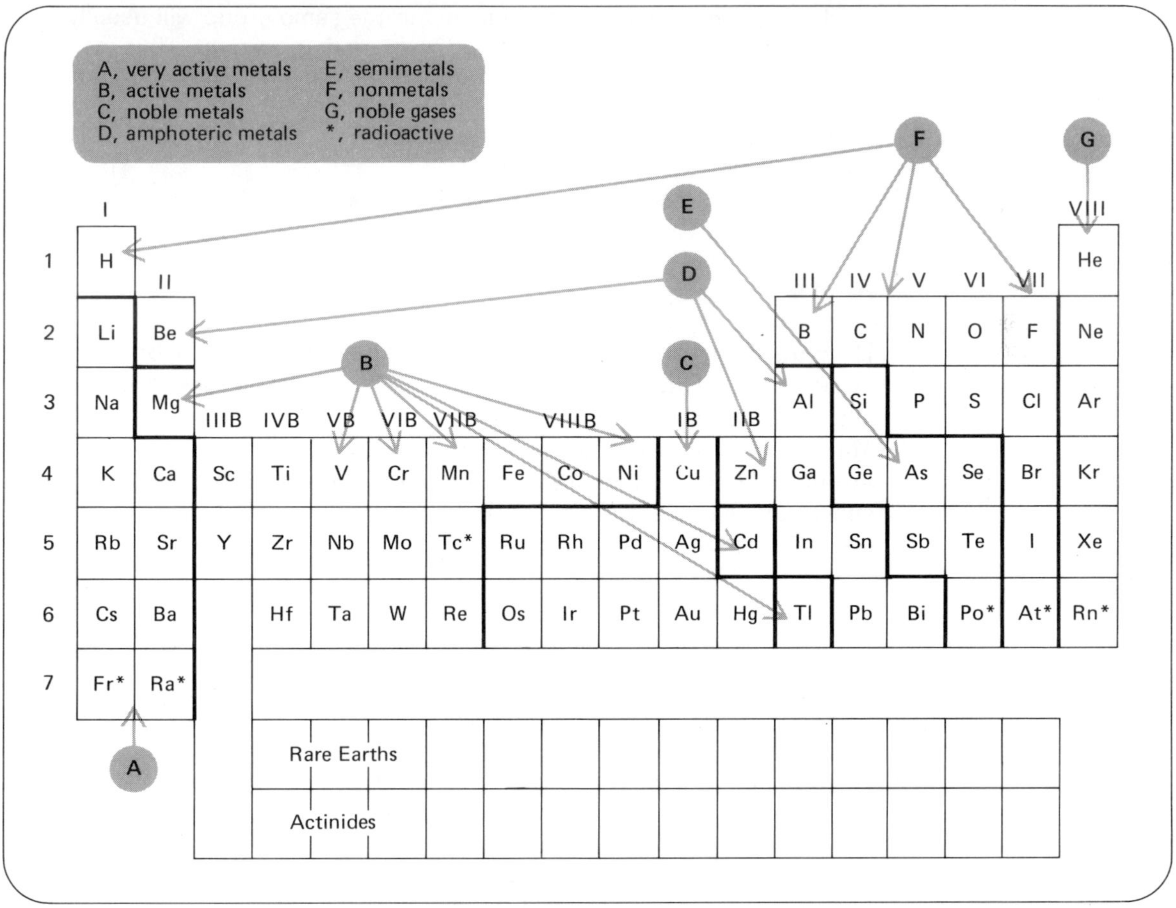

Table 9.1
Chemical properties of the elements

and antimony (Sb) are more metallic than the others; silicon (Si) and selenium (Se) are more like the nonmetals.

EXERCISE 9.2 Compare the reactivity groups (Table 9.1) with electron configuration (Table 6.1) and electronegativity (Table 7.3). What patterns do you see? We might make these observations:

1. Although the elements in any particular group have the same outer electron configuration they do not have the same electronegativity. The elements with higher atomic numbers in a chemical group have a lower electronegativity.
2. The very active metals have the lowest electronegativity, and the supernonmetals have the highest. Elements with electronegativities less than about 2.0 are metallic in nature; those with electronegativities greater than 2.0 are more nonmetallic.
3. The group electron configuration determines the *number* of valence electrons an element has. Elements in the same group will usually have the same valence when combined with another element. Thus we would predict that nitrogen, phosphorus, arsenic, and antimony in Group VA would have the formulas NCl_3, PCl_3, $AsCl_3$, and $SbCl_3$ when combined with chlorine. We might also predict that PCl_5, $AsCl_5$, and $SbCl_5$ are possible. (But NCl_5 is not. Why not?)

Alloys

Humans first broke out of the Stone Age when they discovered that an alloy of tin and copper made a better tool than a rock. Most metals when melted together form alloys or solutions of one in the other. Many alloys have more useful properties than the pure metals. They are not compounds, though, because it is not possible to isolate a pure substance with a definite mole ratio of elements from them. They are mixtures.

9.4 *Chemical reaction: direct combination of elements*

In contrast, if a metal is heated with a nonmetal, the result is often spectacular. Magnesium (Mg) ribbon heated in air burns fiercely with a white hot flame. Zinc dust (Zn) and sulfur (S) combine almost explosively. Iron (Fe) powder sprinkled into chlorine gas sparkles like a thousand meteorites.

The products isolated from these combinations can be purified and are found to have a definite composition, unlike an alloy. The

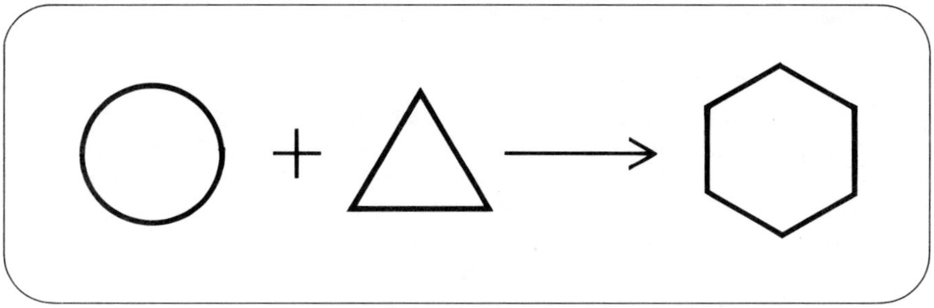

Figure 9.2 Direct combination or addition. Two kinds of reactants form only one kind of product.

physical properties of the products are unlike a metal or a nonmetal too. They are crystalline substances like salt or rocks.

Our first generalization about reactions, then, is that metals tend to combine chemically with nonmetals.

EXAMPLE 9.2

1. $2\,Mg + O_2 \longrightarrow 2\,MgO$
   magnesium and oxygen give magnesium oxide
2. $Zn + S \longrightarrow ZnS$
   zinc and sulfur give zinc sulfide
3. $2\,Fe + 3\,Cl_2 \longrightarrow 2\,FeCl_3$
   iron and chlorine give iron(III) chloride

EXERCISE 9.3

Write balanced equations for the formation of lithium oxide (Li_2O), sodium oxide (Na_2O), aluminum oxide (Al_2O_3), and calcium oxide (CaO) from the elements.

Direct combination of nonmetals 9.5

Nonmetals may combine with each other also. The energy released per mole of product formed is generally lower than that in the reaction between metals and nonmetals, but the rate of the reaction can sometimes be explosive, especially if the reactants are in the gaseous state.

The products of the combination of nonmetals are physically different from the rock-like or salt-like solids obtained from metal/nonmetal reactions. These products are often gases, liquids, or low melting solids like the nonmetals themselves.

EXAMPLE 9.3

1. $2\ H_2$ + O_2 $\longrightarrow$ $2\ H_2O$
hydrogen gas and oxygen gas give water vapor

This reaction gives off a lot of energy as heat and is explosive if the right amounts of oxygen and hydrogen are present. The middle stage of the Saturn rocket uses liquid hydrogen (boiling point −253°C) and liquid oxygen (boiling point −218°C) as its propellant. The elements are mixed and ignited under controlled conditions in the combustion chamber.

2. S + O_2 $\longrightarrow$ SO_2
sulfur and oxygen give sulfur dioxide

The odor of burning match heads is mostly sulfur dioxide. Sulfur impurities in fuel oil can contribute much sulfur dioxide to our atmosphere. It is an irritating, suffocating gas.

3. $2\ SO_2$ + O_2 $\longrightarrow$ $2\ SO_3$
sulfur dioxide and oxygen give sulfur trioxide

An important industrial process converts sulfur dioxide to sulfur trioxide. This product can also form in the atmosphere, and it is even more irritating than sulfur dioxide.

4. N_2 + $3\ H_2$ $\rightleftarrows$ $2\ NH_3$
nitrogen and hydrogen give ammonia (gas)

This reaction is mildly energetic and only occurs at a reasonable rate under high pressure at about 400°C. Ammonia is a valuable fertilizer, as well as a necessary chemical intermediate, so huge tonnages are made this way.

Second generalization: Nonmetals combine with nonmetals, but the products are usually gases, liquids, or low-melting solids. Boron nitride and silicon dioxide are exceptions to this observation. Since oxygen and fluorine will combine with almost everything in sight, we could call them "super nonmetals."

EXERCISE 9.4

Write equations for the formation of carbon dioxide (CO_2), nitric oxide (NO), and phosphorus trichloride (PCl_3) from the elements.

9.6 *Reactions of metal and nonmetal oxides with water*

Earth is the water planet, and the fact that water can exist in all three phases on earth accounts for the evolution of living creatures in the form that they have. It is natural therefore that much of chemistry is concerned with the interactions of substances with water. If we lived on Jupiter, we might be more interested in liquid ammonia.

Bases

Some metal oxides react when dissolved in water to form substances that are bitter to the taste, slippery to the touch, and turn pink litmus blue. Litmus is a naturally occurring organic dye that changes color when water is not neutral. The products of hydration of metal oxides (combination with water) are called *bases.*

EXAMPLE 9.4

1. $CaO + H_2O \longrightarrow Ca(OH)_2$
calcium oxide and water gives calcium hydroxide

2. $Na_2O + H_2O \longrightarrow 2\ NaOH$
sodium oxide and water gives sodium hydroxide

Acids

In a similar manner, oxides of nonmetals react with water, but the products are sour, corrosive to metals, and turn blue litmus pink. These are called *acids.*

EXAMPLE 9.5

1. $SO_2 + H_2O \rightleftarrows H_2SO_3$
sulfur dioxide and water give sulfurous acid

2. $CO_2 + H_2O \rightleftarrows H_2CO_3$
carbon dioxide and water give carbonic acid

3. $P_2O_5 + 3\ H_2O \longrightarrow 2\ H_3PO_4$
phosphorus pentoxide and water give phosphoric acid

4. $SO_3 + H_2O \longrightarrow H_2SO_4$
sulfur trioxide and water give sulfuric acid

The first two reactions are reversible. That is, when a steady state (equilibrium) is reached both products and reactants will be present in solution. The last two reactions go completely from left to right and essentially cannot be reversed.

EXERCISE 9.5

Write the reaction of N_2O_5 with water to form HNO_3.

9.7 *Reactions of metals with water, acids, and bases*

The very active metals (Region A) react vigorously with cold water to liberate hydrogen and form strong bases.

EXAMPLE 9.6

1.	2 Na	+	2 H_2O	$\longrightarrow$	2 NaOH	+	H_2
	sodium	and	water	give	sodium hydroxide	and	hydrogen
2.	Ca	+	2 H_2O	$\longrightarrow$	$Ca(OH)_2$	+	H_2
	calcium	and	water	give	calcium hydroxide	and	hydrogen

The above are examples of *displacement* or substitution reactions. An atom or ion takes the place of another atom or ion in a compound. Displacement (substitution) reactions may occur with changes in oxidation numbers, as in the above example, or they may occur as simple exchanges without changing oxidation numbers. Examples of these exchange reactions are sometimes called *double displacement* reactions (Figure 9.3).

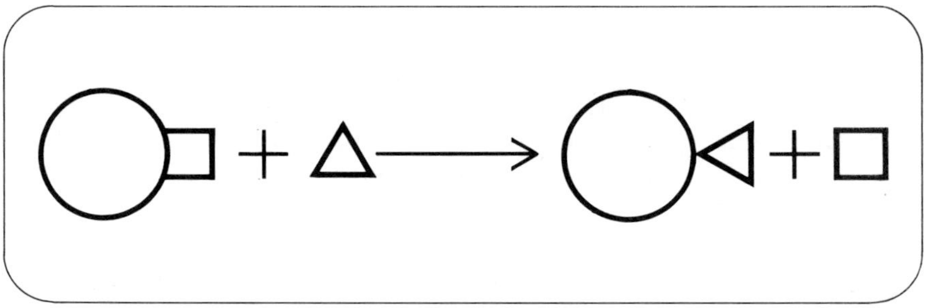

Figure 9.3 Displacement (substitution) reactions. An atom, ion, or group displaces other atoms, ion, or groups from a compound.

Active metals (Region B) react with strong acids in water to liberate hydrogen and form salts in solution.

EXAMPLE 9.7

1.	Mg	+	2 HCl	$\longrightarrow$	$MgCl_2$	+	H_2
	magnesium	and	hydrochloric acid	give	magnesium chloride	and	hydrogen
2.	Fe	+	H_2SO_4	$\longrightarrow$	$FeSO_4$	+	H_2
	iron	and	sulfuric acid	give	iron(II) sulfate	and	hydrogen

The noble metals (Region C) do not react with acids in this way. Combinations of very active nonmetals (aqua regia) are required to make salts of these metals.

Amphoteric metals (Region D) react with strong acids *and* strong bases to form salts and liberate hydrogen.

EXAMPLE 9.8

1. $2\ Al + 3\ H_2SO_4 \longrightarrow Al_2(SO_4)_3 + 3\ H_2$

2. $2\ Al + 2\ NaOH + 6\ H_2O \longrightarrow 2\ NaAl(OH)_4 + 3\ H_2$

EXERCISE 9.6

Write the reactions of zinc (Zn) and of tin (Sn) with hydrochloric acid (HCl). The formulas of the salts are $ZnCl_2$ and $SnCl_2$.

Reactions of charged atoms and molecules: ions 9.8

The product of the reaction between a metal and a nonmetal will not conduct electricity in the solid state, but it will conduct if it is molten or dissolved in water. Electrical conduction requires a mobile charge carrier. Evidently the original reaction has caused these atoms to acquire an electrical charge, so that melting the substance or dissolving it allows these charges to move. Since the substance itself is not charged, the atoms must have an equal number of positive charges and negative charges. As you know from Chapter 7, charged atoms and molecules are called *ions*. Positive ions were named *cations* and negative ions *anions* by Michael Faraday, who first discovered these mobile charge carriers in solutions. The action of water in overcoming the forces holding the ions together so that they can move freely in solution is called *ionization*.

We will come back to this subject in the chapters on Solutions, and Acids and Bases. For now, it is sufficient to know that strong acids and bases and soluble salts are completely ionized in water.

EXAMPLE 9.9

$$HNO_3 + \text{water} \longrightarrow H^+(aq) + NO_3^-(aq)$$
$$HCl + \text{water} \longrightarrow H^+(aq) + Cl^-(aq)$$
$$NaOH + \text{water} \longrightarrow Na^+(aq) + OH^-(aq)$$
$$CuSO_4 + \text{water} \longrightarrow Cu^{2+}(aq) + SO_4^{2-}(aq)$$

These are not balanced equations, because each ion is surrounded by an indeterminate number of water molecules. The (aq) after each ion stands for *aqueous* and means that the ion is in water solution.

Reactions of ions with neutral substances[1]

An ion can do three things when it collides with a neutral molecule or atom:

1. combine with it or add to it, forming a new molecule ion:

$$Cu^{2+}(aq) + 4\ NH_3 \longrightarrow Cu(NH_3)_4{}^{2+}(aq)$$

copper(II) ion and ammonia give tetraammine copper(II) ion

$$H^+(aq) + NH_3 \longrightarrow NH_4{}^+(aq)$$

hydrogen ion and ammonia give ammonium ion

2. displace part of the neutral molecule, forming a new ion:

$$CO_3{}^{2-} + H_2O \rightleftarrows HCO_3{}^- + OH^-$$

carbonate ion and water give bicarbonate ion and hydroxide ion

$$OH^- + HC_2H_3O_2 \rightleftarrows H_2O + C_2H_3O_2{}^-$$

hydroxide ion and acetic acid give water and acetate ion

3. exchange charges:

$$Cu^{2+}(aq) + Zn \longrightarrow Cu + Zn^{2+}(aq)$$

copper(II) ion and zinc give copper and zinc ion

$$2\ Ag^+(aq) + Cu(c) \longrightarrow 2\ Ag(c) + Cu^{2+}(aq)$$

silver ion and copper give silver and copper(II) ion

Reactions of ions with ions

Two species have to collide with each other to react, and since like charges repel each other, it follows that ions won't react with other ions having the same sign. Positive ions won't come near enough to other positive ions to react with them. Oppositely charged ions attract each other and often react to form different products. Reaction is likely to occur if

1. a virtually non-ionized product is formed:

$$HCO_3{}^- + H^+(aq) \rightleftarrows H_2CO_3 \rightleftarrows CO_2 + H_2O$$

bicarbonate ion and hydrogen ion give carbonic acid which gives carbon dioxide and water

$$NH_4{}^+ + OH^- \rightleftarrows NH_3 + H_2O$$

ammonium ion and hydroxide ion give ammonia and water

[1] These equations have been written in the "net ionic" form. Ions which must be present to neutralize the charges of the specified ions are not reacting themselves, so they are left out of the equation.

2. an insoluble product is formed:

$Al^{3+}(aq)$	+	$3\ OH^-$	$\longrightarrow$	$\underline{Al(OH)_3}$
aluminum ion	and	hydroxide ion	give	aluminum hydroxide
$Ba^{2+}(aq)$	+	$SO_4^{2-}(aq)$	$\longrightarrow$	$BaSO_4(s)$
barium ion	and	sulfate ion	give	barium sulfate

3. there is an exchange of atoms and/or charges (change of oxidation state):

$$14\ H^+ + Cr_2O_7^{2-} + 6\ Fe^{2+} \longrightarrow 2\ Cr^{3+} + 6\ Fe^{3+} + 7H_2O$$

Writing and balancing ionic equations involving changes of oxidation state require special methods which will be taken up in Chapter 13.

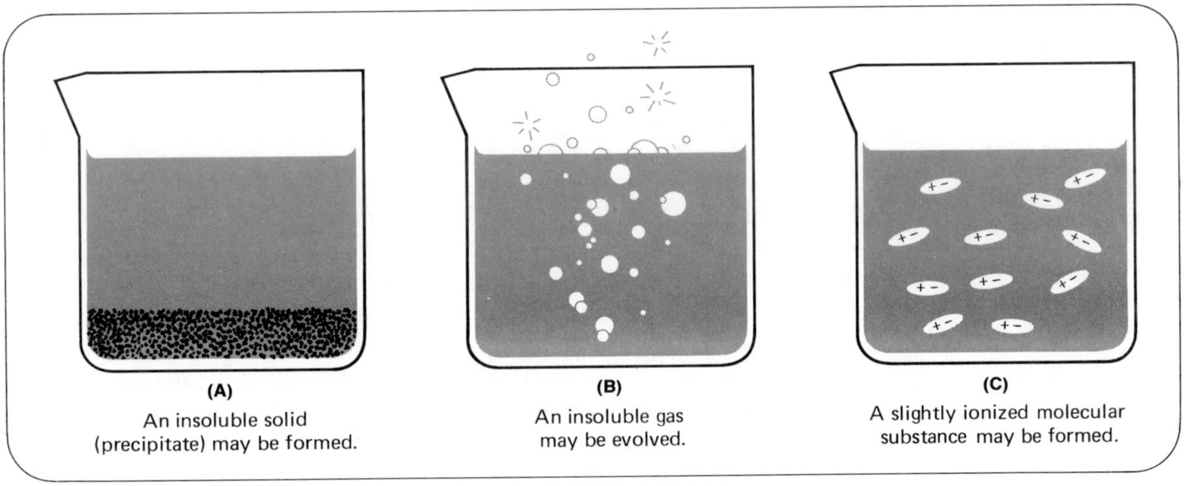

Figure 9.4 Reactions of ions with ions. When oppositely charged ions react, one of three things may happen.

The terms "non-ionized" and "insoluble" are used in a relative sense. Only a tiny fraction of a molecular substance like carbonic acid or water is ionized at any one time. Similarly, the solubilities of substances vary from immeasurably small to very large. In general a substance is considered insoluble if less than 3 g will dissolve in 100 milliliters of water at room temperature. Some "rules of thumb" about the solubilities of common salts are given in Table 9.2.

Table 9.2
Solubility rules

Generalizations	Exceptions
1. Nitrates and acetates are generally soluble.	1. No common ones. Silver acetate is moderately insoluble.
2. Compounds of the alkali metals and ammonium ion are generally soluble.	2. No common ones.
3. Chlorides, bromides, and iodides are generally soluble.	3. Halides of Ag, Hg(I), and Pb(II). HgI_2, BiOCl, and SbOCl.
4. Sulfates are generally soluble.	4. $PbSO_4$, $SrSO_4$, $BaSO_4$. $CaSO_4$ is moderately insoluble.
5. Carbonates and sulfites are generally insoluble.	5. Alkali metals and ammonium.
6. Sulfides are generally insoluble.	6. Alkali metals and ammonium. Sulfides of the alkaline earth metals, Cr_2S_3, and Al_2S_3 are decomposed by water.
7. Hydroxides are generally insoluble.	7. Alkali metals and ammonium. The hydroxides of Ba, Sr, and Ca are moderately soluble.

9.9 *Ionic equations*

The course of a chemical reaction can most easily be predicted if the reactants are written as they exist. Gases and other non-ionic substances are written as molecular uncharged formulas. Insoluble salts are also written as uncharged formulas. Substances that ionize in aqueous solution form a number of charged species that are essentially free to react independently of each other. This state can be indicated in writing a reaction by separating the substance into its ions. Similarly, substances that react with water to form ions can be shown in the ionic form. This form has two advantages. First, it enables you to recognize which species *can* react with each other and, from that, to recognize general class reactions. Second, it helps in balancing the equation, because the polyatomic ions can be treated as units. It is much easier to count groups than it is to count atoms.

EXAMPLE 9.10

$Ca(OH)_2$	$+$	H_3PO_4	$\longrightarrow$	$Ca_3(PO_4)_2$	$+$	H_2O
calcium hydroxide	and	phosphoric acid	give	calcium phosphate	and	water

In ionic form,

$$Ca^{2+} + OH^- + H^+ + PO_4^{3-} \longrightarrow Ca_3(PO_4)_2 + H_2O$$

Calcium phosphate is insoluble, and water is non-ionic, so they are written in molecular form. Now it is easy to see that we need three Ca^{2+} ions and two PO_4^{3-} ions to make one calcium phosphate. We can also see that hydrogen ions and hydroxide ions are combining to form water. Since we must have two hydroxide ions (OH^-) for each calcium ion and three hydrogen ions (H^+) for each phosphate, we will need a total of six each.

$$3\ Ca^{2+} + 6\ OH^- + 6\ H^+ + 2\ PO_4^{3-} \longrightarrow Ca_3(PO_4)_2 + 6\ H_2O$$

EXAMPLE 9.11

$$NaHCO_3(aq) + HCl(aq) \longrightarrow NaCl(aq) + CO_2 + H_2O$$

Carbon dioxide and water are non-ionic, but all the other substances can be written as ions in solution.

$$Na^+ + HCO_3^- + H^+ + Cl^- \longrightarrow Na^+ + Cl^- + CO_2 + H_2O$$

The sodium ions (Na^+) and the chloride ions (Cl^-) do not actually react. As long as they remain in solution, their state is the same before and after the reaction. They are called *spectator ions* and can be left out of the equation:

$$HCO_3^- + H^+ \longrightarrow CO_2 + H_2O$$

The result is a "net ionic equation" representing a general reaction that is true regardless of the identity of the counter-ions. The counter-ions must be present, of course, because we cannot have a bucket of bicarbonate ions or hydrogen ions, but it does not matter what the counter-ions are.

Decomposition (elimination) reactions 9.10

The internal arrangement of atoms in a compound can be changed by supplying energy in the form of heat, electricity, or radiation. New arrangements and pairings can then form, sometimes liberating a considerable amount of energy.

EXAMPLE 9.12

1. $Na_2SO_4 \cdot 10H_2O$	$\underset{\Delta}{\longrightarrow}$	Na_2SO_4	$+$	$10\ H_2O(g)$
sodium sulfate decahydrate	(heated) gives	sodium sulfate	and	water vapor

2. $$CaCO_3 \xrightarrow[\text{heat}]{\text{high}} CaO + CO_2(g)$$

calcium carbonate gives calcium oxide and carbon dioxide

3. $$2\ KClO_3 \xrightarrow[\text{heat}]{\text{cat.}} 2\ KCl + 3\ O_2(g)$$

potassium chlorate gives potassium chloride and oxygen gas

4. $$4\ C_3H_5N_3O_9 \xrightarrow[\text{or shock}]{\text{heat}} 12\ CO_2 + 10\ H_2O + 6\ N_2 + O_2 + \text{energy}$$

glyceryltrinitrate gives carbon dioxide water nitrogen oxygen

Decompositions are sometimes reversible, particularly if heat must be added to effect the decomposition. Both examples **1** and **2** are reversible and go from right to left at low temperatures. Examples **3** and **4** are not reversible. In fact, **4** generally decomposes everything in the neighborhood of the reaction.

9.11 *Polymerization and rearrangements*

Pure substances can also change their chemical nature without breaking into parts. Most of the plastics we use are the result of the addition of molecules to others of the same kind to form giant molecules, called *polymers*. Polystyrene, for example, is a long chain of styrene units joined head-to-tail:

$$n\ CH_2{=}\underset{\displaystyle C_6H_5}{\underset{|}{CH}} \xrightarrow{\text{cat.}} -(CH_2-\underset{\displaystyle C_6H_5}{\underset{|}{CH}})_n-$$

Other substances rearrange their bonds to form different substances with the same number and kind of atoms. One of the first molecules of organic chemistry to be synthesized was made by the rearrangement of ammonium cyanate:

$$NH_4CNO \longrightarrow H_2N-\underset{\displaystyle O}{\underset{\|}{C}}-NH_2$$

ammonium cyanate gives urea

These kinds of reactions are usually studied in more advanced courses in chemistry.

Summary 9.12

1. When arrayed in the form of the periodic table, the elements display a more or less continuous change of properties from metallic to nonmetallic as read from left to right in the table.
2. Elements in any one group (vertical column) characteristically are very similar chemically. The nonmetal character is more pronounced in the first members of each group, but the formulas and the reactions follow the same pattern for the members of each group.
3. Metals rarely combine with each other, but metals and nonmetals often combine to form compounds. When in solution or molten, the compounds behave as if they are made of oppositely charged particles (ions).
4. Compounds formed by combination of nonmetals do not seem to be ionic by themselves. Some of them interact strongly with water, and the products (acids) may become ionized in water.
5. The oxides of the metals often react with water to form hydroxides. The hydroxides of the very active metals are completely ionized in water (strong bases) into cations and hydroxide ions. The hydroxides of the active metals are only weak bases, slightly soluble in water. They are soluble in strong acid solutions. The hydroxides of the amphoteric metals are insoluble in water but soluble in strong acids and bases.
6. The oxides of the nonmetals react with water to form compounds which ionize in water in varying degrees to form hydrated hydrogen ions (hydronium ions, H_3O^+) and oxyanions. These compounds are called acids.
7. Ions in solution may react with neutral substances
   (a) by combination to form new ions; or,
   (b) by displacement of an ion from the neutral compound; or,
   (c) by an exchange of electrons with the neutral substance.
8. Ions in solution may react with ions of opposite charge to form
   (a) an insoluble substance,
   (b) a slightly ionized neutral substance, or
   (c) a gas
9. Pure compounds can change their chemical nature by decomposition, rearrangement, or polymerization.

EXERCISES

I. Balance the following equations:

1. $B + O_2 \longrightarrow B_2O_3$
2. $H_2 + Cl_2 \longrightarrow HCl$
3. $P + O_2 \longrightarrow P_4O_{10}$
4. $K + H_2O \longrightarrow KOH + H_2$

5. $BaO + H_2O \longrightarrow Ba(OH)_2$
6. $NaOH + H_2SO_4 \longrightarrow Na_2SO_4 + H_2O$
7. $NaOH + CO_2 \longrightarrow Na_2CO_3 + H_2O$
8. $CH_4 + Cl_2 \longrightarrow CCl_4 + HCl$
9. $PCl_5 + H_2O \longrightarrow H_3PO_4 + HCl$
10. $(NH_4)_2Cr_2O_7 \longrightarrow Cr_2O_3 + N_2 + H_2O$

II. Balance the following equations:

1. $Fe(OH)_3 + HCl \longrightarrow FeCl_3 + H_2O$
2. $Al(OH)_3 + NaOH \longrightarrow NaAlO_2 + H_2O$
3. $AgNO_3 + BaCl_2 \longrightarrow AgCl + Ba(NO_3)_2$
4. $Fe_2O_3 + Al \longrightarrow Al_2O_3 + Fe$
5. $AgNO_3 + Cu \longrightarrow Ag + Cu(NO_3)_2$
6. $C_8H_{18} + O_2 \longrightarrow CO_2 + H_2O$
7. $CS_2 + Cl_2 \longrightarrow CCl_4 + SCl_2$
8. $NaHCO_3 \longrightarrow Na_2CO_3 + CO_2 + H_2O$
9. $CaCO_3 + HCl \longrightarrow CaCl_2 + CO_2 + H_2O$
10. $KClO_3 \xrightarrow{MnO_2} KCl + O_2$

III. Write chemical equations from these word equations:

1. ammonium sulfate and sodium hydroxide give ammonia, sodium sulfate and water
2. lead(II) nitrate and potassium chromate give lead(II) chromate and potassium nitrate
3. zinc and hydrochloric acid (aqueous hydrogen chloride) give zinc chloride and hydrogen gas
4. copper and oxygen heated, give copper(II) oxide
5. copper(II) oxide and hydrogen give copper and water
6. barium carbonate and hydrochloric acid give barium chloride, carbon dioxide and water
7. mercury(II) nitrate and ammonium sulfide give mercury(II) sulfide and ammonium nitrate
8. magnesium oxide and water give magnesium hydroxide
9. lead(IV) oxide heated gives lead(II) oxide and oxygen
10. sodium carbonate and water (reversibly) give sodium bicarbonate and sodium hydroxide

IV. Complete and Balance

A. Combination (Addition) Reactions

1. **(a)** $Al + O_2 \longrightarrow$
   (b) $S + O_2 \longrightarrow$
   (c) $CaO + SO_3 \longrightarrow$
   (d) $SO_2 + H_2O \longrightarrow$
   (e) $NH_3(g) + HCl(g) \longrightarrow$
   (f) $Mg + Cl_2 \longrightarrow$
   (g) $CaO + H_2O \longrightarrow$
   (h) $C + O_2 \longrightarrow$
   (i) $Na_2O + H_2O \longrightarrow$
   (j) $CaO + CO_2 \longrightarrow$

2. (a) $Cu + Br_2 \longrightarrow$
(b) $Fe + S \longrightarrow$
(c) $H_2 + O_2 \longrightarrow$
(d) $Al + Cl_2 \longrightarrow$
(e) $Ca + O_2 \longrightarrow$
(f) $Mg + N_2 \longrightarrow$
(g) $MgO + H_2O \longrightarrow$
(h) $Si + O_2 \longrightarrow$
(i) $B_2O_3 + H_2O \longrightarrow$
(j) $SO_3 + H_2O \longrightarrow$

B. Displacement Reactions

1. (a) $Na + H_2O \longrightarrow$
(b) $Ca + H_2O \longrightarrow$
(c) $Al + HCl(aq) \longrightarrow$
(d) $Fe + H_2SO_4(aq) \longrightarrow$
(e) $Zn + HCl(aq) \longrightarrow$
(f) $Cu + AgNO_3(aq) \longrightarrow$
(g) $Zn + Cu(NO_3)_2(aq) \longrightarrow$
(h) $Al + Fe_2O_3 \longrightarrow$
(i) $Cl_2 + NaBr(aq) \longrightarrow$
(j) $NH_3 + H_2O \longrightarrow$

2. (a) $K + H_2O \longrightarrow$
(b) $Al + NaOH \longrightarrow$
(c) $Mg + H_3PO_4(aq) \longrightarrow$
(d) $Pb + HCl \longrightarrow$
(e) $Sn + FeCl_2(aq) \longrightarrow$
(f) $Zn + Fe(NO_3)_3(aq) \longrightarrow$
(g) $Br_2 + NaI(aq) \longrightarrow$
(h) $NaH_2PO_4(aq) + Na_2CO_3(aq) \longrightarrow$
(i) $Cd + HgCl_2 \longrightarrow$
(j) $H_2S + NaOH \longrightarrow$

C. Reactions of Ions with Ions

1. (all in aqueous solution)
(a) $NaOH + HCl \longrightarrow$
(b) $NaHCO_3 + H_2SO_4 \longrightarrow$
(c) $Fe(NO_3)_3 + NaOH \longrightarrow$
(d) $AgNO_3 + NaCl \longrightarrow$
(e) $Cd(NO_3)_2 + Na_2S \longrightarrow$
(f) $CuSO_4 + Na_3PO_4 \longrightarrow$
(g) $Fe(NO_3)_3 + KSCN \longrightarrow$
(h) $PbCl_2 + K_2CrO_4 \longrightarrow$
(i) $CaCO_3 + HCl \longrightarrow$
(j) $NH_4Cl + NaOH \longrightarrow$

2. (all in aqueous solution)
(a) $Ca(NO_3)_2 + Na_2C_2O_4 \longrightarrow$
(b) $AlCl_3 + NaOH \longrightarrow$
(c) $NaC_2H_3O_2 + H_2SO_4 \longrightarrow$
(acetate)
(d) $BaCl_2 + Na_2SO_4 \longrightarrow$
(e) $Ca(OH)_2 + H_2SO_4 \longrightarrow$
(f) $Fe(NO_3)_3 + NaCN \longrightarrow$
(g) $AgNO_3 + Na_2S \longrightarrow$
(h) $PbI_2 + Na_2S \longrightarrow$
(i) $BaCl_2 + Na_2CO_3 \longrightarrow$
(j) $FeS + HCl \longrightarrow$

D. Decomposition (Elimination) Reactions

1. (a) $CuSO_4 \cdot 5H_2O \longrightarrow$
(b) $(NH_4)_2Cr_2O_7 \longrightarrow$
(c) $KClO_3 \longrightarrow$
(d) $HgO \longrightarrow$
(e) $NI_3 \longrightarrow$
(f) $BaCl_2 \cdot 2H_2O \longrightarrow$
(g) $NaHCO_3 \longrightarrow$
(h) $CaCO_3 \longrightarrow$
(i) $NH_4NO_3 \longrightarrow$
(j) $H_2O_2 \xrightarrow{\text{cat.}}$

SOLUTIONS

I. Coefficients are

1. 4,3 = 2
3. 4,5 = 1
5. 1,1 = 1
7. 2,1 = 1,1
9. 1,4 = 1,5

II. Coefficients are

1. 1,3 = 1,3
3. 2,1 = 2,1
5. 2,1 = 2,1
7. 1,4 = 1,2
9. 1,2 = 1,1,1

III.

1. $(NH_4)_2SO_4 + 2NaOH \longrightarrow 2NH_3 + Na_2SO_4 + 2H_2O$
3. $Zn + 2HCl \longrightarrow ZnCl_2 + \overline{H}_2$
5. $CuO + H_2 \longrightarrow Cu + H_2O$
7. $Hg(NO_3)_2 + (NH_4)_2S \longrightarrow \underline{HgS} + 2NH_4NO_3$
9. $2PbO_2 \longrightarrow 2PbO + \overline{O}_2$

IVA. **1.**

(a) $4Al + 3O_2 \longrightarrow 2Al_2O_3$
(b) $S + O_2 \longrightarrow SO_2$
(c) $CaO + SO_3 \longrightarrow CaSO_4$
(d) $SO_2 + H_2O \longrightarrow H_2SO_3$
(e) $NH_3(g) + HCl(g) \longrightarrow NH_4Cl(s)$
(f) $Mg + Cl_2 \longrightarrow MgCl_2$
(g) $CaO + H_2O \longrightarrow Ca(OH)_2$
(h) $C + O_2 \longrightarrow CO_2$
(i) $Na_2O + H_2O \longrightarrow 2NaOH$
(j) $CaO + CO_2 \longrightarrow CaCO_3$

IVB. **1.**

(a) $2Na + 2H_2O \longrightarrow 2NaOH + \overline{H}_2$
(b) $Ca + 2H_2O \longrightarrow Ca(OH)_2 + \overline{H}_2$
(c) $2Al + 6HCl \longrightarrow 2AlCl_3 + 3\overline{H}_2$
(d) $Fe + H_2SO_4 \longrightarrow FeSO_4 + \overline{H}_2$
(e) $Zn + 2HCl \longrightarrow ZnCl_2 + \overline{H}_2$
(f) $Cu + 2AgNO_3 \longrightarrow Cu(NO_3) + 2Ag$
(g) $Zn + Cu(NO_3)_2 \longrightarrow Zn(NO_3)_2 + Cu$
(h) $2Al + Fe_2O_3 \longrightarrow Al_2O_3 + 2Fe$
(i) $Cl_2 + 2NaBr \longrightarrow Br_2 + 2NaCl$
(j) $NH_3 + H_2O \longrightarrow NH_4^+ + OH^-$

IVC. **1.**

(a) $NaOH + HCl \longrightarrow NaCl + H_2O$
(b) $2NaHCO_3 + H_2SO_4 \longrightarrow Na_2SO_4 + 2\overline{CO}_2 + 2H_2O$
(c) $Fe(NO_3)_3 + 3NaOH \longrightarrow \underline{Fe(OH)_3} + 3NaNO_3$
(d) $AgNO_3 + NaCl \longrightarrow \underline{AgCl} + NaNO_3$
(e) $Cd(NO_3)_2 + Na_2S \longrightarrow \underline{CdS} + 2NaNO_3$
(f) $3CuSO_4 + 2Na_3PO_4 \longrightarrow Cu_3(PO_4)_2 + 3Na_2SO_4$

(g) $Fe(NO_3)_3 + 3NaSCN \longrightarrow \underline{Fe(SCN)_3} + 3NaNO_3$
(h) $PbI_2 + Na_2S \longrightarrow \underline{PbS} + 2NaI$
(i) $CaCO_3 + 2HCl \longrightarrow CaCl_2 + \overline{CO_2} + H_2O$
(j) $NH_4Cl + NaOH \longrightarrow NH_3 + NaCl + H_2O$

IVD. **1.** **(a)** $CuSO_4 \cdot 5H_2O \longrightarrow CuSO_4 + 5\overline{H_2O}$
(b) $(NH_4)_2Cr_2O_7 \longrightarrow Cr_2O_3 + \overline{N}_2 + 4\overline{H_2O}$
(c) $2KClO_3 \longrightarrow 2KCl + 3\overline{O}_2$
(d) $2HgO \longrightarrow 2Hg + \overline{O}_2$
(e) $2NI_3 \longrightarrow \overline{N}_2 + 3\overline{I}_2$
(f) $BaCl_2 \cdot 2H_2O \longrightarrow BaCl_2 + 2\overline{H_2O}$
(g) $2NaHCO_3 \longrightarrow Na_2CO_3 + CO_2 + H_2O$
(h) $CaCO_3 \longrightarrow CaO + CO_2$
(i) $2NH_4NO_3 \longrightarrow 2\overline{NO} + \overline{N}_2 + 4\overline{H_2O}$
(j) $2H_2O_2 \longrightarrow 2H_2O + \overline{O}_2$

After the "how" and "why" of chemical reactions the question of "how much" is considered.

After reading this chapter you should be able to

1. Use the chemical equation to obtain mole/mole relationships;
2. Use the mole/mole factor with molar weights to calculate the amount of one substance from another substance;
3. Calculate the percent composition of a sample by analysis of the products;
4. Use percent and density relationships in synthesis problems;
5. Find the limiting reagent in a reaction;
6. Find the percent yield of a reaction;
7. Use the molar volume of gases to calculate quantities in a chemical reaction.

calculations from equations

10

Three problems dealing with quantities commonly confront chemists in the practice of their art. The first has to do with *analysis:* how much of a given component is in a sample. The second concerns *synthesis:* how much of each reagent is necessary to produce a given amount of products. The third is the problem of how to get a raise in salary, but that need not concern us here.

To analyze a sample the chemist may physically isolate the component and measure it directly. In most instances this is impossible or impractical so the chemist must measure something *related to* the quantity wanted. One way to do this is to carry out a quantitative chemical reaction that will give products easier to isolate and measure. *The relationship between the products and reactants is given by the chemical equation and it is a mole/mole relationship.*

In a synthesis problem the chemist controls the amounts of reactants. He chooses the amounts necessary to give the desired amount of products. The relationship again depends on the mole relationships of the chemical equation.

The calculation of the relative amounts of the reactants and products of a chemical reaction is called *stoichiometry.* The word looks like it is related to sorcery but it is not all that difficult. We already know two properties that relate weight to moles (molar weight, grams/mole), and volume to moles (molar volume of gases, RT/P liters/mole) for pure substances. Now the mole/mole relationship from the balanced chemical equation allows us to calculate the amount of one substance from the amount of a different substance in the reaction equation.

EXAMPLE 10.1

$$N_2 + 3\ H_2 \rightleftharpoons 2\ NH_3$$

This equation not only says that nitrogen and hydrogen react reversibly to form ammonia but also that for every mole of hydrogen consumed $\frac{2}{3}$ mole of ammonia is produced. The ratio

$$\frac{2\ \text{moles}(NH_3)}{3\ \text{moles}(H_2)}$$

is a factor that relates the quantity of ammonia produced to the quantity of hydrogen consumed. The other components are similarly related:

$$\frac{2\ \text{moles}(NH_3)}{1\ \text{mole}(N_2)} \quad \text{and} \quad \frac{3\ \text{moles}(H_2)}{1\ \text{mole}(N_2)}$$

The mole/mole factor in a chemical equation is the link between two compounds. To find the amount of the compound that reacts with or is produced from another compound we must first convert the amount of the compound we have into moles. The mole/mole factor from the equation can then be used to find the moles of the desired compound.

What weight of hydrogen is consumed when 200 grams of ammonia is produced?

Want: weight of hydrogen in grams

Have: 200 grams of ammonia

Know: The relation of weight to moles of ammonia is the molar weight:

$$\frac{17\ \text{g}}{\text{mol}(NH_3)}$$

The relation of moles of ammonia to moles of hydrogen is the mole/mole factor: $\frac{3\ \text{mol}(H_2)}{2\ \text{mol}(NH_3)}$

The relation of the weight to moles of hydrogen is the molar weight:

$$\frac{2\ \text{g}}{\text{mol}(H_2)}$$

Solution: $200\ \text{g} \times \frac{1\ \text{mol}(NH_3)}{17\ \text{g}} \times \frac{3\ \text{mol}(H_2)}{2\ \text{mol}(N_2)} \times \frac{2\ \text{g}}{\text{mol}(H_2)} = 35\ \text{g of}\ H_2$

EXAMPLE 10.2

How much silver is produced when 0.153 gram of copper dissolves in silver nitrate solution?

$$2\ AgNO_3 + Cu \longrightarrow 2\ Ag + Cu(NO_3)_2$$

Want: grams of silver

Have: 0.153 gram of copper

Know: From the reaction equation the relation of silver to copper is

$$\frac{2\ \text{mol Ag}}{1\ \text{mol Cu}}$$

The relation of weight to moles of Cu $= \frac{63.5\text{ g}}{\text{mol(Cu)}}$

The relation of weight to moles of Ag $= \frac{107.9\text{ g}}{\text{mol(Ag)}}$

Solution: $0.153\text{ g} \times \frac{1\text{ mol(Cu)}}{63.5\text{ g}} \times \frac{2\text{ mol(Ag)}}{1\text{ mol(Cu)}} \times \frac{107.9\text{ g}}{\text{mol(Ag)}} = 0.520\text{ g of silver}$

We know the only connection between silver and copper is a mole/mole factor. We have and want grams of copper and silver, respectively, so we must look up the molar weights to make the proper conversions.

EXERCISE 10.1

Cover the right side of the page. Write out the answers; then check them.

Aluminum reacts with sodium hydroxide and water to produce hydrogen gas according to the following equation:

$$2\text{ Al} + 2\text{ NaOH} + 2\text{ H}_2\text{O} \longrightarrow 2\text{ NaAlO}_2\text{ (aq)} + 3\text{ H}_2\text{(g)}$$

How much aluminum is needed to make 5.0 grams of hydrogen gas?

Want: ______	*x* grams of aluminum
Have: ______	5.0 g of hydrogen
Know: mole/mole factor ______	$\frac{2\text{ mol(Al)}}{3\text{ mol(H}_2\text{)}}$
weight/mole factor(Al) ______	27.0 g/mol(Al)
weight/mole factor(H_2) ______	2.0 g/mol(H_2)
Solution: You want a quantity; Start with the quantity you have and use the chain rule ______	$5.0\text{ g(H}_2\text{)} \times \frac{1\text{ mol(H}_2\text{)}}{2.0\text{ g}} \times \frac{2\text{ mol(Al)}}{3\text{ mol(H}_2\text{)}} \times \frac{27.0\text{ g}}{\text{mole(Al)}} = 45\text{ g (Al)}$

Chemical analysis 10.1

Chemical and medical analysts report information about the composition of pure substances and mixtures. This information is derived from a sampling of the substance to be analyzed, so it is usually reported as a weight percentage. Medical and biochemical analysts often use weight/volume units too.

In either case, the answer is a property characteristic of the substance. To determine an intensive property two measurements must be made: one measurement from which the weight of the component can be inferred, and another measurement of the quantity of sample.

EXAMPLE 10.3 Iron(III) ions in solution react with aqueous ammonia to form insoluble iron(III) hydroxide. When iron(III) hydroxide is heated strongly it forms iron(III) oxide, Fe_2O_3.

$$Fe^{3+}(aq) + 3\ NH_3 + 3\ H_2O \longrightarrow \underline{Fe(OH)_3} + 3\ NH_4^+\ (aq)$$
$$2\ Fe(OH)_3 \xrightarrow[\Delta]{} Fe_2O_3 + 3\ \overline{H_2O}$$

Calculate the weight percent of iron(III) in a 1.50 gram sample if 0.320 grams of iron(III) oxide are obtained from the sample.

Want: percent iron in the sample; i.e., $\frac{\text{weight of iron}}{\text{weight of sample}} \times 100$

Have: weight of sample = 1.50 grams
weight of iron oxide = 0.320 grams

Know: All of the iron(III) in the sample is now present in iron oxide; the mole/mole factor is $\frac{2\ \text{mol(Fe)}}{1\ \text{mol}(Fe_2O_3)}$

The relationship of weight to moles of iron is 55.8 g/mol(Fe).
The relationship of weight to moles of iron oxide is 159.6 g/mol(Fe_2O_3).

Solution: This time we want a *ratio* of two quantities, so we start with the ratio of the two quantities we have. The weight of the sample goes in the denominator because weight percent is

$$\frac{\text{weight of the part}}{\text{weight of the whole}} \times 100$$

Then we convert the weight of iron oxide to the weight of iron using the relationships we know.

$$\frac{0.320\ \text{g}(Fe_2O_3)}{1.50\ \text{g sample}} \times \frac{1\ \text{mol}(Fe_2O_3)}{159.6\ \text{g}} \times \frac{2\ \text{mol(Fe)}}{1\ \text{mol}(Fe_2O_3)} \times \frac{55.8\ \text{g}}{\text{mol(Fe)}} \times 100 = 14.9\%$$

EXAMPLE 10.4 An impure sample of sodium bicarbonate containing some sand was decomposed by heat. The original sample weighed 5.00 grams, and 1.55 grams were driven off as CO_2 and H_2O. What percent of the sample is sodium bicarbonate?

$$2\ NaHCO_3 \xrightarrow[\Delta]{} Na_2CO_3 + H_2CO_3$$
$$H_2CO_3 \xrightarrow[\Delta]{} \overline{CO_2} + \overline{H_2O}$$

Want: $\frac{\text{grams of } NaHCO_3}{\text{grams of sample}} \times 100$

Have: 5.00 grams of sample; 1.55 grams of $CO_2 + H_2O$

Know: The weight of carbonic acid (H_2CO_3) is equal to the weight of $CO_2 + H_2O$, and 2 moles of $NaHCO_3$ are required to produce one mole of H_2CO_3. Molar weight of H_2CO_3 = 62.0 g/mol; $NaHCO_3$ = 84.0 g/mol

Solution: We start with what we have, in the form that we want, and convert the quantities to the desired quantities by the relationships we know:

$$\frac{1.55\text{ g}}{5.00\text{ g}} \times \frac{1\text{ mol}(H_2CO_3)}{62.0\text{ g}} \times \frac{2\text{ mol}(NaHCO_3)}{1\text{ mol}(H_2CO_3)} \times \frac{84.0\text{ g}}{\text{mol}(NaHCO_3)} \times 100$$
$$= 84.0\%\ NaHCO_3$$

EXAMPLE 10.5

Analysis of organic compounds for the elemental composition is usually done by combustion in oxygen. The equation is not known, but all the carbon in the compound goes to CO_2, and all the hydrogen goes to water.

What is the percent elemental composition of a compound consisting of C, H, and O, if 125 milligrams burned in oxygen produced 275 milligrams of CO_2 and 150 milligrams of water?

Want: **(a)** $\frac{\text{mg carbon}}{\text{mg compound}} \times 100$; **(b)** $\frac{\text{mg hydrogen}}{\text{mg compound}} \times 100$;
(c) the percent remainder is oxygen.

(a) Have: 125 mg of compound, from which 275 mg of CO_2 was produced.

Know: Each mole of C in the compound produces one mole of CO_2; C = 12.0 g/mol; CO_2 = 44.0 g/mol.

Solution: $$\frac{275\text{ mg}}{125\text{ mg}} \times \frac{1\text{ mol}(CO_2)}{44\text{ g}} \times \frac{1\text{ mol(C)}}{\text{mol}(CO_2)} \times \frac{12.0\text{ g}}{\text{mol(C)}} \times 100 = 60.0\%\text{ C}$$

(b) Have: 125 mg of compound, from which 150 mg of H_2O was produced.

Know: Two moles of H from the compound produces one mole of water; H_2O = 18.0 g/mol; H = 1.0 g/mol.

Solution: $$\frac{150\text{ mg}}{125\text{ mg}} \times \frac{1\text{ mol}(H_2O)}{18.0\text{ g}} \times \frac{2\text{ mol(H)}}{1\text{ mol}(H_2O)} \times \frac{1.0\text{ g}}{\text{mol(H)}} \times 100 = 13.3\%\text{ H}$$

(c) The percent oxygen is the remainder:

$$100.0 - (60.0 + 13.3) = 26.7\%\text{ O}$$

EXERCISE 10.2

Cover the right side of the page!

The percent chloride in water can be determined by precipitating the chloride as silver chloride, weighing the silver chloride, and then relating the amount of silver chloride to the quantity of chloride and the amount of sample.

$$Ag^+ + Cl^- \longrightarrow AgCl\ (s)$$

Calculate the weight percent of chloride in sea water if a 10.0-gram sample of sea water gave 0.736 gram of silver chloride.

Want:	$\frac{\text{grams chloride}}{\text{grams sea water}} \times 100$
Have:	grams sea water = 10.0 g grams silver chloride = 0.736 g
Know: mole/mole factor	1 mol Cl/mol AgCl
weight/mole factor (Cl)	35.5 g/mol(Cl)
weight/mole factor (AgCl)	143.4 g/mol(AgCl)
Solution: You want a ratio of quantities; start with the ratio you have and use the chain rule to change the quantity of AgCl to the weight of Cl.	$\frac{0.736\ \text{g(AgCl)}}{10.0\ \text{g (sample)}} \times \frac{1\ \text{mol(AgCl)}}{143.4\ \text{g}} \times \frac{1\ \text{mol(Cl)}}{1\ \text{mol(AgCl)}} \times \frac{35.5\ \text{g}}{\text{mol(Cl)}} \times 100 = 1.82\%$

10.2 *Chemical synthesis*

The time is long past when people satisfied all their needs by extracting natural products from roots, herbs, and animals. Most of the manufactured articles in use today are the result of chemical synthesis. Synthesis chemists bring together the proper amounts of starting materials under such conditions of temperature, pressure, and concentration that they get the maximum yield of the desired product.

Before any of those other considerations chemists need a balanced equation or equations that give the mole relationship between reactants and products.

EXAMPLE 10.6 Calcium acetylide (common name calcium carbide) reacts with water to give acetylene and calcium hydroxide.

$$CaC_2 + 2\ H_2O \longrightarrow \overline{C_2H_2} + Ca(OH)_2$$

How much crude carbide that is 90.0% CaC_2 must be taken to produce 500 grams of acetylene?

The amount of acetylene is related to the actual amount of pure CaC_2 by the mole/mole factor from the equation. The amount of crude carbide is obtained from the relation of CaC_2 to crude:

$$\frac{90.0\ \text{g}\ (CaC_2)}{100\ \text{g crude}}$$

Want: grams crude carbide salt

Have: 500 grams acetylene (C_2H_2)

Know: weight to moles for acetylene; $C_2H_2 = 26.0$ g/mol

$$\text{moles } C_2H_2 \text{ to moles } CaC_2 = \frac{1 \text{ mol}(CaC_2)}{1 \text{ mol}(C_2H_2)}$$

weight to moles $CaC_2 = 64.1$ g/mol(CaC_2)

$$\text{weight crude to weight } CaC_2 = \frac{100 \text{ g crude}}{90.0 \text{ g } CaC_2}$$

Solution: $$500 \text{ g} \times \frac{1 \text{ mol}(C_2H_2)}{26.0 \text{ g}} \times \frac{1 \text{ mol}(CaC_2)}{1 \text{ mol}(C_2H_2)} \times \frac{64.1 \text{ g}}{\text{mol}(CaC_2)} \times \frac{100 \text{ g crude}}{90.0 \text{ g } CaC_2}$$

= 1370 g of crude carbide salt

Starting materials are often impure, as in the above example, or in the form of liquid mixtures. The chemist must then know the relationship of the measurable amount of starting materials (weight or volume) to the weight of the pure component. The mole/mole relationship from the equation refers, of course, only to the pure substances.

EXAMPLE 10.7

What volume of 37% hydrochloric acid, density 1.18 grams/milliliter, must be added to calcium carbonate to obtain 1.00 kilogram of carbon dioxide?

$$CaCO_3 + 2\ HCl \longrightarrow CaCl_2 + CO_2 + H_2O$$

The weight of pure hydrogen chloride can be obtained from the mole/mole relationship and the molar weights. The relationship between the weight of pure HCl and the volume of hydrochloric acid is given by the weight/weight property (percent) and the weight/volume property (density).

Want: volume of hydrochloric acid (ml)

Have: 1000 grams of CO_2

Know: $$\frac{1 \text{ mol}(CO_2)}{2 \text{ mol}(HCl)};\ \frac{44 \text{ g}}{\text{mol}(CO_2)};\ \frac{36.5 \text{ g}}{\text{mol}(HCl)};\ \frac{37 \text{ g}(HCl)}{100 \text{ g acid}};\ \frac{1 \text{ ml acid}}{1.18 \text{ g acid}}$$

Solution:

$$1000 \text{ g} \times \frac{\text{mol}(CO_2)}{44 \text{ g}} \times \frac{2 \text{ mol}(HCl)}{1 \text{ mol}(CO_2)} \times \frac{36.5 \text{ g}(HCl)}{\text{mol}(HCl)} \times \frac{100 \text{ g acid}}{37 \text{ g}(HCl)} \times \frac{1 \text{ ml}}{1.18 \text{ g}}$$

= 3800 ml of acid

EXERCISE 10.3

Cover the right side of the page!

Nitrogen(IV) oxide reacts with water to give nitric acid and nitrogen(II) oxide, thusly,

$$3\ NO_2(g) + H_2O(l) \longrightarrow 2\ HNO_3(aq) + NO(g)$$

How much NO_2 is necessary to produce 1000 grams of a 60.0% by weight solution of nitric acid?

Want:	grams (NO_2)
Have:	1000 grams acid
Know: $\frac{\text{weight of } HNO_3}{\text{weight of acid}}$	$\frac{60.0 \text{ g } (HNO_3)}{100 \text{ g acid}}$
weight/mole (HNO_3)	$\frac{63.0 \text{ g}}{\text{mol}(HNO_3)}$
$\frac{\text{moles } NO_2}{\text{moles } HNO_3}$	$\frac{3 \text{ mol}(NO_2)}{2 \text{ mol}(HNO_3)}$
weight/mole(NO_2)	$\frac{46.0 \text{ g}}{\text{mol } NO_2}$

Solution:

$$1000 \text{ g (acid)} \times \frac{60.0 \text{ g}(HNO_3)}{100 \text{ g (acid)}} \times \frac{1 \text{ mol}(HNO_3)}{63.0 \text{ g}} \times \frac{3 \text{ mol}(NO_2)}{2 \text{ mol}(HNO_3)} \times \frac{46.0 \text{ g}}{\text{mol}(NO_2)} = 657 \text{ g of } NO_2$$

In synthesis reactions one reagent may be present in greater amounts than called for by the chemical equation. This may be done to make the reaction go faster or more toward completion, or the excess reagent may serve as a solvent or diluent to repress unwanted side reactions. The amount of product obtained, of course, is limited by the minimum number of moles of reactant needed to make it.

EXAMPLE 10.8 How much carbon tetrachloride will be formed from the complete reaction of the limiting reagent when 100 grams of CS_2 and 200 grams of Cl_2 react?

$$CS_2 + 3\ Cl_2 \xrightarrow[\Delta]{} CCl_4 + S_2Cl_2$$

Carbon tetrachloride will be produced until one reagent or the other is consumed. To find out which is used up first, we convert each of the two quantities to moles of carbon tetrachloride.

$$100 \text{ g}(CS_2) \times \frac{1 \text{ mol}(CS_2)}{76 \text{ g}} \times \frac{1 \text{ mol}(CCl_4)}{1 \text{ mol}(CS_2)} = 1.3 \text{ mol}(CCl_4)$$

$$200 \text{ g}(Cl_2) \times \frac{1 \text{ mol}(Cl_2)}{71 \text{ g}} \times \frac{1 \text{ mol}(CCl_4)}{3 \text{ mol}(Cl_2)} = 0.94 \text{ mol}(CCl_4)$$

Chlorine is the limiting reagent, and there is enough to make 0.94 moles of CCl_4.

$$0.94 \text{ mol}(CCl_4) \times \frac{154 \text{ g}}{\text{mol}(CCl_4)} = 145 \text{ g of } CCl_4$$

The actual amount of products isolated is often less than the amount predicted from the equation because of losses encountered in isolating and purifying the product, losses due to side reactions, or reactions that do not go to completion. The *theoretical yield* is the amount of product predicted from the number of moles of the limiting reagent used. The *actual yield* is the amount obtained after isolation and purification. The *percentage yield* is the ratio of the actual yield to the theoretical yield.

The chemist who wants to isolate a particular amount of product must calculate the amounts of starting materials on the basis of an expected percentage yield.

EXAMPLE 10.9

How much salicylic acid should be taken to produce 500 grams of acetylsalicylic acid (aspirin) if the expected yield is 80.0%?

$$\underset{\text{salicylic acid}}{C_7H_6O_3} + (C_2H_3O)_2O \longrightarrow \underset{\text{aspirin}}{C_9H_8O_4} + HC_2H_3O_2$$

Want: grams salicylic acid

Have: 500 grams aspirin (actual)

Know: The ratio of actual to theoretical is

$$\frac{80.0\ \text{g(act)}}{100\ \text{g(th)}};\ \frac{180\ \text{g}}{\text{mol}(C_9H_8O_4)};\ \frac{138\ \text{g}}{\text{mol}(C_7H_6O_3)};\ \frac{1\ \text{mol}(C_7H_6O_3)}{1\ \text{mol}(C_9H_8O_4)}$$

Solution:

$$500\ \text{g} \times \frac{100\ \text{g(th)}}{80.0\ \text{g(act)}} \times \frac{1\ \text{mol}(C_9H_8O_4)}{180\ \text{g}} \times \frac{1\ \text{mol}(C_7H_6O_3)}{1\ \text{mol}(C_9H_8O_4)} \times \frac{138\ \text{g}}{\text{mol}(C_7H_6O_3)}$$

$= 479$ g of salicylic acid

Calculations involving gases 10.3

Gases are usually measured by volume because it is inconvenient to weigh them. That should be no problem for us, because we know a property that changes the volume of a gas to moles, $\frac{P}{RT}\frac{\text{moles}}{\text{liter}}$.

EXAMPLE 10.10

How much potassium chlorate is necessary to produce 500 milliliters of dry oxygen gas measured at 745 torr and 25°C?

$$2\ KClO_3 \xrightarrow[\Delta]{} 2\ KCl + 3\ \overline{O}_2$$

Want: grams potassium chlorate

Have: 500 ml of O_2 gas

Know: $\frac{P}{RT} = \frac{745}{(62.4)(298)}\ \frac{\text{mol}(O_2)}{\text{liter}};\ \frac{2\ \text{mol}(KClO_3)}{3\ \text{mol}(O_2)};\ \frac{122.5\ \text{g}}{\text{mol}(KClO_3)}$

Solution: $0.500\ \text{liter} \times \frac{745}{(62.4)(298)}\ \frac{\text{mol}(O_2)}{\text{liter}} \times \frac{2\ \text{mol}(KClO_3)}{3\ \text{mol}(O_2)} \times \frac{122.5\ \text{g}}{\text{mol}(KClO_3)}$

$= 1.64$ g of $KClO_3$

10.4 *Summary*

1. The quantities of products and reactants in a chemical reaction are related by the ratio of the molar coefficients in the chemical equation.
2. Chemical analysts compare the amount of a component to a measured sample of a mixture or a compound. The amount of the component is usually calculated from a measurement of a related substance.
3. The weights of products and reactants in chemical synthesis are calculated from the mole/mole relationships and the molar weights.
4. Mole/mole factors apply only to pure substances. If reactants or products are measured as impure mixtures, the amounts of pure substances must first be calculated from weight/weight or weight/volume relationships.
5. The maximum moles of product obtainable is limited by the minimum moles of reactants available.
6. The actual amount of product obtained may be less than that predicted because of mechanical losses, side reactions, and incomplete reaction. The ratio of the actual yield to the theoretical yield ($\times$ 100) is the percent yield.
7. Quantities of gases in chemical reactions are usually measured by volume. The volume is related to the moles by the molar volume, RT/P liters/mole, and the moles are related by the mole/mole factor of the chemical equation.

PROBLEMS

1. The combustion of ethanol proceeds according to the following equation:

$$C_2H_5OH + 3\ O_2 \longrightarrow 2\ CO_2 + 3\ H_2O$$

   (a) How many moles of oxygen are needed to consume 2.0 moles of ethanol?
   (b) How many moles of water are produced from 1.5 moles of ethanol?
   (c) How many moles of CO_2 are produced from 100 grams of ethanol (molar weight 46 grams/mole)?
   (d) What weight of ethanol is needed to produce 100 grams of CO_2 (molar weight 44 grams/mole)?

2. A reaction useful as a rocket propellant occurs when hydrazine and hydrogen peroxide are mixed.

$$N_2H_4 + 2\ H_2O_2 \longrightarrow \overline{N_2} + 4\ \overline{H_2O}$$

   (a) How many moles of gas are produced by the combustion of 2.0 moles of N_2H_4?

(b) How many moles of H_2O_2 are needed to react with 5.0 moles of N_2H_4?

(c) How many moles of H_2O_2 are needed to react with 100 grams of N_2H_4 (molar weight 32 grams/mole)?

(d) How many grams of H_2O_2 (molar weight 34 grams/mole) are needed to react with 500 grams of N_2H_4?

(e) How many grams of 90% H_2O_2 solution are needed to react with 500 grams of N_2H_4?

3. The overall photosynthesis reaction is

$$6\ CO_2 + 6\ H_2O \xrightarrow{\text{light}} C_6H_{12}O_6 + 6\ O_2$$

What weight of sugar (molar weight 180 grams/mole) could be produced from 440 grams of CO_2, assuming 100% conversion?

4. Ammonia is made from nitrogen and hydrogen. Ammonium sulfate is made from ammonia and sulfuric acid.

$$N_2 + 3\ H_2 \longrightarrow 2\ NH_3$$
$$2\ NH_3 + H_2SO_4 \longrightarrow (NH_4)_2SO_4$$

(a) How many moles of hydrogen gas are consumed per mole of ammonium sulfate produced?

(b) How many grams of ammonia react with one gram of sulfuric acid (molar weight 98.0 grams/mole)?

(c) If 600 grams of hydrogen are converted completely to ammonia, and the ammonia is neutralized with sulfuric acid, how much ammonium sulfate is produced?

5. How much nitric acid containing 63% by weight HNO_3 can be made from 46 grams of NO_2 according to the equation:

$$3\ NO_2 + H_2O \longrightarrow 2\ HNO_3 + NO$$

Molar weights: $NO_2 = 46$ grams/mole; $HNO_3 = 63$ grams/mole

6. What weight of 10% hydrochloric acid will just dissolve 2.0 grams of zinc?

$$Zn + 2\ HCl \longrightarrow ZnCl_2 + H_2$$

Molar weights: $Zn = 65$ grams/mole; $HCl = 36.5$ grams/mole

7. 0.380 gram of a compound containing only carbon and hydrogen was burned in oxygen to give 1.272 grams of CO_2 and 0.297 gram of H_2O. What is the weight percent composition of the compound?

8. When an excess of barium chloride solution was added to 10.0 grams of a sulfuric acid solution, 0.378 gram of barium sulfate was obtained. What percent H_2SO_4 is in the solution?

$$BaCl_2 + H_2SO_4 \longrightarrow \underline{BaSO_4} + 2\ HCl$$

Molar weights: $H_2SO_4 = 98.0$ grams/mole; $BaSO_4 = 233$ grams/mole

9. 1.000 gram of a sample containing oxalic acid was treated with an excess of calcium chloride to precipitate calcium oxalate. The calcium oxalate was decomposed by heat to give 0.440 gram of calcium oxide. What was the percent oxalic acid in the sample?

$$H_2C_2O_4 + CaCl_2 \longrightarrow CaC_2O_4 + 2\ HCl$$
$$CaC_2O_4 \longrightarrow CaO + CO_2 + CO$$

Molar weights: $CaO = 56.1$ grams/mole; $H_2C_2O_4 = 90.0$ grams/mole

10. How much chlorine is needed to make 200 grams of potassium chlorate, assuming 100% yield of products?

$$2\ KOH + Cl_2 \longrightarrow KClO + KCl + H_2O$$
$$3\ KClO \longrightarrow KClO_3 + 2\ KCl$$

Molar weights: $Cl_2 = 71.0$ grams/mole; $KClO_3 = 122.5$ grams/mole

11. What volume of 95.0% sulfuric acid, density 1.84 grams/milliliter is necessary to produce 100 grams of hydrogen chloride gas?

$$2\ NaCl + H_2SO_4 \longrightarrow 2\ HCl + Na_2SO_4$$

12. How much 85% phosphoric acid can be produced from 100 grams of P_4O_{10}?

$$P_4O_{10} + 6\ H_2O \longrightarrow 4\ H_3PO_4$$

13. **(a)** What is the theoretical yield of nitrobenzene ($C_2H_5NO_2$) from 100 grams of benzene (C_6H_6)?

$$C_6H_6 + HNO_3 \xrightarrow{H_2SO_4} C_6H_5NO_2 + H_2O$$

Molar weights: benzene = 78.0 grams/mole; nitrobenzene = 123 grams/mole

(b) If the actual yield of nitrobenzene is 120 grams what is the percent yield?

14. What volume of air at 298°K and 740 torr containing 20% oxygen by volume is necessary to burn 2.5 kilograms (about 1 gallon) of octane to CO_2 and water?

$$2\ C_8H_{18} + 25\ O_2 \longrightarrow 16\ CO_2 + 18\ H_2O$$

Molar weight of octane = 114 grams/mole

15. What weight of magnesium liberates 450 milliliters of hydrogen gas measured at 750 torr and 298°K?

$$Mg + 2\ HCl \longrightarrow MgCl_2 + \overline{H_2}$$

16. How much sodium nitrite is necessary to produce 500 milliliters of nitrogen(I) oxide measured at 720 torr and 288°K?

$$NH_4Cl + NaNO_2 \longrightarrow \overline{N_2O} + NaCl + H_2O$$

ADDITIONAL PROBLEMS

17. The neutralization of soda lime ($NaOH \cdot CaO$) with sulfuric acid proceeds according to the following equation:

$$2\ NaOH \cdot CaO + 3\ H_2SO_4 \longrightarrow Na_2SO_4 + 2\ CaSO_4 + 4\ H_2O$$

(a) How many moles of H_2SO_4 are needed to make 4 moles of $CaSO_4$ from soda lime?

(b) How many moles of Na_2SO_4 are produced from 6 moles of H_2SO_4 and sufficient soda lime?

(c) How many moles of H_2SO_4 are needed to react with 192 grams of soda lime (molar weight = 96.0 grams/mole)?

(d) What weight of H_2SO_4 (molar weight = 98 grams/mole) is needed to react with 100 grams of soda lime?

18. The combustion of natural gas (mostly methane) can be represented by the equation:

$$CH_4 + 2\ O_2 \longrightarrow CO_2 + 2\ H_2O$$

(a) How many moles of oxygen are consumed for each mole of CO_2 produced?

(b) How many moles of oxygen are needed to burn 5.0 moles of methane?

(c) If 0.132 gram of CO_2 (molar weight = 44 grams/mole) are produced, what weight of methane (molar weight = 16 grams/mole) was burned?

(d) What weight of air, 20% oxygen by weight, is needed to burn 16 grams of methane?

(e) If the density of air is 1.29 grams/liter, what volume of air is required to burn 16 grams of methane?

19. Carbon dioxide is taken up by lithium oxide according to the equation:

$$\overline{CO_2} + Li_2O \longrightarrow Li_2CO_3(s)$$

If an astronaut exhales 1.0 kilogram of CO_2 per day how much lithium oxide (molar weight 30 grams/mole) is necessary to take up the CO_2 for a seven day flight?

20. How many grams of ammonium sulfate are necessary to produce 80 grams of ammonia according to the equation

$$(NH_4)_2SO_4 + CaO \longrightarrow 2\ \overline{NH_3} + CaSO_4 \cdot H_2O$$

21. How much limestone which is 85% $CaCO_3$ is necessary to produce 200 grams of CO_2 according to the equation

$$CaCO_3 + 2\ HCl \longrightarrow \overline{CO_2} + CaCl_2 + H_2O$$

22. A compound containing carbon, hydrogen, oxygen, and nitrogen was submitted for analysis. 0.250 gram of the compound gave 0.293 gram of CO_2 and 0.151 gram of H_2O when burned in oxygen. In a separate reaction 0.750 gram of the compound gave 0.140 gram of elemental nitrogen. What is the percent elemental composition.

23. 10.00 grams of witherite rock (mostly $BaCO_3$) gave 8.90 grams of barium sulfate when treated with sulfuric acid. What percent of the rock is $BaCO_3$?

$$BaCO_3 + H_2SO_4 \longrightarrow \underline{BaSO_4} + CO_2 + H_2O$$

24. 20.0 milliliters of an iron salt solution, density 1.10 grams/milliliter, gave 1.430 grams of Fe_2O_3 after precipitating the iron as the hydroxide and converting to the oxide with heat.

$$Fe^{3+} + 3\ OH^- \longrightarrow \underline{Fe(OH)_3}$$
$$2\ Fe(OH)_3 \longrightarrow Fe_2O_3 + 3\ \overline{H_2O}$$

What is the percent iron in the sample?

25. $$Fe_2O_3 + CH_4 \longrightarrow 2\ Fe + CO + 2\ H_2O$$
How many tons of methane are necessary to reduce one ton of iron(III) oxide?

26. $$2\ Al + 3\ H_2SO_4 \longrightarrow Al_2(SO_4)_3 + 3\ \overline{H_2}$$
What weight of aluminum will produce 5.0 grams of hydrogen?

27. $$2\ C_2H_6 + 7\ O_2 \longrightarrow 4\ CO_2 + 6\ H_2O$$
$$CO_2 + 2\ NaOH \longrightarrow Na_2CO_3 + H_2O$$
How much solid sodium carbonate can be produced by burning 100 grams of ethane and absorbing the CO_2 on solid sodium hydroxide?

28. $$\underset{\text{sugar}}{C_6H_{12}O_6} \xrightarrow[\text{fermentation}]{\text{yeast}} \underset{\text{alcohol}}{2\ C_2H_5OH} + 2\ CO_2$$
What weight of molasses, which is 55% sugar, is necessary to produce 1000 grams of alcohol, assuming the fermentation is 100% efficient?

29. $$2\ KMnO_4 + 16\ HCl \longrightarrow 2\ KCl + 2\ MnCl_2 + 5\ \overline{Cl_2} + 8\ H_2O$$
How much $KMnO_4$ (molar weight 158 grams/mole) is necessary to produce 2.0 liters of chlorine gas at 750 torr and 298°K?

30. $$NaOH \cdot CaO + 2\ CO_2 \longrightarrow NaHCO_3 + CaCO_3$$
What weight of soda lime is necessary to completely react with 100 liters of CO_2 at 0°C and 1.0 atmosphere pressure?

31. $$(NH_4)_2Cr_2O_7 \longrightarrow Cr_2O_3(s) + N_2(g) + 4\ H_2O(g)$$
What volume of gas (nitrogen and water vapor) measured at 373°K and 1 atmosphere is produced by the decomposition of 1.0 gram of ammonium dichromate?

SOLUTIONS

The formula of a compound in these solution sets will stand for 1 mole of the substance so identified.

1. (a) $$2.0\ C_2H_5OH \times \frac{3\ O_2}{C_2H_5OH} = 6.0\ O_2$$

(b) $$1.5\ C_2H_5OH \times \frac{3\ H_2O}{C_2H_5OH} = 4.5\ H_2O$$

(c) $$100\ \text{g} \times \frac{C_2H_5OH}{46.0\ \text{g}} \times \frac{2\ CO_2}{C_2H_5OH} = 4.35\ CO_2$$

(d) $$100\ \text{g} \times \frac{CO_2}{44.0\ \text{g}} \times \frac{C_2H_5OH}{2\ CO_2} \times \frac{46.0\ \text{g}}{C_2H_5OH} = 52.3\ \text{g ethanol}$$

2. (a) $$2.0\ N_2H_4 \times \frac{1\ N_2 + 4\ H_2O}{N_2H_4} = 10.0\ \text{mol gas}$$

(b) $$5.0\ N_2H_4 \times \frac{2\ H_2O_2}{1\ N_2H_4} = 10.0\ H_2O_2$$

(c) $$100\ \text{g} \times \frac{N_2H_4}{32.0\ \text{g}} \times \frac{2\ H_2O_2}{1\ N_2H_4} = 6.25\ H_2O_2$$

(d) $$500\ \text{g} \times \frac{N_2H_4}{32.0\ \text{g}} \times \frac{2\ H_2O_2}{1\ N_2H_4} \times \frac{34.0\ \text{g}}{H_2O_2} = 1060\ \text{g hydrogen peroxide}$$

(e) $500\text{ g} \times \frac{N_2H_4}{32.0\text{ g}} \times \frac{2\ H_2O_2}{1\ N_2H_4} \times \frac{34.0\text{ g}}{H_2O_2} \times \frac{100\text{ g}}{90\text{ g}} = 1180\text{ g solution}$

3. $440\text{ g} \times \frac{CO_2}{44.0\text{ g}} \times \frac{1\ C_6H_{12}O_6}{6\ CO_2} \times \frac{180\text{ g}}{C_6H_{12}O_6} = 300\text{ g sugar}$

4. (a) $1\ (NH_4)_2SO_4 \times \frac{2\ NH_3}{(NH_4)_2SO_4} \times \frac{3\ H_2}{2\ NH_3} = 3\ H_2$

(b) $1\text{ g} \times \frac{H_2SO_4}{98.0\text{ g}} \times \frac{2\ NH_3}{H_2SO_4} \times \frac{17.0\text{ g}}{NH_3} = 0.35\text{ g ammonia}$

(c) $600\text{ g} \times \frac{H_2}{2\text{ g}} \times \frac{1\ (NH_4)_2SO_4}{3\ H_2} \times \frac{130\text{ g}}{(NH_4)_2SO_4} = 13{,}000\text{ g}$ ammonium sulfate

5. $46\text{ g} \times \frac{NO_2}{46.0\text{ g}} \times \frac{2\ HNO_3}{3\ NO_2} \times \frac{63.0\text{ g}}{HNO_3} \times \frac{100\text{ g acid}}{63\text{ g}} = 67\text{ g acid}$

6. $2.0\text{ g} \times \frac{Zn}{65.0\text{ g}} \times \frac{2\ HCl}{1\ Zn} \times \frac{36.5\text{ g}}{HCl} \times \frac{100\text{ g acid}}{10\text{ g}} = 22\text{ g acid}$

7. $\frac{1.272\text{ g}}{0.380\text{ g}} \times \frac{CO_2}{44.0\text{ g}} \times \frac{1\ C}{CO_2} \times \frac{12.0\text{ g}}{C} \times 100 = 91.3\%\text{ carbon}$

$\frac{0.297\text{ g}}{0.380\text{ g}} \times \frac{H_2O}{18\text{ g}} \times \frac{2\ H}{H_2O} \times \frac{1.0\text{ g}}{H} \times 100 = 8.7\%\text{ hydrogen}$

8. $\frac{0.378\text{ g}}{10.0\text{ g}} \times \frac{BaSO_4}{233\text{ g}} \times \frac{H_2SO_4}{BaSO_4} \times \frac{98.0\text{ g}}{H_2SO_4} \times 100 = 1.59\%$

9. $\frac{0.440\text{ g}}{1.000\text{ g}} \times \frac{CaO}{56.1\text{ g}} \times \frac{CaC_2O_4}{CaO} \times \frac{H_2C_2O_4}{CaC_2O_4} \times \frac{90.0\text{ g}}{H_2C_2O_4} \times 100 = 70.6\%$

10. $200\text{ g} \times \frac{KClO_3}{122.5\text{ g}} \times \frac{3\ KClO}{KClO_3} \times \frac{1\ Cl_2}{KClO} \times \frac{71.0\text{ g}}{Cl_2} = 348\text{ g chlorine}$

11. $100\text{ g} \times \frac{HCl}{36.5\text{ g}} \times \frac{1\ H_2SO_4}{2\ HCl} \times \frac{98.0\text{ g}}{H_2SO_4} \times \frac{100\text{ g}}{95.0\text{ g}} \times \frac{1.0\text{ ml}}{1.84\text{ g}} = 76.8\text{ ml acid}$

12. $100\text{ g} \times \frac{P_4O_{10}}{284\text{ g}} \times \frac{4\ H_3PO_4}{P_4O_{10}} \times \frac{98.0\text{ g}}{H_3PO_4} \times \frac{100\text{ g}}{85.0\text{ g}} = 162\text{ g acid}$

13. (a) $100\text{ g} \times \frac{C_6H_6}{78.0\text{ g}} \times \frac{1\ C_6H_5NO_2}{C_6H_6} \times \frac{123\text{ g}}{C_6H_5NO_2} = 158\text{ g nitrobenzene}$

(b) $\frac{120\text{ g}}{158\text{ g}} \times 100 = 75.9\%\text{ yield}$

14. $2500\text{ g} \times \frac{C_8H_{18}}{114\text{ g}} \times \frac{25\ O_2}{2\ C_8H_{18}} \times \frac{(62.4)(298)\text{ liter}}{740\quad O_2} \times \frac{100\text{ liter air}}{20\text{ liter } O_2} = 34{,}400$ liters air

15. $0.450\text{ liter} \times \frac{750\quad H_2}{(62.4)(298)\text{ liter}} \times \frac{1\ Mg}{H_2} \times \frac{24.3\text{ g}}{Mg} = 0.441\text{ g magnesium}$

16. $0.500\text{ liter} \times \frac{720\quad N_2O}{(62.4)(288)\text{ liter}} \times \frac{1\ NaNO_2}{N_2O} \times \frac{69.0\text{ g}}{NaNO_2} = 1.38\text{ g}$ sodium nitrite

The chemical and physical properties of solutions are introduced.

After reading this chapter you should be able to

1. Define the relationship between the volume of a solution and the number of moles of solute (molarity);
2. Describe how to prepare solutions of known molarity;
3. Describe the operation of titration, and define equivalence point and end point;
4. Describe how to standardize a solution;
5. Use molarity and the mole/mole relationship to calculate quantities of reactant from the volume of titrant;
6. Describe the process of dynamic equilibrium;
7. Explain why molecular substances have a vapor pressure;
8. Explain the process of dissolving;
9. (optional) Define Raoult's Law;
10. (optional) Define molality, and calculate molalities of solutions;
11. (optional) Perform calculations involving boiling point elevation constants and freezing point depression constants;
12. (optional) Describe the phenomenon of osmotic pressure, and calculate the osmotic pressure of solutions.

11 solutions

A *solution* is a mixture of two or more substances blended so completely that it appears homogeneous. When one substance disperses throughout a different substance, it is said to *dissolve*. The dispersing medium is the *solvent*; that which disperses is the *solute*.

Substances in solution are so common that it is almost correct to say that everything that happens involves solutions at one stage or another. Plants absorb nutrients in solution. The food you eat must be dissolved to be assimilated. The oxygen you breathe must be dissolved in the blood before the hemoglobin can transport it to the cells. Ocean water is dispersed in the air and is carried to the land where lifting and cooling processes condense the water vapor and precipitate it as rain.

Solid substances must be in solution in order to react with each other. The reason of course is that atoms and molecules have to collide with each other in order to react, and if the reaction is confined to the surfaces of solids nothing much happens. Baking powder, for example, is made of potassium acid tartrate (a salt) and sodium bicarbonate (a salt), which will react in solution to form CO_2 and cause bubbles in the biscuits. In the solid state the mixture will keep indefinitely.

Chemical concentration; molarity 11.1

The general term *concentration* in chemistry refers to the relationship between the amount of dispersed substance and the volume or weight of the medium in which it is dispersed. Weight percent, for example, is a concentration term that relates the weight of the solute to the weight of solution. There are many different concentration terms for use in various circumstances.

One of the most useful and important concentration terms is *molarity* (M), the relationship between the *moles* of solute and the *volume* of solution.

$$\text{Molarity } (M) = \frac{n \text{ moles of solute}}{V \text{ liters of solution}} \tag{11.1}$$

Molarity is used to find the number of moles of solute by measuring the volume of the solution. In fact, molarity is the same property for solutions that P/RT is for gases.

$$\text{For solutions: } n \text{ moles} = M \frac{\text{moles}}{\text{liter}} \times V \text{ liters} \tag{11.2}$$

$$\text{For gases: } n \text{ moles} = \frac{P}{RT} \frac{\text{moles}}{\text{liter}} \times V \text{ liters} \tag{11.3}$$

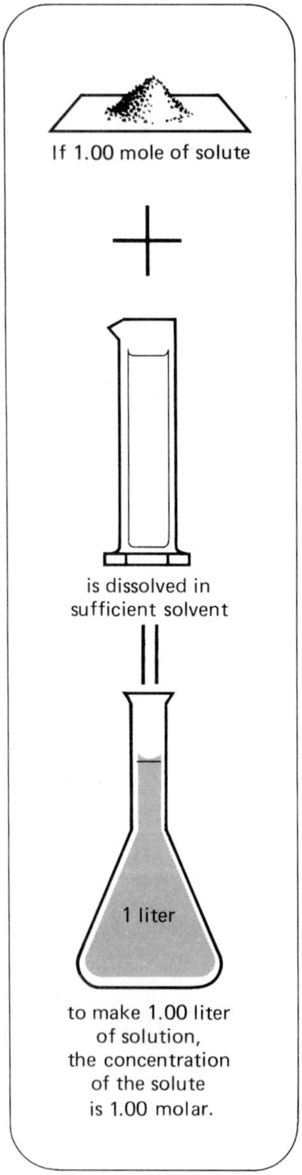

Figure 11.1

EXAMPLE 11.1

1. How many moles of solute are in 30.0 milliliters of 0.50M H_2SO_4?

Solution: $n \text{ mol} = 0.030 \text{ liter} \times 0.50 \frac{\text{mol}(H_2SO_4)}{\text{liter}} = 0.015 \text{ mol}$

2. What volume of 6.0M ammonia contains 0.20 moles of ammonia?

Solution: $\text{volume} = 0.20 \text{ mol}(NH_3) \times \frac{1 \text{ liter}}{6.0 \text{ mol}(NH_3)} = 0.033 \text{ liter}$

EXERCISE 11.1

Cover the right side of the page!

1. How many moles of solute are in 25.7 milliliters of 0.350M HCl?

Want: ______________	moles of HCl
Have: ______________	25.7 ml of solution
Know: relation of moles solute to volume	$\frac{0.350 \text{ mol(HCl)}}{\text{liter solution}}$
relation of milliliters to liters	$1000 \frac{\text{ml}}{\text{liter}}$
Solution: ______________	$25.7 \text{ ml} \times \frac{1 \text{ liter}}{1000 \text{ ml}} \times 0.350 \frac{\text{mol (HCl)}}{\text{liter}} = 9.00 \times 10^{-3} \text{ mol(HCl)}$

2. What volume of 0.500M acetic acid contains 0.150 mole of acetic acid?

Want: ______________	volume of solution
Have: ______________	0.150 mol acetic acid
Know: relation of moles solute to volume	$0.500\ \frac{\text{mol(acetic acid)}}{\text{liter solution}}$
Solution: ______________	$0.150\ \text{mol} \times \frac{1\ \text{liter}}{0.500\ \text{mol}}$
	$= 0.300$ liter

Volumes of unlike substances are not additive when dissolved, so the volume of the solution cannot be found by adding the volumes of the solute and solvent. Solutions of known molarity are prepared by adding sufficient solvent to the solute to make a measured volume of solution.

EXAMPLE 11.2

1. Describe exactly how to prepare 500 milliliters of 0.50*M* glucose ($C_6H_{12}O_6$) solution from solid glucose.
How many moles of glucose are needed?

$$0.500\ \text{liter} \times \frac{0.50\ \text{mol}(C_6H_{12}O_6)}{\text{liter}} = 0.25\ \text{mol}$$

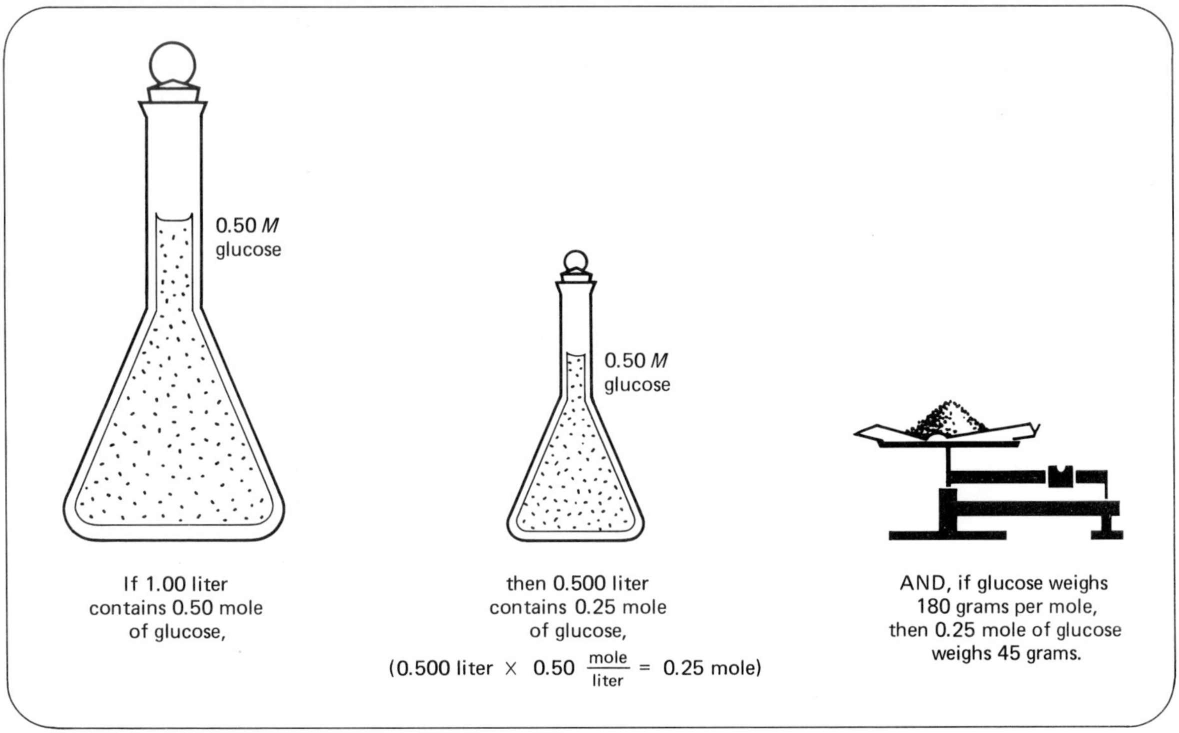

Figure 11.2

How are the moles of glucose related to weight?

$$0.25\ \text{mol}(C_6H_{12}O_6) \times \frac{180\ \text{g}}{\text{mol}(C_6H_{12}O_6)} = 45\ \text{g}$$

By the chain rule,

$$0.500\ \text{liter} \times \frac{0.50\ \text{mol}(C_6H_{12}O_6)}{\text{liter}} \times \frac{180\ \text{g}}{\text{mol}(C_6H_{12}O_6)} = 45\ \text{g}$$

Solution: Take 45 grams of solid glucose and dissolve in sufficient water to make 500 milliliters of solution in a volumetric flask.

EXERCISE 11.2

Cover the right side of the page!

How many grams of sodium chloride (NaCl) are in 200 milliliters of 0.100*M* NaCl solution?

Want: ______________	grams of NaCl
Have: ______________	200 ml of solution
Know: the relation of volume to moles	$0.100\ \frac{\text{mol(NaCl)}}{\text{liter solution}}$
the relation of weight to moles	$58.5\ \frac{\text{g}}{\text{mol(NaCl)}}$
Solution: ______________________	$0.200\ \text{liter} \times 0.100\ \frac{\text{mol(NaCl)}}{\text{liter}} \times 58.5\ \frac{\text{g}}{\text{mol(NaCl)}} = 1.17\ \text{g of NaCl}$

Solutions often must be prepared from stock solutions or from impure mixtures. In each case there must be some known relationship between a measurable quantity of the mixture and the number of moles of solute needed.

EXAMPLE 11.3

1. Describe exactly how to prepare 2.0 liters of 0.50*M* HCl from 37% hydrochloric acid stock solution, density 1.18 grams/milliliter.
How many moles of HCl are needed?

$$2.0\ \text{liters} \times \frac{0.50\ \text{mol(HCl)}}{\text{liter}} = 1.0\ \text{mol(HCl)}$$

How is that related to milliliters of stock solution?

$$1.0\ \text{mol(HCl)} \times \frac{36.5\ \text{g}}{\text{mol(HCl)}} \times \frac{100\ \text{g(stock)}}{37\ \text{g(HCl)}} \times \frac{1\ \text{ml}}{1.18\ \text{g}} = 84\ \text{ml}$$

Solution: Take 84 milliliters of stock solution and dilute with sufficient water to make 2.0 liters in a volumetric flask.

2. Describe exactly how to prepare 500 milliliters of 0.30*M* NaOH from 6.0*M* NaOH stock solution.
How many moles of solute are needed?

$$0.500 \text{ liter} \times \frac{0.30 \text{ mol(NaOH)}}{\text{liter}} = 0.15 \text{ mol(NaOH)}$$

How are the moles of NaOH related to the volume of stock solution?

$$0.15 \text{ mol(NaOH)} \times \frac{1 \text{ liter}}{6.0 \text{ mol(NaOH)}} = 0.025 \text{ liter}$$

By the chain rule,

$$0.500 \text{ liter} \times \frac{0.30 \text{ mol(NaOH)}}{\text{liter}} \times \frac{1 \text{ liter(stock)}}{6.0 \text{ mol(NaOH)}} = 0.025 \text{ liter}$$

Solution: Take 25 milliliters of stock solution and dilute with sufficient water to make 500 milliliters in a volumetric flask.

Cover the right side of the page!

EXERCISE 11.3

Describe how to prepare 100 milliliters of 3.00*M* hydrogen peroxide (H_2O_2) from a stock solution that is 30.0% H_2O_2, density 1.13 grams/milliliter.

Want: ______ | volume of H_2O_2 stock solution

Have: ______ | 100 ml of dilute solution

Know: (a) relation of moles H_2O_2 to volume of dilute solution ______ | $3.00 \frac{\text{mol}(H_2O_2)}{\text{liter(dil)}}$

(b) relation of moles(H_2O_2) to weight of H_2O_2 ______ | $34.0 \frac{\text{g}}{\text{mol}(H_2O_2)}$

(c) relation of weight of H_2O_2 to weight of stock solution ______ | $\frac{30.0 \text{ g}(H_2O_2)}{100 \text{ g stock}}$

(d) relation of weight of stock solution to volume of stock solution ______ | $1.13 \frac{\text{g}}{\text{ml}}$

Solution: ______

Solution:

$$0.100 \text{ liter} \times \frac{3.00 \text{ mol}(H_2O_2)}{\text{liter}} \times \frac{34.0 \text{ g}}{\text{mol}(H_2O_2)} \times \frac{100 \text{ g(stock)}}{30 \text{ g}(H_2O_2)} \times \frac{1.00 \text{ ml}}{1.13 \text{ g}} = 23.8 \text{ ml of stock solution}$$

Take 23.8 milliliters of stock solution and dilute with water to 100 milliliters.

11.2 *Titrations*

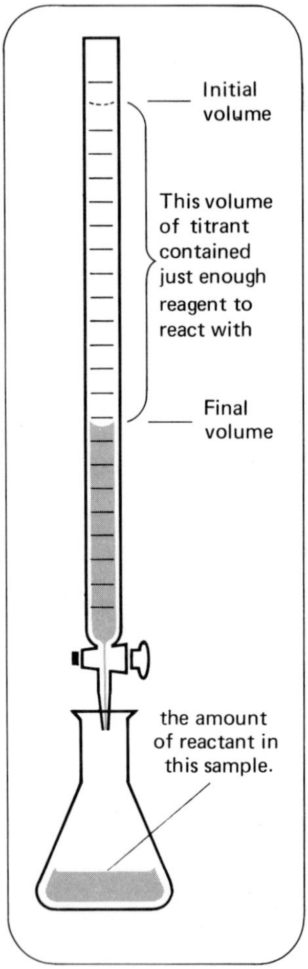

Figure 11.3 Titration.

The word "titrate" means to add a solution of one reagent (the titrant) to another until the amount of titrant added is just sufficient to react with all the second reactant, and no excess of either reagent is present. The titration is at the *equivalence point* when just enough titrant has been added to react with all of the reactant in the sample. The purpose is to determine the quantity of one of the reactants by relating it chemically to a known quantity of the other. To be practical the reagent must react quickly and completely, and the reaction must be free of side reactions. In other words, the stoichiometry must be completely predicted by the chemical equation.

It is also necessary that something signal when the equivalence point is reached, so that the analyst knows when to stop titrating. In some cases the electrical properties of the solution are measured. More often an indicator dye is added that changes color when the reactant to be analyzed is used up. The point in the titration at which the indicator changes color is called the "end point." Indicators must be chosen with some care so that the end point and the equivalence point are at the same place.

The quantity of reactant is related to the titrant by the mole/mole relationship of the chemical equation. Since the titrant is always a liquid solution, the quantity measured is usually the volume, and it is related to the moles of solute by the molarity.

EXAMPLE 11.4

How much sodium bicarbonate is in a sample that requires 35.0 milliliters of 0.50*M* HCl to titrate it?

$$NaHCO_3 + HCl \longrightarrow NaCl + CO_2 + H_2O$$

Titration implies that an amount of hydrogen chloride just sufficient to react with all the sodium bicarbonate was added to the sample. The equation specifies one mole of HCl per mole of $NaHCO_3$. Find the moles of HCl added, convert to moles of $NaHCO_3$, and find the weight from the molar weight.

Want: grams of $NaHCO_3$

Have: 35.0 ml of hydrochloric acid

Know: $\dfrac{0.50 \text{ mol(HCl)}}{\text{liter}}; \dfrac{1 \text{ mol } NaHCO_3}{1 \text{ mol HCl}}; \dfrac{84 \text{ g}}{\text{mol}(NaHCO_3)}$

Solution:

$$0.035 \text{ liter} \times \frac{0.50 \text{ mol(HCl)}}{\text{liter}} \times \frac{1 \text{ mol}(NaHCO_3)}{1 \text{ mol(HCl)}} \times \frac{84 \text{ g}}{\text{mol}(NaHCO_3)}$$

$$= 1.5 \text{ g of } NaHCO_3$$

The same relationships are often used to find the volume of a solution that will react with a known quantity of reagent.

EXAMPLE 11.5

What volume of $0.500M$ HCl is necessary to completely react with 0.375 gram of magnesium?

$$Mg + 2\,HCl \longrightarrow MgCl_2 + H_2$$

Want: volume of HCl solution

Have: 0.375 g of Mg

Know: $\frac{24.3\text{ g}}{\text{mol(Mg)}}$; $\frac{2\text{ mol(HCl)}}{1\text{ mol(Mg)}}$; $\frac{0.500\text{ mol(HCl)}}{\text{liter solution}}$

Solution:

$$0.375\text{ g} \times \frac{1\text{ mol(Mg)}}{24.3\text{ g}} \times \frac{2\text{ mol(HCl)}}{1\text{ mol(Mg)}} \times \frac{1\text{ liter}}{0.500\text{ mol(HCl)}}$$

$= 0.0617$ liter of HCl

EXERCISE 11.4

Cover the right side of the page!

How much chloride is in a sample that requires 35.2 milliliters of $0.0102M$ $AgNO_3$ solution to titrate it?

Want: ______ — grams of chloride

Have: ______ — 35.2 ml of $AgNO_3$ solution

Know: $Ag^+ + Cl^- \longrightarrow AgCl$;

(a) the mole/mole factor is ______ — $\frac{1\text{ mole }Cl^-}{1\text{ mole }AgNO_3}$

(b) the relation of moles to volume of $AgNO_3$ ______ — $\frac{0.0102\text{ moles}(AgNO_3)}{\text{liter}}$

(c) the relation of weight to moles of Cl^- ______ — $\frac{35.5\text{ g}}{\text{mol Cl}}$

Solution: ______

$$0.0352\text{ liter} \times \frac{0.0102\text{ mole}(AgNO_3)}{\text{liter}} \times \frac{1\text{ mol}(Cl^-)}{1\text{ mol}(AgNO_3)} \times \frac{35.5\text{ g}}{\text{mol(Cl)}}$$

$= 1.27 \times 10^{-2}$ g of Cl^-

11.3 *Standard solutions*

Any solution of accurately known concentration is a standard solution. To know the concentration accurately, the number of moles of solute and the volume must be known accurately. Most reagents are impure enough so that the exact number of moles cannot be known by weighing the substance. Sodium hydroxide, for example, absorbs moisture and carbon dioxide from the atmosphere so that it is continuously changing its composition. Once it is in solution, it can be protected from impurities. The answer is to "standardize" the solution; that is, to find its concentration by titration of a primary standard compound. A primary standard is a compound of known purity that can be weighed accurately. One such primary standard is potassium acid phthalate, an acid salt used for standardizing bases.

EXAMPLE 11.6 What is the molarity of a sodium hydroxide solution if 23.4 milliliters of it is required to titrate 1.020 grams of potassium hydrogen phthalate ($KHC_8H_4O_4$, MW 204)?

$$NaOH + KHC_8H_4O_4 \longrightarrow NaKC_8H_4O_4 + H_2O$$

Molarity is an intensive property.

Want: $\dfrac{\text{moles of NaOH}}{\text{volume of solution}}$

Have: 1.020 g of potassium hydrogen phthalate (KHPh)
23.4 ml of sodium hydroxide solution

Know: by titration and the mole ratio the moles of sodium hydroxide equal the moles of KHPh; molar weight $= \dfrac{204\text{ g}}{\text{mol(KHPh)}}$

Solution: $\dfrac{1.020\text{ g}}{0.0234\text{ liter}} \times \dfrac{1\text{ mol(KHPh)}}{204\text{ g}} \times \dfrac{1\text{ mol(NaOH)}}{1\text{ mol(KHPh)}} = 0.214M$

Solutions may also be standardized by titration with other standard solutions.

EXAMPLE 11.7 What is the molarity of a sulfuric acid solution if 25.0 milliliters of it requires 35.2 milliliters of 0.250*M* NaOH to titrate it?

$$H_2SO_4 + 2\ NaOH \longrightarrow Na_2SO_4 + 2\ H_2O$$

Want: $\dfrac{\text{moles of } H_2SO_4}{\text{volume of acid solution}}$

Have: 35.2 ml of sodium hydroxide solution
25.0 ml of sulfuric acid solution

Know: The moles of sodium hydroxide are related to the moles of sulfuric acid by the titration and the mole ratio $\frac{1\ \text{mole}(H_2SO_4)}{2\ \text{moles(NaOH)}}$; there are $\frac{0.250\ \text{mole(NaOH)}}{\text{liter}}$

Solution: $\frac{0.0352\ \text{liter}}{0.0250\ \text{liter}} \times \frac{0.250\ \text{mol(NaOH)}}{\text{liter}} \times \frac{1\ \text{mol}(H_2SO_4)}{2\ \text{mol(NaOH)}} = 0.176M$

Equilibrium and intensity factors 11.4

Up to this point we have assumed that all chemical events have a "before" condition or state represented by the left side of the equation and an "after" condition represented by the right side. We have not considered any "in-between" states. However, "in-between" or equilibrium processes are occurring all the time, everywhere, and one of the main concerns of chemistry is the study of the factors that determine how far a process goes before it seems to stop. A system is said to be "in equilibrium" when the intensity factors that make the system change in a forward or reverse direction are balanced. When the intensity factors are balanced, no more change is observed in the process or reaction. What are the intensity factors?

EXAMPLE 11.8

1. A mass balance is in equilibrium when the forces that make the balance arm rotate are equal and opposite. Balances are constructed so that the torque (twisting force) decreases on the side going down and increases on the side going up. If the balance arm were not constructed in this way,

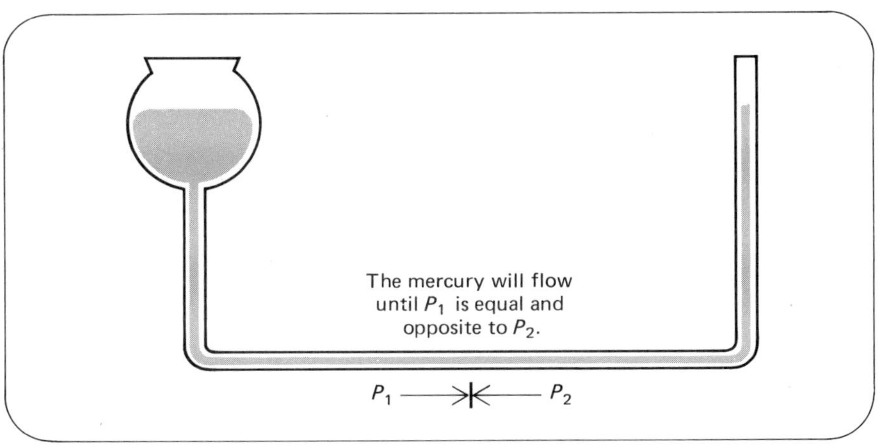

Figure 11.4 Static equilibrium.

it would continue to rotate until it fell off the fulcrum. The intensity forces are the weights times the moment arms. It is a necessary condition for equilibrium that the intensity factor in one direction decreases as the opposing factor increases.

2. Two objects at different temperatures in contact will exchange heat until they are at the same temperature. The intensity factor is the difference in temperatures. The hot object will decrease and the cool one increase in temperature until the intensity factors are equal.
3. If a connection exists between two reservoirs, water will flow from one to the other until the surfaces are at the same height. It does not matter if one reservoir is Lake Mead and the other is a teacup. The intensity factor is the difference in pressures. If the pressures (height times density) are the same, there will be no flow.

The above are examples of static equilibria. Something moves or is exchanged until forces that stop the action build up. In chemistry we often encounter *dynamic equilibrium*. The forward and reverse processes continue to go, but *they go at the same rate* so that no net change is observed.

EXAMPLE 11.9

1. Mike earns all his money working at a gas station, but he spends all he earns on gas. The money is in a state of dynamic equilibrium, because as fast as it goes to Mike it returns to the gas station. The intensity factor here depends on how much Mike needs gas.
2. If a yellow gas and a green gas are placed in two different containers with a passageway between them, the yellow will migrate into the green, and the green will migrate into the yellow until both containers are the same color. The process does not stop, but neither container ever becomes yellow or green again. Yellow molecules migrate back and forth but at a rate such that the *concentration* of yellow is the same in both containers. The green molecules do the same thing. The intensity factors here are the differences in concentration (measured as the partial pressures) of each gas in the two containers.

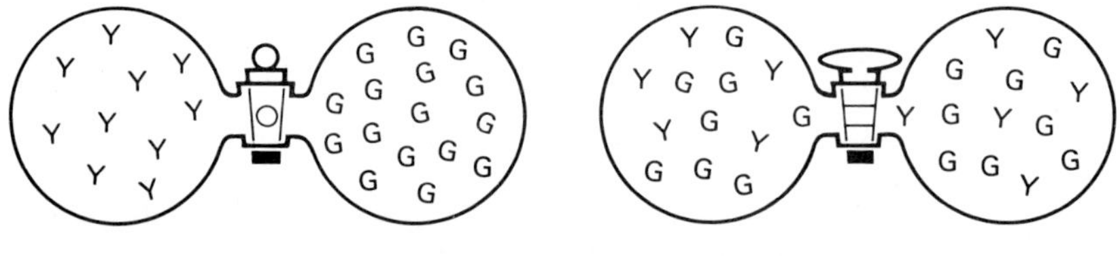

When the stopcock is opened, Y will flow to the region of lower concentration of Y, and G will flow to the region of lower concentration of G until the concentration of each in the two flasks is equal.

Figure 11.5 Dynamic equilibrium.

3. A solution of salt and a solution of sugar placed in contact will blend and diffuse until the two solutions are uniformly salty and sweet. The salt ions will diffuse into the region of lower salt concentration, and the sugar molecules will diffuse into the region of lower sugar concentration. Diffusion continues on both sides, but the two solutions never separate into salt and sweet again by themselves. The intensity factors that cause diffusion in the first place are the differences in concentration between the two solutions.
4. Hydrogen gas and nitrogen gas will combine at high temperature and pressure to form ammonia. Unfortunately, ammonia will decompose at the same temperature and pressure to form nitrogen and hydrogen. Eventually the rate of formation of ammonia and the rate of decomposition become equal, and the reaction seems to stop. The intensity factors for the forward and reverse reactions are proportional to the partial pressures of the gases.

Equilibrium and energy 11.5

Differences in intensity factors such as force, temperature, pressure, and concentration tell us the direction that change will take, but they do not tell us *how much* change will occur.

Water will flow from one reservoir to another until the surface levels are equal, but what will the level be when they are equal?

A hot object will warm a cool object until the temperatures are equal, but what will the final temperatures be?

Hydrogen and nitrogen form an equilibrium mixture with ammonia, but what fraction of the mixture is ammonia?

"Well," you say, "the amount of change must depend on the sizes of the changing systems. A twenty-five pound block of ice will not cool the ocean very much, but that much ice will cool a case of Coke very nicely."

The amount of change that occurs depends on the sizes (capacities) of the interacting systems *and* on the differences in intensity factors. The parts of a system come to equilibrium when all the intensity factors are balanced, but the ways in which the intensities change depend on the sizes or capacities of the various parts.

This very abstract concept, the capacity to cause change, is called *energy*. It is measured as the *product* of a capacity factor and an intensity factor. In Chapter 14 we will explore the various ways that systems can change energy and how to calculate the changes. Mechanical systems possess energy in the form of energy of motion (kinetic energy) and energy of position in a force field (potential energy). It is the experience of mankind that, *by itself*, a mechanical system will always change by converting its potential energy into kinetic energy when restraints are removed.

A mass balance will initially swing when the weights are unequal, but after a while it will settle at an equilibrium position with the minimum potential energy. The excess energy has been dissipated as random kinetic energy of air motion and friction.

A clock with a mainspring converts the potential energy of its spring into motion of cogs, wheels, and hands. Eventually the potential energy is gone, converted into random motion.

A space ship in orbit must convert the potential energy of height into heat in order to return to earth. The heat is dissipated to the atmosphere by the heat shield.

A rubber ball dropped from a height will bounce and convert some of its kinetic energy back into height. But each bounce also converts some energy into *random* kinetic energy of heat, until the ball has a minimum potential energy and comes to rest.

To summarize, every mechanical system, left to itself, will change to an equilibrium position with the minimum potential energy, and the excess potential energy will appear as random kinetic energy of heat.

Chemical systems possess potential energy as the energy stored in chemical bonds and other forces between atoms and molecules. Do chemical systems always change to an equilibrium condition of lowest potential energy when left to themselves? In other words, do chemical systems always give off heat to the surroundings when they spontaneously go to equilibrium? The next two sections will explore these questions.

11.6 *Solution processes and phase equilibria*

Almost everyone knows that a partially filled container of gasoline is more dangerous than a completely filled container. The space above the liquid is a potentially explosive mixture of gasoline vapor and air. If any liquid is in the tank, some of it will evaporate until the vapor space is saturated. If the tank is sealed with a pressure-tight cap, the pressure will increase to some value above atmospheric pressure and then stop.

If the air is removed with a vacuum pump, it is found that the pressure cannot be reduced below a certain value. As long as any liquid remains, the liquid will boil to maintain that pressure. Evidently gasoline molecules have a certain tendency to escape from the liquid into the vapor phase. The escaping tendency depends on the temperature and is measured by the partial pressure of the vapor.

Experiments with other liquids and many molecular solids show the same general behavior. Each molecular substance[1] has a *vapor pressure* characteristic of the substance and the temperature but independent of the amount of substance or vapor present. If the vapor is not removed, the liquid will evaporate until that pressure is reached and evaporation apparently ceases. The vapor and liquid are said to be in *dynamic equilibrium*, which implies that the rates of evaporation and condensation are in balance.

If the system is insulated from its surroundings, the liquid will cool because it requires energy to separate a molecule from its neighbors and put it in the gas phase (Figure 11.6). Since heat energy is kinetic energy, and a molecule in the vapor phase has a higher potential energy than one in the liquid phase, we are forced to conclude that a substance exerting a vapor pressure is spontaneously increasing its potential energy at the expense of its kinetic energy.

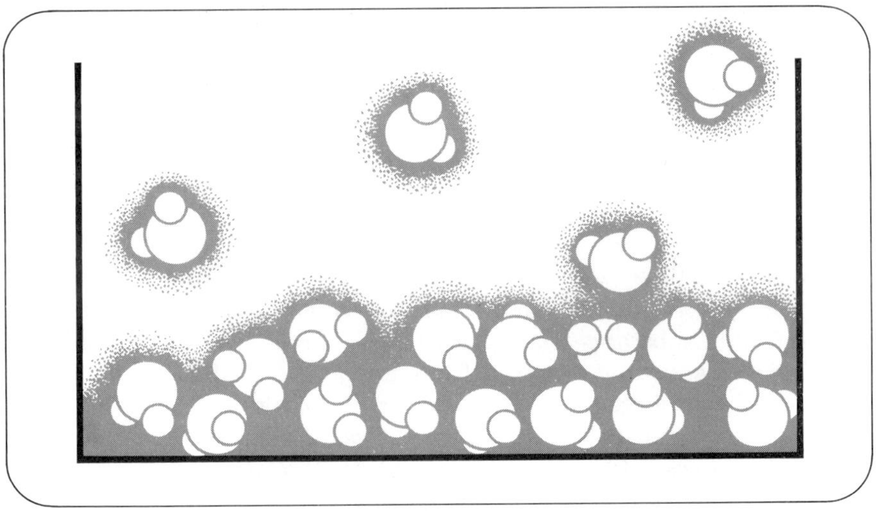

Figure 11.6 Evaporation. The attractive force between molecules is pictured as an "aura" that fades rapidly with distance. Molecules with kinetic energy much greater than average may escape the attractive force of the surface. In doing so they are slowed down by the attractive force to the average speed of all the molecules.

Table 11.1

Vapor pressure vs. temperature of some substances

Substance	1 torr	10 torr	100 torr	760 torr
water	− 17°C	11°C	52°C	100°C
carbon dioxide	−134°C	−119°C	−100°C	− 78°C
octane	− 14°C	19°C	66°C	126°C
ethyl ether	− 74°C	− 48°C	− 11°C	35°C

[1] All substances, even metals and ionic substances, either vaporize or decompose if the temperature is high enough. At ordinary temperatures the vapor pressure of these substances is vanishingly small.

A very similar situation is observed when a solute is placed in a solvent. If a quantity of salt is placed in water, some of the salt will dissolve or disperse in the water until a certain concentration is reached and then no more salt dissolves. The solution is saturated with salt. The saturation concentration depends on the temperature and the nature of the solvent, and like vapor pressure is characteristic of each substance.

There are some differences, however. Whereas the vaporization of a substance *always* absorbs heat from the system or the surroundings, the solution of a substance may absorb or evolve heat. The mixing of sulfuric acid in water, for example, may evolve enough heat to boil the solution, whereas the dissolving of sodium nitrate causes the solution to become cold. On the proposition that an evolution of heat means the system is going to a lower potential energy and an absorption of heat means a higher potential energy, we must conclude that a solution process can go either way! Evidently another factor, one that influences the normal tendency to seek the lowest potential energy, is at work.

11.7 *Entropy*

All substances have appreciable attractive forces between atoms and molecules when they are close enough to touch each other. The force between ionic substances is the coulombic force. The attractive force of molecular substances arises from unequal distribution of charge (polar or Debye forces) and from distortion of electron clouds due to neighboring atoms. Taken as a class, attractive forces are called "van der Waals forces." It requires work to remove a molecule from its neighbors and that energy must come from heat (kinetic energy). This means a substance that spontaneously evaporates or dissolves with an absorption of heat is moving toward a higher potential energy, and one does not expect that any more than one expects a boulder to fall uphill.

A system must gain some other advantage when it spontaneously changes in the "wrong" direction toward higher potential energy. The answer lies in the degree of disorder available to the system. When a molecule goes from the liquid phase into the gas phase, it automatically gains a large free volume in which to move. In a similar way an ion or molecule freed from an ordered crystal to the disordered liquid phase has gained a large measure of disorder.

"Freedom" and "disorder" in a system containing a very large number of molecules is a calculable quantity, not the subjective thing it is with us. It can be calculated as the work necessary to restore order, or as a quantity based on probability and statistics. Both methods require more mathematics than we have at our disposal at the moment so we will not show them in this book.

The change in disorder occurring in a system when it spontaneously changes to an equilibrium condition is called the *entropy* change. The natural tendency of everything is to become disordered, mixed, chaotic, and jumbled. Entropy is a property of systems containing very large numbers of particles, and the entropy increases as the disorder in the possible arrangements of the particles increases. The two laws of experience are

- A system will change to a condition of lowest potential energy when restraints are removed; and,
- A system will change to a condition of maximum entropy when restraints are removed.

As we have seen, sometimes these two laws are in opposition. That is, sometimes an increase in entropy requires an increase in potential energy. When that happens, the system will change to an equilibrium

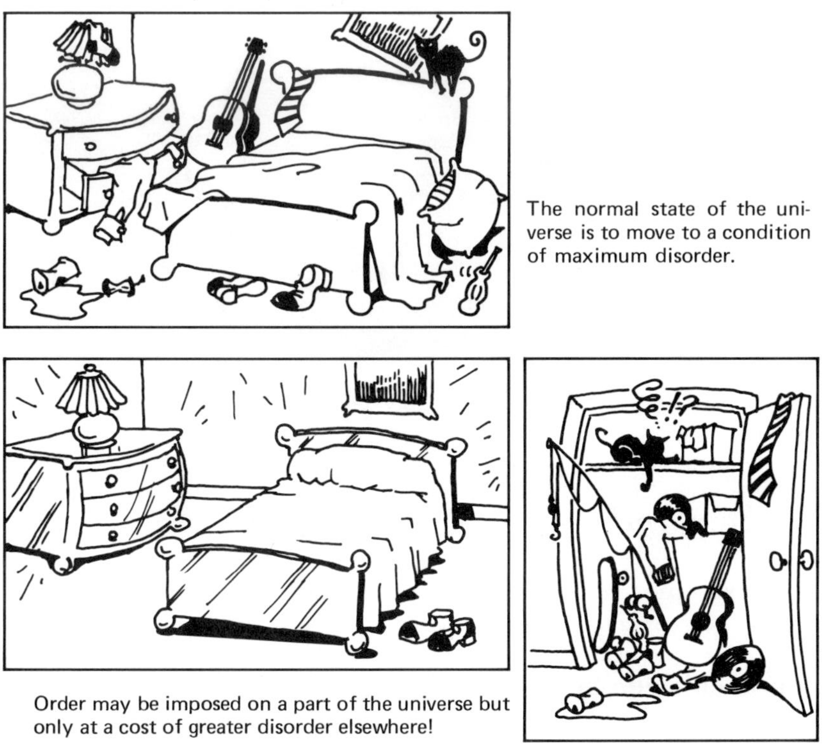

The normal state of the universe is to move to a condition of maximum disorder.

Order may be imposed on a part of the universe but only at a cost of greater disorder elsewhere!

Figure 11.7 Entropy.

condition in which the entropy gain and the potential energy gain are equal.

The working of these two laws is observed in all kinds of equilibrium processes. Entropy is the "spreading factor"; it always increases in the direction of expanding, vaporizing, dissolving, mixing, melting, and decomposing. Whether a process will increase or decrease in potential energy is more difficult to predict on a molecular basis than the direction of entropy change, but the energy change is easy to measure. If a process gives off heat to the surroundings, it is going to a lower potential energy. If it absorbs heat without increasing its temperature, it is going to a higher potential energy.

The ammonia reaction (given in Example 11.9) is an example of the working of entropy in a chemical reaction. The reaction of nitrogen and hydrogen to form ammonia is mildly exothermic; that is, the reaction gives up heat to the surroundings. Ammonia is therefore lower in potential energy than nitrogen and hydrogen. However, the ammonia molecule is more ordered than nitrogen and hydrogen (four molecules combine to form two molecules), so the entropy factor is unfavorable—it is opposed to the energy factor. The reaction comes to equilibrium at the point where the energy factor and the entropy factor are balanced. (Figure 11.8)

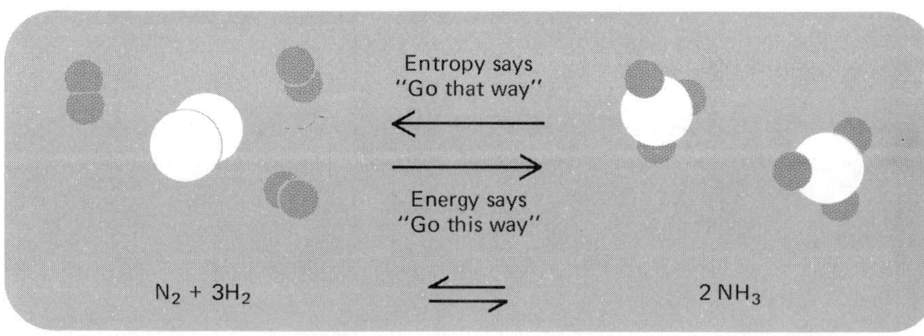

Figure 11.8 The ammonia synthesis.

11.8 *The how and why of dissolving*

Let us examine the working of the two factors, energy and entropy, in the physical process of dissolving.

Sodium chloride in water

When the ionic crystal of sodium chloride dissolves in water, the ions of sodium and chloride become free to wander throughout the solution. The energy necessary to separate the sodium ions and the chloride ions

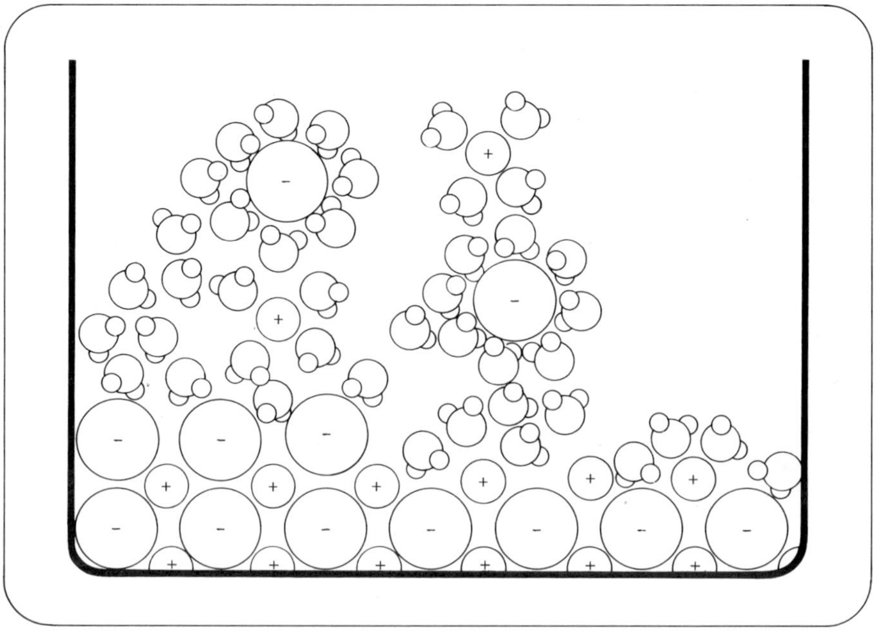

Figure 11.9 Sodium chloride dissolving in water. The sodium chloride crystal is much lower in entropy than the disordered ions in solution. The energy required to disrupt the crystal is supplied by the clustering of polar water molecules around the charged ions.

is supplied by the energy of interaction with the water. Water molecules are attracted to the charged ions more strongly than water is to itself. In this case the energy required to disrupt the crystal and the energy supplied by the interaction with water are about equal, so very little heat is required.

There is an immediate large gain in the spreading factor, entropy, because the disorder of the solution is much greater than that of the ordered crystals. As the salt dissolves, it causes a small but increasing amount of ordering in the water molecules, because so many of them are stuck to the ions in solution. Eventually, when large amounts of salt are in solution, the loss of entropy by the water balances the gain by the salt, and the dissolving process appears to stop. The solution is saturated with salt.

Alcohol in water

Alcohol and water interact with each other to about the same degree that they interact with other molecules of their own kind, so when alcohol and water are mixed the solution neither heats nor cools. Both liquids gain entropy on mixing, so the spreading factor is unlimited. Alcohol and water will mix in any proportion.

Carbon tetrachloride and water

Carbon tetrachloride is typical of many organic liquids in that it does not have enough polarity (electrical lopsidedness) to interact with water at all. Water has less attraction for carbon tetrachloride than it has for itself. This means that energy must be supplied for carbon tetrachloride and water to mix.

The entropy factor is not very large, because both substances are liquids and already disordered. The net result is that the entropy factor cannot overcome the large positive energy factor, and carbon tetrachloride and water do not dissolve in each other to any appreciable extent.

The effect of temperature on solubility

In most cases an increase in temperature increases the entropy factor, so that substances become more soluble at higher temperatures. In a few cases, the entropy factor is negative and the contrary is true. The substance becomes less soluble when heated. Table 11.2 shows the effect of temperature on the solubility of a few representative substances.

Table 11.2

Solubility vs. temperature of some substances (grams per 100 grams of water)

Substance	0°C	30°C	100°C
ammonium nitrate	118	242	871
barium hydroxide	1.67	5.59	100
potassium dichromate	5	20	80
silver chloride	9×10^{-5}	2×10^{-4}	2×10^{-3}
sodium chloride	35.7	36.3	39.8

We have only touched upon the many and varied aspects of solubility in this book. In subsequent courses you will explore this fascinating subject at more depth.

*11.9 *Colligative properties of solutions*

The colligative properties of solutions are those changes in physical properties of solvents dependent on the numbers of solute particles present. Vapor pressure, boiling point, freezing point, and osmotic

* Optional section.

pressure are the properties in this category. The laws describing the colligative properties are derived from the assumption that there is no potential energy change when the solute and solvent mix and that their physical states depend only on the entropy gain on mixing. A solution like this is called an "ideal solution." It is doubtful that any ideal solutions exist, but very dilute solutions approach ideal behavior. The following "laws" therefore apply only to very dilute solutions.

Vapor pressure of solutions

The vapor pressure of a solvent depends on the number of molecules with sufficient kinetic energy to leave the surface. If the number of solvent molecules in the surface is reduced by the presence of a solute, the vapor pressure will be reduced. *Raoult's Law* states that the vapor pressure (P) of a solvent is equal to the normal vapor pressure of the pure solvent (P_0) times the mole fraction (X) of the solvent in the solution.

$$P = XP_0 = \frac{n \text{ moles solvent}}{n \text{ moles solute} + n \text{ moles solvent}} \times P_0 \qquad (11.4)$$

Since colligative properties depend on numbers of particles and not on their identity, the number of moles of solute is taken as the total number of moles of particles produced by the solute.

EXAMPLE 11.10

1. If the vapor pressure of water is 760 torr at 100°C, what is the vapor pressure of a solution of 50 moles of water and 0.5 moles of sugar? Sugar does not ionize in water.

Solution: $P = \frac{50 \text{ mol } H_2O}{0.5 \text{ mol sugar} + 50 \text{ mol } H_2O} \times 760 \text{ torr} = 752 \text{ torr}$

2. What is the vapor pressure at 100°C of a solution of 50 moles of water and 0.5 mole of sodium chloride? Sodium chloride ionizes to $Na^+(aq)$ and $Cl^-(aq)$ in water.

Solution: $P = \frac{50 \text{ mol}(H_2O)}{50 \text{ mol } H_2O + 0.5 \text{ mol } Na^+ + 0.5 \text{ mol } Cl^-} \times 760 \text{ torr} = 745 \text{ torr}$

Boiling point and freezing point changes *11.10

The boiling point of a liquid is the temperature at which its vapor pressure equals the atmospheric pressure. From Example 11.10 it is

* Optional section.

seen that the presence of a solute decreases the vapor pressure and therefore elevates the boiling point.

The graph of the vapor pressure versus temperature of most substances looks like the solid line in Figure 11.10. The break in the curve is the freezing point and values to the left of the break represent the vapor pressure of the solid. If a solute is present, the vapor pressure is diminished somewhat and follows the dashed curve. It can be seen from the figure that the dashed curve intercepts the solid/vapor equilibrium line at a lower temperature, and therefore the solution has a lower freezing point than the pure solvent.

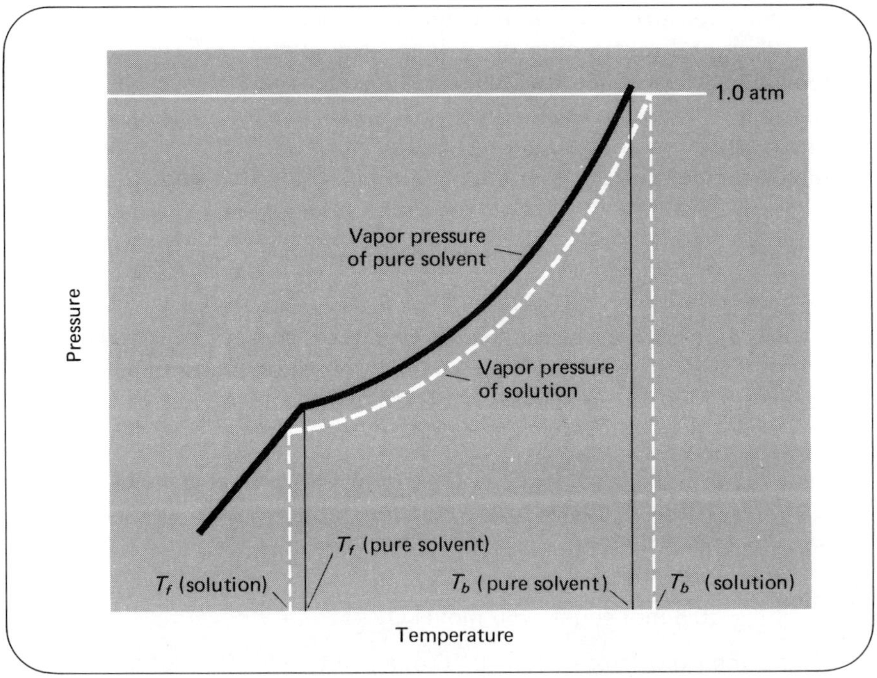

Figure 11.10 Vapor pressure vs. temperature.

*11.11 *Molality*

The vapor pressure change and, therefore, the change in the boiling point and freezing point depend on the nature of the solvent and its mole fraction (X) in the solution. The mole fraction of the *solute* is

$$(1 - X) \quad \text{or} \quad \frac{\text{moles solute}}{\text{moles solute} + \text{moles solvent}}$$

* Optional section.

This fraction is small for dilute solutions, so dropping the number of moles of solute in the denominator does not introduce much error.

$$\frac{0.1}{50 + 0.1} \simeq \frac{0.1}{50}$$

The moles of solvent in the denominator is related to the weight of the solvent by the molar weight. If we express the concentration of solute in terms of $\frac{\text{moles of solute}}{\text{weight of solvent}}$, we have a concentration term related to the mole fraction (approximately) and independent of the volume. When the weight of solvent is in kilograms, this concentration term is called the *molality* (m).

$$\text{molality } (m) = \frac{\text{moles of solute}}{\text{kilograms of solvent}} \tag{11.5}$$

Molality is independent of the density of a solution and does not change with temperature. From its derivation, we can see that this concentration term is useful for physical changes of solutions and is not used for stoichiometry.

EXAMPLE 11.11

What is the molality of a solution of 2.0 grams of naphthalene ($C_{10}H_8$) in 100 grams of benzene?

Solution: $2.0 \text{ g} \times \frac{1 \text{ mol}(C_{10}H_8)}{128 \text{ g}} = 0.016 \text{ mol}$

$$\frac{0.016 \text{ mol}}{0.100 \text{ kg solvent}} = 0.16 \text{ molal } (m)$$

Molecular weights from colligative properties *11.12

The boiling point elevation or the freezing point depression of dilute or ideal solutions depends on the solvent and the molality only. The factor that relates the change of temperature to the molality is called the boiling point elevation constant (K_b) or the freezing point depression constant (K_f).

K_b is the change in the boiling point of a solvent at one atmosphere pressure when one mole of a solute is dissolved in one kilogram of solvent. The solute should be molecular (not ionic) and should have no vapor pressure of its own at that temperature. K_f is the corresponding change in the freezing point of the solvent for a one molal solution.

The change in the boiling point or the freezing point is calculated

* Optional section.

by multiplying the corresponding constant, K_b or K_f, by the molality of the solution.

$$\Delta T_b \text{ degrees} = K_b \frac{\text{deg-kg}}{\text{mol}} \times m \frac{\text{mol}}{\text{kg}} \tag{11.6}$$

$$\Delta T_f \text{ degrees} = K_f \frac{\text{deg-kg}}{\text{mol}} \times m \frac{\text{mol}}{\text{kg}} \tag{11.7}$$

Table 11.3

Boiling point elevation constants

	$K_b \frac{\text{deg-kg}}{\text{mol}}$
acetic acid	3.07
carbon tetrachloride	5.03
benzene	2.53
water	0.512

K_b and K_f for some representative substances are given in Tables 11.3 and 11.4.

Table 11.4

Freezing point depression constants

	$K_f \frac{\text{deg-kg}}{\text{mol}}$
acetic acid	3.90
benzene	4.90
camphor	40.0
water	1.86

Soluble impurities in a substance always depress the melting point of the substance, so a melting point determination is a measure of the purity of a substance. Camphor has one of the highest melting point depression constants and may be used to determine molecular weights.

EXAMPLE 11.12

1. The freezing point of pure benzene is 5.5°C. What would the freezing point of benzene containing 1% of a soluble impurity of molecular weight 100 be?

Solution: The molality of the impurity is

$$\frac{1 \text{ g} \times \frac{1 \text{ mol}}{100 \text{ g}}}{0.099 \text{ kg benzene}} = 0.10\ m$$

The melting point depression is

$$\Delta T_f = -4.9 \frac{\text{deg-kg}}{\text{mol}} \times 0.10 \frac{\text{mol}}{\text{kg}} = -0.49 \text{ deg}$$

$$T_f = 5.5 - 0.49 = 5.0 \text{ deg}$$

2. What is the molecular weight of a substance if 0.250 gram dissolved in 10.0 grams of camphor depresses the melting point 10 degrees?

Want: molecular weight $\frac{\text{grams}}{\text{mole}}$

Have: 0.250 g of compound; 10 g of camphor

Know: **(a)** the melting point is depressed 10 degrees

(b) $-40\ \frac{\text{deg-kg}}{\text{mole}}$ is the freezing point depression constant for camphor

Solution: $\text{MW} = \frac{0.250\text{ g}}{0.010\text{ kg}} \times \frac{-40\text{ deg-kg}}{\text{mol}} \times \frac{1}{-10.0\text{ deg}} = 100\ \frac{\text{g}}{\text{mol}}$

Osmotic pressure *11.13

The cell walls of plants and animals have an amazing faculty for selectively allowing molecules and ions to diffuse through them. The walls act as barriers to some substances, while allowing others such as water to pass freely as through a sieve. For survival the cells must be able to keep nutrients inside them and pass waste products out.

The free diffusion of water, however, presents a problem to the cell. The law of entropy that says molecules will diffuse from the more concentrated region to the less concentrated region when restraints are removed applies to the water molecules. If a difference in solute concentration exists, water will pass through the membrane in the direction of the greater concentration of solute. This may cause the cell to burst or shrivel up like a dried pea. Whichever happens, it is decidedly unhealthy for the cell; so the organism must maintain the same concentration of dissolved particles in the fluid surrounding the cell as inside it.

If a difference in concentration of solutes on the two sides of the membrane exists, water can be prevented from diffusing by increasing the potential energy of the side with the greater concentration of solute. This can be done by increasing the external pressure on the fluid until the rate of diffusion due to the pressure difference is equal to the rate due to the concentration difference.

The diffusion process through a semipermeable membrane is called *osmosis*, and the pressure necessary to restore equilibrium is the *osmotic pressure*. The osmotic pressure can be calculated simply from the equation

$$P = MRT \tag{11.8}$$

where M is the molarity of the nondiffusing solute, R is the gas constant, and T is the absolute temperature.

* Optional sections.

EXAMPLE 11.13 What is the osmotic pressure of a 0.50*M* sugar solution at 298°K?

Solution: $P = \frac{0.50 \text{ mol}}{\text{liter}} \times \frac{0.0821 \text{ liter-atm}}{\text{mol-deg}} \times 298 \text{ deg} = 12 \text{ atm}$

11.14 *Summary*

1. Solutions are homogeneous mixtures, consisting of a dispersing phase, the solvent, and a dispersed component, the solute.
2. Molarity (M) is the number of moles of solute dissolved in one liter of solution. Molarity is used to compute the moles of solute in a given volume of solution.
3. Solutions of known molarity may be prepared by adding sufficient solvent to a measured quantity of solute to make a measured volume of solution. This method is possible only if there is a known relationship between a measurable quantity (weight or volume) of solute and the required number of moles.
4. Titration is the addition of one reagent (the titrant) to another until the quantity of titrant added is just sufficient to react with all of the second reactant, and no excess of either reagent is present. The purpose is to determine the quantity of one of the reactants by relating it to a known quantity of the other through the mole/mole relationship of the chemical equation.
5. Solutions of accurately known concentration are standard solutions. Solutions may be standardized by titration against primary standard compounds or other standard solutions.
6. The causes of physical and chemical changes are differences in intensity factors (force, pressure, temperature, concentration, and the like). When intensity factors are balanced, the system is in equilibrium, and no net changes are observed. Static equilibrium implies no change is occurring at all. Dynamic equilibrium implies that opposing processes are occurring at the same rate.
7. The amount of physical or chemical change that may occur depends on the sizes or capacities of the interacting parts as well as differences in intensity factors. The product of the intensity factor and capacity factor is called *energy*. Systems normally change to a condition of minimum potential energy when restraints are removed.
8. Systems containing large numbers of particles will change to a condition of maximum disorder when restraints are removed. The

measure of this change is called *entropy*. The equilibrium condition of solid/liquid/vapor phases, saturated solutions, and chemical reactions depend upon a balance of energy and entropy factors.

*9. Changes in vapor pressure, boiling point, freezing point, and osmotic pressure of solvents due to the presence of a solute are called colligative properties.

*10. Raoult's law relates the change in vapor pressure of a solvent to its mole fraction in the solution. The boiling point elevation constant (K_b) and the freezing point depression constant (K_f) relate the change in those properties to the molality of the solute. The molality is the number of moles of solute dissolved in a kilogram of solvent.

*11. Osmosis is the selective diffusion of molecules through a semipermeable membrane. The osmotic pressure of a solution is the pressure necessary to equalize the rates of diffusion of water in opposing directions through a membrane, when a difference in concentration exists between the solutions on each side of the membrane.

REVIEW QUESTIONS

1. Define the following terms:

- **(a)** solute
- **(b)** solvent
- **(c)** solution
- **(d)** molarity
- *(e) molality
- *(f) mole fraction
- **(g)** entropy
- **(h)** vapor pressure
- **(i)** boiling point
- **(j)** freezing point
- *(k) Raoult's law
- *(l) ideal solution
- *(m) osmosis

PROBLEMS

I. How many moles of solute are in

A. 1. 225 ml of 0.3*M* $Ba(OH)_2$
2. 125 ml of 1.5*M* $Al(NO_3)_3$
3. 25 ml of 0.10*M* H_2SO_4

B. 1. 150 ml of 0.10*M* glucose
2. 27.5 ml of 0.3*M* HCl
3. 39.1 ml of 0.295*M* NaOH

II. What volume of solution contains

A. 1. 0.015 mole of $BaCl_2$ if the molarity is 0.10*M*
2. 0.0050 mole of NaOH if the molarity is 0.30*M*
3. 0.020 mole of KCl if the molarity is 0.50*M*

B. 1. 0.15 mole of acetic acid if the molarity is 3.0*M*

* Optional sections.

2. 0.002 mole of $FeCl_3$ if the molarity is 0.5*M*
3. 0.018 mole of silver ion if the molarity is 0.1*M*

III. Describe exactly how to prepare aqueous solutions of the following. Assume pure compounds are available.

A. 1. 500 ml of 0.20*M* $CaCl_2$ solution
2. 200 ml of 0.10*M* KOH solution
3. 100 ml of 1.0*M* $C_6H_{12}O_6$ (glucose) solution
4. 100 ml of 0.10*M* glycine ($C_2H_5NO_2$) solution
5. 250 ml of 0.20*M* $CuSO_4$ solution from $CuSO_4{\cdot}5H_2O$

B. 1. 1.0 liter of 0.10*M* NaH_2PO_4
2. 5.0 liters of 0.50*M* acetic acid ($C_2H_4O_2$)
3. 1.5 liters of 0.10*M* monosodium citrate ($NaH_2C_6H_7O_7$)
4. 100 ml of 0.20*M* $Al(NO_3)_3$ from $Al(NO_3)_3{\cdot}9H_2O$
5. 250 ml of 0.10*M* Na_2CO_3

IV. Describe exactly how to prepare aqueous solutions from the specified reagents.

A. 1. 500 ml of 0.20*M* NaOH from 6.0 *M* NaOH
2. 200 ml of 1.0*M* NH_3 from 14*M* NH_3
3. 500 ml of 1.0*M* HCl from 37% HCl, density 1.18 g/ml
4. 500 ml of 1.0*M* H_2SO_4 from 96% H_2SO_4, density 1.84 g/ml
5. 250 ml of 3.0*M* HNO_3 from 68% HNO_3

B. 1. 250 ml of 0.10*M* NaH_2PO_4 from 5*M* NaH_2PO_4
2. 500 ml of 0.30*M* HCl from 12*M* HCl
3. 100 ml of 0.010*M* $PbCl_2$ solution from a solution containing 4.50 g of $PbCl_2$ per liter
4. 250 ml of 0.2*M* H_3PO_4 from 85% H_3PO_4
5. 100 ml of 0.1*M* KOH from 10% KOH solution

V. Stoichiometry

A. 1. $NaHCO_3 + HCl = NaCl + H_2O + \overline{CO_2}$
What weight of sodium bicarbonate is equivalent to 30.0 ml of 0.30*M* HCl?

2. $FeCl_3 + 3NaOH = \underline{Fe(OH)_3} + 3NaCl$
What weight of iron(III) hydroxide could be precipitated by 20.0 ml of 0.50*M* NaOH?

3. $AgNO_3 + HCl = \underline{AgCl} + HNO_3$
What weight of silver chloride would precipitate if 20.0 ml of 0.10*M* $AgNO_3$ and 30 ml of 0.075*M* HCl were mixed?

4. $H_2SO_4 + 2NaOH = Na_2SO_4 + 2H_2O$
What volume of 0.25*M* NaOH will convert 500 ml of 0.30*M* H_2SO_4 to sodium sulfate?

5. $C_6H_5CO_2CH_3 + KOH = C_6H_5CO_2K + CH_3OH$
Methyl benzoate was hydrolized with 100 ml of 1.0*M* KOH. The

excess KOH required 25.0 ml of 1.0M HCl to titrate it. How much methyl benzoate was in the sample?

B. **1.** $Mg + H_2SO_4 = MgSO_4 + \overline{H_2}$
What weight of magnesium is equivalent to 20 ml of 0.5M H_2SO_4?

2. $BaCl_2 + Na_2SO_4 = \underline{BaSO_4} + 2NaCl$
What weight of barium sulfate would be precipitated by 20.0 ml of 0.5M Na_2SO_4?

3. $CH_3CO_2C_2H_5 + KOH = CH_3CO_2K + C_2H_5OH$
What weight of ethyl acetate would be hydrolized by 75.0 ml of 2.0M KOH?

4. $2CaO{\cdot}NaOH + 3H_2SO_4 = 2CaSO_4 + Na_2SO_4 + 4H_2O$
What would it cost to neutralize 1000 kg of 10% sulfuric acid with soda lime if soda lime costs $0.30 per kilogram?

5. $Na_2CO_3 + H_2SO_4 = Na_2SO_4 + CO_2 + H_2O$
What volume of 0.2M H_2SO_4 will neutralize 400 ml of 0.5M Na_2CO_3?

VI. Standard Solutions

A. **1.** $KHC_8H_4O_4 + NaOH = KNaC_8H_4O_4 + H_2O$
Find the molarity of the NaOH solution if 27.5 ml is equivalent to 1.300 g of potassium acid phthalate.

2. $$5Na_2C_2O_4 + 2KMnO_4 + 8H_2SO_4 = 10\overline{CO_2} + 2MnSO_4 + K_2SO_4 + 5Na_2SO_4 + 8H_2O$$
Find the molarity of the potassium permanganate solution if 35.0 ml is equivalent to 2.95 g of sodium oxalate.

3. $CaCO_3 + H_2SO_4 = \underline{CaSO_4} + H_2O + \overline{CO_2}$
Find the weight percent calcium carbonate in a 2.00 g sample of rock if it reacts with 30.0 ml of 0.60M sulfuric acid.

4. $NH_3 + HCl = NH_4Cl$
The nitrogen in 5.0 g of a food sample was converted to ammonia, and this required 20.0 ml of 0.105M HCl to titrate it. What was the percent *nitrogen* in the food sample?

B. **1.** $NaOH + HCl = NaCl + H_2O$
What is the molarity of a sodium hydroxide solution if 25.0 ml requires 30.0 ml of 0.500M HCl to titrate it?

2. What is the percent of ammonia in household ammonia if 10.0 g of solution requires 5.9 ml of 0.500M HCl to titrate it?

3. $H_3C_6H_5O_7 + 3NaOH = Na_3C_6H_5O_7 + 3H_2O$
What is the percent of citric acid in lemon juice if 20.0 g of lemon juice requires 5.0 ml of 0.100M NaOH to titrate it?

4. 2.00 g of a solid acid (molar weight = 200 g/mol) requires 25.0 ml of a base solution to titrate it. One mole of acid reacts with two moles of base. What is the molarity of the base solution?

VII. Physical Concentration Terms

A. **1.** Describe how to prepare 200 g of a 20% NaOH solution.

2. What weight of 3.0% salt solution contains 10 g of salt?

3. What weight of $Cr(NO_3)_3$ is needed to prepare 250 ml of solution containing 0.10 mg Cr^{3+} ion per milliliter of solution?

4. A hydrochloric acid solution has a density of 1.18 g/ml and is 37.0% HCl by weight. What is the weight of HCl per liter of solution?

5. How much 15% sodium hypochlorite solution (so called "liquid chlorine") is necessary to establish a concentration of 4 ppm of available chlorine in a 30,000 gallon swimming pool? Assume 1 NaOCl is chemically equivalent to 1 Cl_2 and the density of water is 8.34 lb/gallon.

6. What is the mole fraction of a solution of 1.0 g of acetic acid in 99.0 g of water?

*7. What is the molality of the above solution?

*8. Calculate the molality and the molarity of a 10% ethylene glycol ($C_2H_6O_2$) solution, density 1.00 g/ml.

B. 1. Describe how to prepare 20 liters of a 23% sugar solution, density 1.095 kg/liter.

2. What weight of a 5% acetic acid solution contains 3.0 g of acetic acid?

3. What weight of $AlCl_3$ is needed to prepare 250 ml of solution containing 10 mg per ml of chloride ion?

4. Sea water is 3% sodium chloride. Its density is 1.025 kg/liter. What is the weight of sodium chloride per liter of sea water?

5. The formula of a popular water treatment chemical is $NaC_3N_3Cl_2O_3$. If one mole of the chemical is chemically equivalent to two moles of chlorine (Cl_2) and if the density of water is 3.79 kg/gallon, how much of this chemical should be added to a 20,000 gallon pool so that there are 1.5 ppm of available chlorine?

*6. What is the mole fraction of ethanol (C_2H_6O) in a 12% solution of ethanol?

*7. What is the molality of the above solution?

*8. Calculate the molality and molarity of a 30% ammonia solution, density 0.91 kg/liter.

VIII. *Colligative Properties

A. 1. The vapor pressure of water at 25°C is 23.8 torr. Calculate the vapor pressure of a solution of 8.55 grams of sucrose ($C_{12}H_{22}O_{11}$) in 100 grams of water.

2. The freezing point depression constant for water is $-1.86 \frac{\text{deg-kg}}{\text{mol}}$. What is the freezing point of the solution in problem 1?

3. The freezing point depression constant for benzene is $-4.9 \frac{\text{deg-kg}}{\text{mol}}$. The freezing point of pure benzene is 5.5°C. What is the molecular weight of a chlorinated hydrocarbon if a solution containing 0.90 g in 10.0 g of benzene freezes at 2.5°C?

* Optional sections.

B. 1. The vapor pressure of trichloroethylene (C_2HCl_3) is 400 torr at 67°C. What is the vapor pressure of a solution of 10 grams of fat (molar weight $C_3H_5(C_{18}H_{35}O_2)_3$ = 890 g/mole) in 90 grams of trichloroethylene at 67°C?

2. The boiling point elevation constant for water is 0.513 $\frac{\text{deg-kg}}{\text{mol}}$

What is the boiling point of a 3% sodium chloride solution under 1 atmosphere pressure? Remember sodium chloride ionizes.

3. A solution of 3.6 grams of a substance in 100 g of water freezes at −0.372°C. What is the molar weight of the solute?

SOLUTIONS

I.A. 1. $0.225 \text{ liter} \times 0.3 \frac{\text{mol}}{\text{liter}} = 0.07 \text{ mol}$

2. $0.125 \text{ liter} \times 1.5 \frac{\text{mol}}{\text{liter}} = 0.19 \text{ mol}$

3. $0.025 \text{ liter} \times 0.10 \frac{\text{mol}}{\text{liter}} = 0.0025 \text{ mol}$

II.A. 1. $0.015 \text{ mol} \times \frac{1 \text{ liter}}{0.10 \text{ mol}} = 0.15 \text{ liter}$

2. $0.0050 \text{ mol} \times \frac{1 \text{ liter}}{0.30 \text{ mol}} = 0.017 \text{ liter}$

3. $0.020 \text{ mol} \times \frac{1 \text{ liter}}{0.50 \text{ mol}} = 0.040 \text{ liter}$

III.A. 1. $0.500 \text{ liter} \times \frac{0.20 \text{ CaCl}_2}{\text{liter}} \times \frac{111 \text{ g}}{\text{CaCl}_2} = 11 \text{ g}$

Take 11 g of calcium chloride and dissolve in sufficient water to make 500 ml.

2. $0.200 \text{ liter} \times \frac{0.10 \text{ KOH}}{\text{liter}} \times \frac{56 \text{ g}}{\text{KOH}} = 1.1 \text{ g}$

Take 1.1 g of potassium hydroxide and dissolve in sufficient water to make 200 ml.

3. $0.100 \text{ liter} \times \frac{1.0 \text{ C}_6\text{H}_{12}\text{O}_6}{\text{liter}} \times \frac{180 \text{ g}}{\text{C}_6\text{H}_{12}\text{O}_6} = 18 \text{ g}$

Take 18 g of glucose and dissolve in sufficient water to make 100 ml.

4. $0.100 \text{ liter} \times 0.10 \frac{\text{C}_2\text{H}_5\text{NO}_2}{\text{liter}} \times \frac{75 \text{ g}}{\text{C}_2\text{H}_5\text{NO}_2} = 0.75 \text{ g}$

Take 0.75 g of glycine and dissolve in sufficient water to make 100 ml.

5. $0.250 \text{ liter} \times 0.20 \frac{\text{CuSO}_4 \cdot 5\text{H}_2\text{O}}{\text{liter}} \times \frac{249 \text{ g}}{\text{CuSO}_4 \cdot 5\text{H}_2\text{O}} = 12.5 \text{ g}$

Take 12.5 g of copper sulfate pentahydrate and dissolve in sufficient water to make 250 ml.

IV.A. **1.** $0.500 \text{ liter} \times \frac{0.20 \text{ NaOH}}{\text{liter}} \times \frac{1 \text{ liter}}{6.0 \text{ NaOH}} = 0.017 \text{ liter}$

To 17 ml of 6M NaOH add sufficient water to make 500 ml.

2. $0.200 \text{ liter} \times \frac{1.0 \text{ NH}_3}{\text{liter}} \times \frac{1 \text{ liter}}{14 \text{ NH}_3} = 0.014 \text{ liter}$

To 14 ml of 14M NH_3 add sufficient water to make 200 ml.

3. $0.500 \text{ liter} \times \frac{1.0 \text{ HCl}}{\text{liter}} \times \frac{36.5 \text{ g}}{\text{HCl}} \times \frac{100 \text{ g acid}}{37 \text{ g}} \times \frac{1 \text{ ml}}{1.18 \text{ g}} = 42 \text{ ml}$

To 42 ml of concentrated acid add sufficient water to make 500 ml.

4. $0.500 \text{ liter} \times \frac{1.0 \text{ H}_2\text{SO}_4}{\text{liter}} \times \frac{98 \text{ g}}{\text{H}_2\text{SO}_4} \times \frac{100 \text{ g}}{96 \text{ g}} \times \frac{1 \text{ ml}}{1.84 \text{ g}} = 28 \text{ ml}$

To 28 ml of concentrated acid add sufficient water to make 500 ml.

5. $0.250 \text{ liter} \times \frac{3.0 \text{ HNO}_3}{\text{liter}} \times \frac{63 \text{ g}}{\text{HNO}_3} \times \frac{100 \text{ g}}{68 \text{ g}} = 69 \text{ g}$

To 69 grams of concentrated acid add sufficient water to make 250 ml.

V.A. **1.** $.030 \text{ liter} \times \frac{0.30 \text{ HCl}}{\text{liter}} \times \frac{1}{1} \frac{\text{NaHCO}_3}{\text{HCl}} \times \frac{84 \text{ g}}{\text{NaHCO}_3} = 0.76 \text{ g}$

2. $.020 \text{ liter} \times \frac{0.50 \text{ NaOH}}{\text{liter}} \times \frac{\text{Fe(OH)}_3}{3 \text{ NaOH}} \times \frac{107 \text{ g}}{\text{Fe(OH)}_3} = 0.36 \text{ g}$

3. $.020 \text{ liter} \times \frac{0.10 \text{ Ag}^+}{\text{liter}} \times \frac{\text{AgCl}}{\text{Ag}^+} = .0020 \text{ AgCl}$

$.030 \text{ liter} \times \frac{0.075 \text{ Cl}^-}{\text{liter}} \times \frac{\text{AgCl}}{\text{Cl}^-} = .0022 \text{ AgCl}$

The silver ion is the limiting reagent, therefore:

$$.0020 \text{ AgCl} \times \frac{143 \text{ g}}{\text{AgCl}} = 0.29 \text{ g of silver chloride}$$

4. $0.500 \text{ liter} \times \frac{0.30 \text{ H}_2\text{SO}_4}{\text{liter}} \times \frac{2 \text{ NaOH}}{\text{H}_2\text{SO}_4} \times \frac{1 \text{ liter}}{0.25 \text{ NaOH}} = 1.2 \text{ liters}$

5. The methyl benzoate is related to the moles of KOH that reacted. That is the initial moles minus the excess moles.

$$\left[0.100 \text{ liter} \times \frac{1.0 \text{ KOH}}{\text{liter}}\right] - \left[.025 \text{ liter} \times \frac{1.0 \text{ HCl}}{\text{liter}} \times \frac{\text{KOH}}{\text{HCl}}\right] = .075 \text{ KOH}$$

$$.075 \text{ KOH} \times \frac{1 \text{ MB}}{\text{KOH}} \times \frac{136 \text{ g}}{\text{MB}} = 10 \text{ g of methyl benzoate}$$

VI.A. **1.** $\frac{1.300 \text{ g}}{.0275 \text{ liter}} \times \frac{\text{KHPh}}{204 \text{ g}} \times \frac{\text{NaOH}}{\text{KHPh}} = 0.232M \text{ NaOH}$

2. $\frac{2.92 \text{ g}}{.035 \text{ liter}} \times \frac{\text{Na}_2\text{C}_2\text{O}_4}{134 \text{ g}} \times \frac{2 \text{ KMnO}_4}{5 \text{ Na}_2\text{C}_2\text{O}_4} = 0.25M \text{ KMnO}_4$

3. $\frac{.030 \text{ liter}}{2.00 \text{ g}} \times \frac{0.60 \text{ H}_2\text{SO}_4}{\text{liter}} \times \frac{\text{CaCO}_3}{\text{H}_2\text{SO}_4} \times \frac{100 \text{ g}}{\text{CaCO}_3} \times 100$

$= 90\%$ calcium carbonate

4. $\frac{.020\text{ liter}}{5.0\text{ g}} \times \frac{0.105\text{ HCl}}{\text{liter}} \times \frac{1\text{ NH}_3}{\text{HCl}} \times \frac{1\text{ N}}{\text{NH}_3} \times \frac{14\text{ g}}{\text{N}} \times 100 = 0.59\%$

VII.A. 1. $200\text{ g} \times \frac{20\text{ g}}{100\text{ g}} = 40$ g of sodium hydroxide

Take 40 g of sodium hydroxide and dissolve in 160 g of water.

2. $10\text{ g} \times \frac{100\text{ g solution}}{3.0\text{ g salt}} = 330$ g of solution

3. $250\text{ ml} \times \frac{0.10\text{ mg}}{\text{ml}} \times \frac{\text{Cr}^{3+}}{52\text{ mg}} \times \frac{\text{Cr(NO}_3)_3}{\text{Cr}^{3+}} \times \frac{238\text{ mg}}{\text{Cr(NO}_3)_3}$

$= 110$ mg of chromium(III) nitrate

4. $\frac{37.0\text{ g}}{100\text{ g}} \times \frac{1.18\text{ g}}{\text{ml}} \times \frac{1000\text{ ml}}{\text{liter}} = 437$ g of hydrogen chloride

5. $30{,}000\text{ gal} \times \frac{8.34\text{ lb}}{\text{gal}} \times \frac{4\text{ lb chlorine}}{10^6\text{ lb water}} \times \frac{1\text{ Cl}_2}{71\text{ lb}}$

$\times \frac{1\text{ NaOCl}}{\text{Cl}_2} \times \frac{74.5\text{ lb}}{\text{NaOCl}} \times \frac{100\text{ lb}}{15\text{ lb}} = 7.0\text{ lb}$

6. $1.0\text{ g} \times \frac{\text{HOAc}}{60\text{ g}} = .017\text{ HOAc}$

$99 \times \frac{\text{H}_2\text{O}}{18} = 5.5\text{ H}_2\text{O}$

$X = \frac{.017}{5.5 + .017} = 3.1 \times 10^{-3}$

7. $\frac{.017\text{ mole}}{.099\text{ kg}} = 0.17m$

8. (a) $\frac{10\text{ g}}{.090\text{ kg}} \times \frac{\text{C}_2\text{H}_6\text{O}_2}{62\text{ g}} = 1.8m$

(b) $\frac{10\text{ g}}{100\text{ g}} \times \frac{1.0\text{ g}}{1\text{ ml}} \times \frac{\text{C}_2\text{H}_6\text{O}_2}{62\text{ g}} \times \frac{1000\text{ ml}}{\text{liter}} = 1.6M$

VIII.A. 1. $8.55\text{ g} \times \frac{\text{C}_{12}\text{H}_{22}\text{O}_{11}}{342\text{ g}} = .0250$ mol of solute

$100\text{ g} \times \frac{\text{H}_2\text{O}}{18.0\text{ g}} = 5.55$ mol of solvent

$P = XP_0 = \frac{5.55}{5.55 + 0.025} \times (23.8) = 23.7\text{ torr}$

2. $\text{molality} = \frac{.025}{0.100} = 0.25m$

$\Delta T = -1.86\,\frac{\text{deg-kg}}{\text{mol}} \times 0.25\,\frac{\text{mol}}{\text{kg}} = -0.47\text{ deg°C}$

$T_f = 0.00°\text{C} - 0.47\text{ deg°C} = -0.47°\text{C}$

3. $\Delta T = 2.5°\text{C} - 5.5°\text{C} = -3.0\text{ deg}$

$\text{Molar weight} = \frac{0.90\text{ g}}{.010\text{ kg}} \times -4.9\,\frac{\text{kg-deg}}{\text{mol}} \times \frac{1}{-3.0\text{ deg}} = 150\text{ g/mol}$

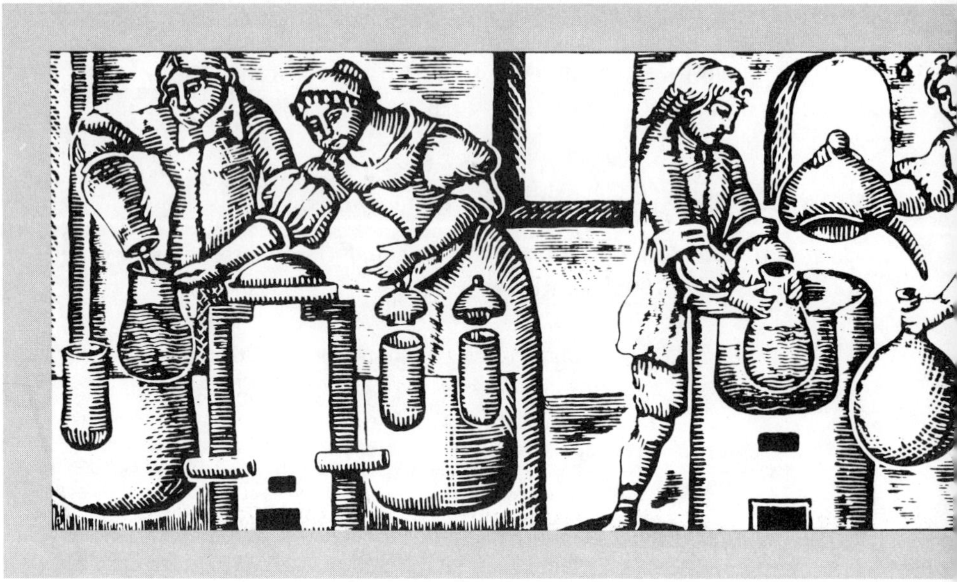

Proton combination and displacement reactions that do not involve changes in oxidation numbers are acid-base reactions.

After reading this chapter you should be able to

1. Classify compounds as Arrhenius acids and bases;
2. Classify molecules and ions as Brønsted acids and bases;
3. Write general chemical equations for Brønsted acid-base reactions of all possible combinations of neutral and charged species;
4. List the relative strengths of some common acids and bases;
5. Describe the factors that make acids and bases "strong" and "weak";
6. Define pH;
7. Calculate pH from the hydrogen ion concentration, and calculate the hydrogen ion concentration from the pH;
8. Write equilibrium equations from chemical equations;
9. Calculate the hydrogen ion concentration in solutions of weak acids;
10. Define Lewis acids.

acids and bases 12

The number of possible chemical reactions is certainly huge if not infinite, so the only way we can hope to understand what reactions *may* occur is to look for similarities in the reactions we know *do* occur. We can then generalize that all compounds possessing these similarities will behave in a similar way. One of the most important generalizations chemists have made about reactions is called acid-base theory.

Like all theories it began modestly enough with the observations noted in Chapter 9. As more observations were made the essential character of the reactions became more evident, so the theory became simpler and included many more compounds.

Arrhenius acids and bases 12.1

The first generalization was that acids taste sour, will react with active metals to release hydrogen gas, and will turn blue litmus pink. Bases are bitter, feel slippery, and turn pink litmus blue. The first thing any budding chemist learned was that acids react with bases to form salts and water.

Aqueous solutions of acids and bases conduct electric current, although the pure anhydrous compounds do not. The reactions that occurred during electrolysis[1] together with knowledge of the com-

[1] Discussed in Chapter 13.

position led Svante Arrhenius to propose that acids are substances that ionize in water to produce hydrated positive hydrogen ions (hydronium ions, H_3O^+) and an anion. Bases dissolve in water to produce hydroxyl ions (OH^-) and a cation. The reaction of acids and bases therefore was simply the combination of hydronium ions and hydroxyl ions to produce non-ionized water and a solution of anions and cations.

Arrhenius Acids: $HA + H_2O = H_3O^+ + A^-(aq)$
Arrhenius Bases: $MOH + H_2O = M^+(aq) + OH^-(aq)$

EXAMPLE 12.1

$$H_2SO_4 + \text{water} = 2\ H_3O^+ + SO_4^{2-}(aq)$$
$$Ca(OH)_2 + \text{water} = Ca^{2+}(aq) + 2\ OH^-(aq)$$
$$H_2SO_4 + Ca(OH)_2 + \text{water} = Ca^{2+}(aq) + SO_4^{2-}(aq) + 2\ H_2O$$

12.2 *Brønsted-Lowry acids and bases*

It was quickly evident that some compounds without hydroxyl ions behaved like bases. Ammonia was a base by all the criteria, but the only way it could be fitted into the Arrhenius scheme was to postulate a molecule NH_4OH (contrary to all the rules of valence!), which ionized in water into NH_4^+ ions and OH^- ions.

A broader classification was therefore proposed to include those substances that had no hydroxyl ions but acted like bases. The Brønsted-Lowry scheme (1923) defined an acid as any substance that could donate a positive hydrogen ion to a base. A base was any substance that could accept a positive hydrogen ion. The structural requirement for a Brønsted acid is the same as that for an Arrhenius acid: it must have one or more polar hydrogens that can be displaced. The structural requirement for a base is different. The only requirement for a Brønsted base is that the molecule or ion have at least *one pair of unshared electrons* that can be used to bond a hydrogen ion.

The *conjugate base* of the acid is the structure that remains when the proton (H^+) is lost. The *conjugate acid* of the base is that structure bonded to the proton.

Figure 12.1 Acids taste sour, turn blue litmus red, and make your teeth fuzzy.

$$H{:}A + {:}B \longrightarrow A{:}^- + H{:}B^+ \qquad (12.1)$$

acid plus base gives conjugate base of HA plus conjugate acid of B

Instead of the Arrhenius rule that an acid and a base give a salt and water, the Brønsted-Lowry rule says an acid plus a base gives the *conjugate base of the acid* plus the *conjugate acid of the base.* The reaction is a *displacement* of the negative part of the acid from its hydrogen by an ion or molecule possessing a pair of nonbonding electrons.

attack by base dissociation

Equation 12.1 can be written in either direction; the products present in the greater concentration at equilibrium depend on the relative strength of the bond to hydrogen of the two bases. Ammonia shares its electrons with hydrogen ion more strongly than chloride ion, so the reaction of ammonia and hydrogen chloride lies almost completely on the side of the ionized products.

Brønsted acids are covalent compounds; the fact that their aqueous solutions conduct electricity must be due to water acting as a base to cause ionization.

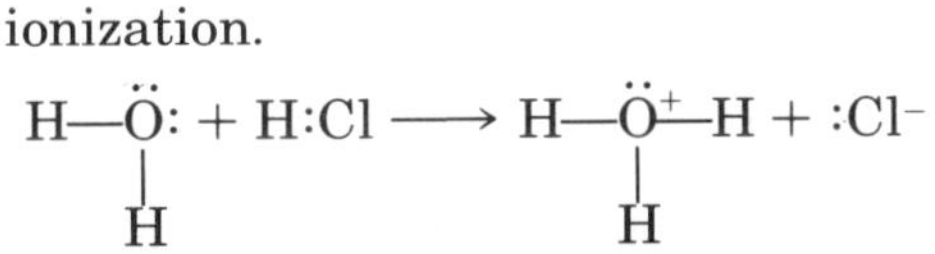

Water is also a stronger base than chloride. A dilute solution of hydrogen chloride is almost completely ionized. Concentrated hydrochloric acid must contain some molecular hydrogen chloride though, because it has a strong odor of HCl.

Acetic acid (abbreviated HOAc) does not give up its hydrogen to water as easily as HCl. Acetate ion is a stronger base than chloride. If the equation for the reaction of acetic acid with water is written in the direction of the greater concentration of products it would be

$$H_3O^+ + OAc^- \rightleftharpoons HOAc + H_2O$$

Both ammonia, $:NH_3$, and water, 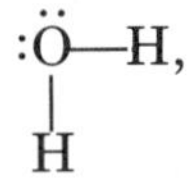, have nonbonding electrons and polar hydrogens. Nitrogen is less electronegative than oxygen (see Table 7.3, page 154), so it is better able to lend its electrons and its N—H bonds are not as polar. Ammonia displaces some OH^- ions from water, whereas water does not displace any NH_2^- ions from ammonia. The reaction of water and ammonia, with water acting as a Brønsted acid, comes to equilibrium after only a small amount of ionized products form. Ammonia is a "weak" base.

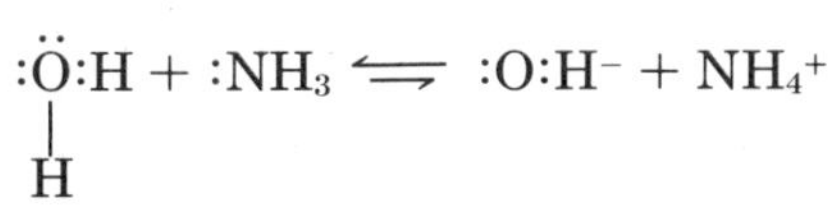

Acid-base reactions are not restricted to neutral reactants and charged products. A Brønsted base may be neutral or negatively charged. A Brønsted acid may be positive, neutral, or negative.

$$HA + :B^- \rightleftharpoons A:^- + HB \qquad (12.2)$$
$$HA^+ + :B \rightleftharpoons A: + HB^+ \qquad (12.3)$$
$$HA^- + :B \rightleftharpoons A:^{2-} + HB^+ \qquad (12.4)$$

EXAMPLE 12.2

1. Neutral acid and charged base
   - **(a)** $HOAc + OH^- \rightleftharpoons OAc^- + H_2O$
   - **(b)** $H_2O + CN^- \rightleftharpoons HCN + OH^-$
   - **(c)** $H_2O + CO_3^{2-} \rightleftharpoons OH^- + HCO_3^-$
2. Charged acid and neutral base
   - **(a)** $NH_4^+ + H_2O \rightleftharpoons NH_3 + H_3O^+$
   - **(b)** $HSO_4^- + H_2O \rightleftharpoons SO_4^{2-} + H_3O^+$

12.3 *Strengths of acids and bases*

Brønsted acids are classified as strong or weak on the basis of their interaction with water, as in equation 12.1:

$$HA + H_2O \rightleftharpoons A^- + H_3O^+$$

If a dilute solution (1M or less) of the acid interacts strongly with water so that the composition of the solution is described by the right side of the equation, it is a strong acid. In other words, dilute solutions of strong acids ionize completely.

The composition of an aqueous solution of a weak acid is not the right side of the equation 12.1, because only a small fraction of the acid is ionized. The concentration of hydronium ion is much less than the concentration of molecular acid. Table 12.1 lists some common Brønsted acids in the order of their strengths as proton donors.

The relative strengths of solutions of bases can be judged by comparing the concentration of hydroxide ion produced to the concentration of the base. For substances that contain hydroxide ion (Arrhenius bases), the strength of the base depends on the degree of ionization.

$$NaOH + \text{water} \longrightarrow Na^+(aq) + OH^-(aq)$$
$$Ca(OH)_2 + \text{water} \rightleftharpoons Ca^{2+}(aq) + 2\ OH^-(aq)$$
$$Fe(OH)_2 + \text{water} \rightleftharpoons Fe^{2+}(aq) + 2\ OH^-(aq)$$

Sodium hydroxide is very soluble in water and is completely ionized; it is a strong Arrhenius base. Calcium hydroxide is moderately soluble in water and is a moderately strong base. Iron(II) hydroxide is only slightly soluble in water. The hydroxide concentration of most of these substances is limited by their solubility.

Table 12.1

Relative strengths of acids

	Acid		Conjugate Base
STRONG	perchloric	$HClO_4$	ClO_4^-
	sulfuric	H_2SO_4	HSO_4^-
	hydriodic	HI	I^-
	hydrobromic	HBr	Br^-
	hydrochloric	HCl	Cl^-
	nitric	HNO_3	NO_3^-
	hydronium ion	H_3O^+	H_2O
MODERATELY STRONG	chromic	H_2CrO_4	$HCrO_4^-$
	periodic	HIO_4	IO_4^-
	bisulfate ion	HSO_4^-	SO_4^{2-}
	phosphoric	H_3PO_4	$H_2PO_4^-$
	hydrofluoric	HF	F^-
WEAK	acetic	$HOAc$	OAc^-
	carbonic	H_2CO_3	HCO_3^-
	hydrosulfuric	H_2S	HS^-
	hydrocyanic	HCN	CN^-
	water	H_2O	OH^-

The concentration of hydroxide ion of Brønsted bases depends on the degree of interaction with water acting as an acid.

$$H_2O + :B \rightleftharpoons OH^- + HB^+$$

and

$$H_2O + :B^- \rightleftharpoons OH^- + HB$$

A strong base has a strong affinity for a hydrogen ion; the conjugate acid is therefore very weak. Conversely a weak base has little tendency to share its electrons with hydrogen; its conjugate acid is strong. Table 12.2 lists some bases in the order of increasing proton-accepting power. The strongest base is at the bottom of the list.

Table 12.2

Relative strengths of bases

	Base		Conjugate Acid
WEAK	ammonia	NH_3	NH_4^+
	acetate ion	OAc^-	$HOAc$
	bicarbonate ion	HCO_3^-	H_2CO_3
	carbonate ion	CO_3^{2-}	HCO_3^-
STRONG	phosphate ion	PO_4^{3-}	HPO_4^{2-}
	hydroxide ion	OH^-	H_2O
	methoxide ion	OCH_3^-	CH_3OH
	amide ion	NH_2^-	NH_3
	oxide ion	O^{2-}	OH^-
	hydride ion	H^-	H_2

No stronger acid than hydronium ion can exist in molecular form in dilute water solutions, because the stronger acid would immediately react with water to form hydronium ions and an anion. For the same reason no stronger base than hydroxide can exist in water. If stronger acids or bases are needed the chemist must carry out the reactions in solvents other than water.

12.4 *Factors affecting base strength*

A study of the structures of the bases in Tables 12.1 and 12.2 shows two independent factors affecting the tendency of a base to share its nonbonding electron pair with a hydrogen ion. The principal factor is charge density. A large negative charge on an atom or molecule predisposes it to neutralize that charge with a proton. The size of the atom or group also plays a part. A small atom or group with many electrons around it has a tendency to spread the charge by sharing a pair of electrons with a proton.

Resonance

Certain molecules and polyatomic ions with multiple bonds have a mechanism for delocalizing or spreading the charge on an atom to the whole molecule. This very much reduces the charge density on any particular atom and therefore decreases the base strength.

EXAMPLE 12.3

1. We can write the structure of nitrate ion three ways:

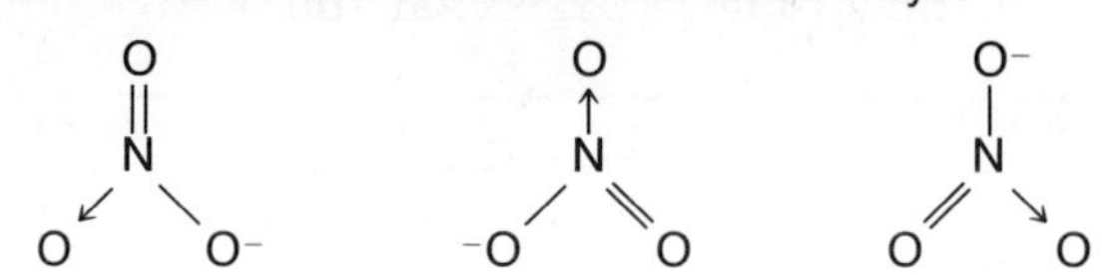

None of the three is wholly correct. The actual ion is a hybrid of the three forms. Each oxygen will have a partial negative charge, and the strength of the ion as a base is less than it would be if the charge were localized on one oxygen.

2. We can write the structure of acetate ion in two ways without moving any atoms, just electrons:

The acetate ion has two equivalent oxygens which share one negative charge. The charge density is less by this means and the base strength of the ion is less than it would be if the charge were on one atom.

Electronegativity

The electronegativity of the atoms comprising a base counteracts the effect of charge density to some extent. Given the same electron density the more electronegative atom will be the weaker base.

EXAMPLE 12.4

	F^- is weaker than	OH^- is weaker than	NH_2^-
Electronegativity	4.0	3.5	3.0
	Cl^- is weaker than	SH^- is weaker than	PH_2^-
Electronegativity	3.0	2.5	2.1

These observations can be summarized by three rules of thumb:

1. The greater the negative charge the stronger the base.

$$H_2O << OH^- << O^{2-}$$
$$H_2PO_4^- < HPO_4^{2-} < PO_4^{3-}$$

2. The greater the size of the ion the weaker the base.

$$ClO_4^- < Cl^- << NH_2^-$$

Corollary: Resonance smears out the charge, making the effective size of the ion greater.

$$NO_3^- < CH_3CO_2^- << CH_3O^-$$

3. The greater the electronegativity the weaker the base, if charge, size, and resonance factors are similar.

$$NO_3^- < HCO_3^-$$

pH and the water equilibrium 12.5

Water having polar hydrogens and nonbonding electrons in its structure is capable of undergoing self-ionization.

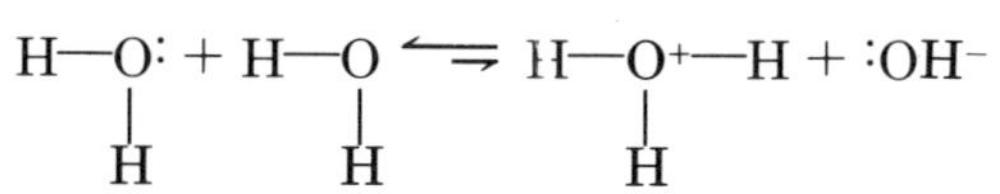

The gain of entropy in this process is so small and the increase of potential energy so large that one does not expect this process to go very far to the right. In fact, only about two water molecules in each billion are in the ionized state at any instant.

The addition of either hydrogen ions or hydroxyl ions from any other source will increase the probability that hydrogen ions and hydroxyl ions will recombine to form water. For example, the addition of sodium hydroxide to water will decrease the concentration of hydronium ions present, whereas the addition of hydrochloric acid will decrease the concentration of hydroxide ion.

The hydronium ion concentration and the hydroxide ion concentration depend on an equilibrium constant (K_w). The water equilibrium follows the law

$$(\text{Molarity of } H_3O^+)(\text{Molarity of } OH^-) = 10^{-14}$$

or

$$[H_3O^+]\,[OH^-] = K_w = 10^{-14} \tag{12.5}$$

The symbol of ions and molecules enclosed in brackets conventionally means its molar concentration. This expression says that the concentration of hydronium ions from any source times the concentration of hydroxyl ions from any source always equals 10^{-14} (at room temperature).

It is reasonable to ask what significance such extremely small concentrations of hydronium ion and hydroxyl ion have. The answer is that the rates of many reactions, particularly biological reactions, and the concentrations of acids and their conjugate bases depend on the concentration of hydronium ion to a great degree, so much so that a change from $10^{-7}M$ to $10^{-6}M$ in the blood causes serious disorders.

pH

Expressing a hydrogen ion concentration as $4.0 \times 10^{-8}M$ is certainly awkward, even though the exponential notation is not as clumsy as $0.00000004M$. The logarithm of a number is a more convenient expression especially if the number is very large or very small.[2] Almost all hydrogen ion concentrations of interest are less than $1.0M$, so that the logarithm of the concentration is a negative number. We can make the expression still more convenient by multiplying the logarithm by minus one to make it a positive number. This method of expressing the hydrogen ion concentration is called pH.[3]

Dilute solutions of strong acids ionize completely in water, so the hydrogen ion concentration is the molarity of the acid times the

[2] How to find the logarithm of a number and vice versa is given in the Appendix.

[3] More precisely, pH is defined as the negative logarithm of the hydrogen ion activity compared to unit activity. Logarithms have no units so a logarithm must be taken of a pure number. We ignore the distinction in this book.

number of hydrogens donated by one molecule of the acid. To calculate the pH from the hydrogen ion concentration, find the logarithm of the hydrogen ion concentration, and multiply by minus one.

$$pH = -\log [H_3O^+] \tag{12.6}$$

EXAMPLE 12.5

1. What is the pH of a 0.10*M* HCl solution?
   (a) A solution of 0.10*M* HCl is 0.10*M* in hydrogen ion.
   (b) The logarithm of 0.10 is -1.0
   (c) The pH is $-(-1) = 1.0$
2. A solution of a weak acid has a hydrogen ion concentration of $5.0 \times 10^{-4}M$. What is its pH?
   (a) The logarithm of 5.0×10^{-4} is $0.7 + (-4) = -3.3$
   (b) The pH is $-(-3.3) = 3.3$
3. A 0.001*M* solution of sodium hydroxide has a hydroxide ion concentration of 0.001*M* ($1. \times 10^{-3}M$). What is its pH?
   (a) The hydrogen ion concentration depends on the hydroxide ion concentration through the water equilibrium law:

$$[H_3O^+]\,[OH^-] = 10^{-14}$$
$$[H_3O^+]\,[1 \times 10^{-3}] = 10^{-14}$$
$$[H_3O^+] = 1.0 \times 10^{-11}$$

   (b) The logarithm of 1.0×10^{-11} is -11.0
   (c) The pH is $-(-11.0) = 11.0$

The pH of basic solutions can be found more quickly by observing that the negative logarithm of the hydroxide ion concentration (pOH) and the pH always add to 14:

$$pH + pOH = 14 \tag{12.7}$$

A shortcut way to solve Example 12.5(c) is as follows:

$$[OH^-] = 1.0 \times 10^{-3}$$
$$pOH = -\log [OH^-] = -\log(1.0 \times 10^{-3}) = 3.0$$
$$pH = 14 - pOH = 14 - 3.0 = 11.0$$

Table 12.3 shows the useful range of pH. Notice that the acidity decreases as the pH increases. A pH of 0 or 1 is strongly acid, whereas a pH of 13 or 14 is strongly basic.

Table 12.3

pH	
0	1*M* strong acid
1	
2	1*M* acetic acid
3	
4	
5	
6	
7	neutral
8	normal blood pH
9	
10	
11	1*M* ammonia
12	
13	
14	1*M* strong base

Scale of pH

Acid-base equilibrium 12.6

An aqueous solution of a weak acid has hydronium ion, molecular acid, water, the conjugate base of the acid, and minute amounts of hydroxide present. These species are constantly interacting with each other according to the equations

$$HA + H_2O \rightleftharpoons H_3O^+ + A^-$$
$$2\ H_2O \rightleftharpoons H_3O^+ + OH^-$$

The two equilibria are working simultaneously. The water equilibrium was discussed in the previous section, and we find that we can write a similar equilibrium equation for the interaction of the weak acid and water.

$$\frac{[H_3O^+]\,[A^-]}{[HA]} = K_a \tag{12.8}$$

where K_a is an equilibrium constant. K_a is called the acid dissociation constant because it is a measure of the degree of ionization of the acid in water.

The explanation for the origin and form of the equilibrium equation comes from the study of the relationships of energy and entropy in chemical systems. These relationships will be discussed in a later course. For now, accept the equilibrium equation as a stone tablet.

The explanation of the origin of the equilibrium equation may be complicated but the construction of one of them is ridiculously easy. From the balanced chemical equation write the concentrations of the product species in the numerator and the concentrations of the reactant species in the denominator. The coefficients in the balanced chemical equation go into the equilibrium equation as powers (exponents) of the concentrations. The ratio of the numerator and the denominator is a constant characteristic of the reaction. The concentration (or activity) of water in dilute solutions is considered constant, and its concentration is not included in the left side of the equation.

EXAMPLE 12.6

Chemical Equation	Equilibrium Equation
1. $HOAc + H_2O = H_3O^+ + OAc^-$	$\frac{^*[H^+]\,[OAc^-]}{[HOAc]} = 1.8 \times 10^{-5}$
2. $H_2S + H_2O \rightleftharpoons H_3O^+ + HS^-$	$\frac{[H^+]\,[HS^-]}{[H_2S]} = 9.1 \times 10^{-8}$
3. $H_2C_2O_4 + H_2O \rightleftharpoons H_3O^+ + HC_2O_4^-$	$\frac{[H^+]\,[HC_2O_4^-]}{H_2C_2O_4} = 5.9 \times 10^{-2}$
4. $NH_4^+ + H_2O \rightleftharpoons H_3O^+ + NH_3$	$\frac{[H^+]\,[NH_3]}{[NH_4^+]} = 5.5 \times 10^{-10}$
5. $H_2PO_4^- + H_2O \rightleftharpoons H_3O^+ + HPO_4^{-2}$	$\frac{[H^+]\,[HPO_4^{2-}]}{[H_2PO_4^-]} = 6.2 \times 10^{-8}$

* Hydronium ion is often written H^+ for short, with the understanding that it is hydrated.

You will find tables of acid dissociation constants in reference books. The examples in Table 12.4 are typical of moderate to weak acids. The size of the constant is related to the degree to which a reaction proceeds from left to right. Very weak acids have very small constants; strong acids have constants much greater than unity. The second and third acid dissociation constants for acids with more than one hydrogen are always much smaller than the first. This follows from the rule that an increase of negative charge on an ion greatly increases its base strength.

Table 12.4

Dissociation constants of acids

Acid	Formula	K_a
acetic acid	$HC_2H_3O_2$	1.8×10^{-5}
ascorbic acid (Vitamin C)	$H_2C_6H_6O_6$	7.9×10^{-5}
carbonic acid	H_2CO_3	4.3×10^{-7}
bicarbonate ion	HCO_3^-	5.6×10^{-11}
hydrocyanic acid	HCN	4.9×10^{-10}
hypochlorous acid	$HOCl$	2.9×10^{-8}
phosphoric acid	H_3PO_4	7.5×10^{-2}
dihydrogen phosphate ion	$H_2PO_4^-$	6.2×10^{-8}
monohydrogen phosphate ion	HPO_4^{2-}	2.2×10^{-13}
bisulfate ion	HSO_4^-	1.2×10^{-2}

Use of dissociation constants to find pH 12.7

We will treat only one example of the use of dissociation constants. Subsequent courses in chemistry will cover acid-base equilibrium in much more depth.

An equilibrium equation relates the *concentrations* of the species in a chemical reaction to each other. A chemical equation relates the stoichiometry of the reactants and products. The equilibrium equation and the chemical equation can be used to find the concentrations of all the species in the solution, if the initial conditions are known, or if some final measurement is made.

Find the pH of a solution that is 0.50*M* in acetic acid. **EXAMPLE 12.7**

Chemical equation: $HOAc + H_2O = H_3O^+ + OAc^-$

Equilibrium equation: $\dfrac{[H^+]\,[OAc^-]}{[HOAc]} = 1.8 \times 10^{-5}$

The chemical equation states that one mole of hydronium ion and one mole of acetate ion are formed from each mole of acetic acid *that reacts with water.* The term 0.50*M* refers to the moles of molecular acetic acid dissolved in a liter of solution. Some of that, say *x* moles per liter, has reacted with water to form *x* moles per liter of hydronium ion and acetate ion. The concentration of molecular acetic acid left is therefore $(0.50 - x)M$.

The equilibrium equation relates these concentrations to each other:

$$\frac{[H^+]\,[OAc^-]}{[HOAc]} = \frac{(x)(x)}{0.50 - x} = 1.8 \times 10^{-5}$$

The exact solution of this equation requires a quadratic form. Luckily it is seldom necessary to solve the quadratic equation for solutions of weak acids that are not too dilute (greater than 10^{-2} *M*), because the concentration of hydronium ion and acetate ion (x) is small compared to the molarity of the acid. That is, we may ignore the x in the denominator, because leaving it out will not change the answer appreciably. The equation in approximate form is

$$\frac{x^2}{0.50} = 1.8 \times 10^{-5}$$

$$x^2 = 0.50(1.8 \times 10^{-5})$$

$$x = \sqrt{9.0 \times 10^{-6}}$$

$$x = 3.0 \times 10^{-3}$$

The concentration of hydrogen ion and acetate ion are each 3.0×10^{-3}. The logarithm of 3.0×10^{-3} is $0.48 - 3$. The pH is $-(0.48 - 3) = 2.52$.

12.8 *Lewis acids*

The beauty of a scientific theory is that it leads to understanding a particular happening in a way that leads to greater generalization and a greater understanding of a broad variety of happenings. The Brønsted-Lowry theory of acid-base reactions focuses attention on the proton and pictures the reaction as a substitution or displacement of one proton-base bond by a stronger base.

$$\text{Base}_1\text{—H}^+ + \text{Base}_2 = \text{Base}_1 + \text{H}^+\text{—Base}_2$$

Having this picture of the reaction in mind, one wonders what is so special about a proton. Cannot the same mechanism occur with other positive atoms or groups attached to the base? Even more generally, are there neutral molecules with the capacity to combine with a base?

The proton is tiny and has a high positive charge density. It might be expected that other small positive atoms with a high charge density would also form bonds to bases.

EXAMPLE 12.8

1. $AlCl_3 + 3\ OH^- = Al(OH)_3 + 3\ Cl^-$
2. $Cu(H_2O)_4^{2+} + 4\ NH_3 = Cu(NH_3)_4^{2+} + 4\ H_2O$

These are substitution or displacement reactions of the same type as equation 12.8, but a metal atom is the positive part instead of the proton.

$$M\text{—}Base_1 + Base_2 = M\text{—}Base_2 + Base_1$$

Here are compounds that react with bases but that have no protons in them! G. N. Lewis proposed that an acid be defined as any ion or molecule that can accept a pair of electrons to form a bond with a base.

Water will displace any weaker base from a cation. If the hydrated cation is small and highly charged, it can react further with water to form hydronium ions.

EXAMPLE 12.9

$$FeCl_3 + 6\ H_2O = Fe(H_2O)_6^{3+} + 3\ Cl^-$$
$$Fe(H_2O)_6^{3+} + 3\ H_2O = Fe(OH)_3(H_2O)_3 + 3\ H_3O^+$$

Certain neutral compounds may combine directly with bases to form coordinate covalent bonds. Since a base by definition is supplying a pair of electrons to a bond, it is only necessary that the acid have an empty orbital to put them in.

EXAMPLE 12.10

1. $F{:}\overset{F}{\underset{F}{B}} + {:}NH_3 = F_3B{:}NH_3$

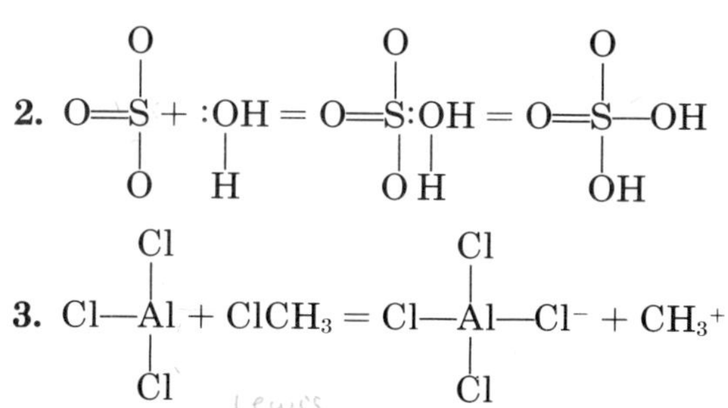

The most general picture of acid-base reactions now emerges. An electron-rich ion or molecule (base) attacks an electron-poor site on another ion or molecule (acid). If the acid has an empty orbital to accept the electrons from the base, they simply combine. If the acid does not have an empty orbital but instead is combined with a weaker base, the attacking base displaces (substitutes) the weaker base from the acid. In this view the proton is simply one of a variety of possible acids.

The strengths of Lewis acids follow the same principles as the strengths of Brønsted bases, but with the signs reversed.

12.9 *Summary*

1. Arrhenius acid-base theory was concerned only with the chemistry of substances which produced hydronium ions and hydroxyl ions in water.
2. Brønsted and Lowry expanded the idea of a base to include not only hydroxyl ions but all structures possessing at least one pair of unshared electrons.
3. Lewis changed the idea of an acid from a proton-donating structure to an electron-pair accepting structure. The theory is now focused on the formation of bonds between strongly positive centers and strongly negative centers on molecules.
4. The strength of a Brønsted acid is measured by the degree to which it transfers a proton to water. The strength of a Brønsted base is measured by the degree to which it accepts a proton from water.
5. A qualitative comparison of base strength can be made by considering the quantity of charge, the volume in which the charge is contained, and the electronegativity of the constituent atoms.
6. Water self-ionizes to a small extent. The measure of its self-ionization is given by an equilibrium equation:

$$[H^+]\,[OH^-] = 10^{-14}$$

7. pH is defined as the negative logarithm of the hydrogen ion concentration. pOH is the corresponding expression for the hydroxide ion.
8. The relationship between the concentrations of ionic and molecular species in aqueous solutions of weak acids is given by an equilibrium equation:

$$\frac{[H^+]\,[A^-]}{[HA]} = K_a$$

QUESTIONS

I. Define the following terms.

1. hydronium ion
2. conjugate acid-base pair
3. strong acids
4. weak acids
5. pH
6. equilibrium
7. equilibrium equation

II.

1. State the Brønsted-Lowry rule.
2. What must a molecule or ion have to act as a Brønsted base?
3. What three factors influence the strength of a Brønsted base?
4. How does a Lewis acid differ from a Brønsted acid?

EXERCISES

III. Arrange each series in order of decreasing base strength.

1. F^-, O^{2-}, Cl^-, H^-
2. SO_3^{2-}, NO_2^-, $CH_3CO_2^-$
3. NH_3, H_2O, OH^-
4. PO_4^{3-}, HPO_4^{2-}, $H_2PO_4^-$
5. CN^-, OH^-, HS^-
6. ClO_4^-, ClO_3^-, ClO^-
7. HCO_3^-, HSO_3^-
8. CO_3^{2-}, OAc^-, CH_3O^-

IV. Complete these equations. Label those components acting as Brønsted acids and those acting as bases.

1. $HCl + H_2O =$
2. $HOAc + H_2O =$
3. $NH_3 + H_2O =$
4. $OAc^- + H_2O =$
5. $HCO_3^- + H_2O =$
6. $HPO_4^{2-} + H_2O$
7. $HSO_4^- + OH^- =$
8. $CN^- + H_3O^+ =$
9. $HCO_3^- + OH^- =$
10. $H^- + H_2O =$
11. $NH_2^- + H_2O =$
12. $HS^- + HCl =$
13. $Cl^- + H_2SO_4 =$
14. $H_2CrO_4 + HCO_3^- =$
15. $H_3PO_4 + OAc^- =$
16. $NH_4^+ + OH^- =$

17. $PO_4^{3-} + H_2PO_4^- =$
18. $O^{2-} + H_2O =$
19. $OCl^- + HCl =$
20. $CH_3O^- + H_2O =$

V. 1. Find the pH when the hydrogen ion concentration is

(a) $0.002M$	(e) $4.2 \times 10^{-3}M$
(b) $5 \times 10^{-5}M$	(f) $0.05M$
(c) $6 \times 10^{-8}M$	(g) $9 \times 10^{-4}M$
(d) $1.3 \times 10^{-4}M$	(h) $7.2 \times 10^{-6}M$

2. Find the hydrogen ion concentration when the pH is

(a) 7.3	(d) 2.7
(b) 11.2	(e) 1.3
(c) 4.5	(f) 8.4

3. What is the pOH when the pH is

(a) 6.0	(d) 9.0
(b) 11.2	(e) 5.3
(c) 3.5	(f) 1.3

VI. Find the pH of $0.10M$ solutions of
1. sodium dihydrogen phosphate
2. hypochlorous acid
3. ascorbic acid
4. acetic acid

SOLUTIONS

III. 1. H^- O^{2-} F^- Cl^-
3. OH^- NH_3 H_2O
5. OH^- CN^- HS^-
7. HCO_3^- HSO_3^-

IV.

	Acid	*Base*	*Conj. base*	*Conj. acid*
1.	HCl	H_2O	Cl^-	H_3O^+
3.	H_2O	NH_3	OH^-	NH_4^+
5.	H_2O	HCO_3^-	OH^-	H_2CO_3
7.	HSO_4^-	OH^-	SO_4^{2-}	H_2O
9.	HCO_3^-	OH^-	CO_3^{2-}	H_2O
11.	H_2O	NH_2^-	OH^-	NH_3
13.	H_2SO_4	Cl^-	HSO_4^-	HCl
15.	H_3PO_4	OAc^-	$H_2PO_4^-$	$HOAc$
17.	$H_2PO_4^-$	PO_4^{3-}	HPO_4^{2-}	HPO_4^{2-}
19.	HCl	OCl^-	Cl^-	$HOCl$

V. **1.** **(a)** $-\log(2 \times 10^{-3}) = -\log 2 - \log 10^{-3} = -0.3 + 3 = 2.7$
(c) $-\log(6 \times 10^{-8}) = -\log 6 - \log 10^{-8} = -0.8 + 8 = 7.2$
(e) $-\log(4.2 \times 10^{-3}) = 3 - .62 = 2.38$
(g) $-\log(9 \times 10^{-4}) = 4 - 0.95 = 3.05$

2. **(a)** $-\text{antilog}(7.3) = \text{antilog}(0.7 - 8) = 5.0 \times 10^{-8}$
(c) $-\text{antilog}(4.5) = \text{antilog}(0.5 - 5) = 3.2 \times 10^{-5}$
(e) $-\text{antilog}(1.3) = \text{antilog}(0.7 - 2) = 5.0 \times 10^{-2}$

3. **(a)** 8.0
(c) 10.5
(e) 8.7

VI. **1.** $K_a = 6.2 \times 10^{-8}$; $c = 0.10M$

$$H^+ = \sqrt{K_a c} = [(6.2 \times 10^{-8})(0.10)]^{1/2}$$
$$= 8 \times 10^{-5}$$
$$-\log(8 \times 10^{-5}) = 4.1$$

Note that $-\log(\sqrt{K_a c}) = -\frac{1}{2}\log K_a - \frac{1}{2}\log c$. The shortcut method of finding the pH then is to add the logarithms of K_a and c, divide by 2, and multiply by -1.

Chemical reactions involving changes in oxidation numbers and the use of these reactions in electrochemistry are introduced.

After reading this chapter you should be able to

1. Designate the species being oxidized and reduced in an oxidation-reduction reaction;
2. Write a net ionic equation given the identity of the reactants and products;
3. Compute oxidation number changes for each species;
4. Balance oxidation-reduction equations by relating oxidation number changes;
5. Balance oxidation-reduction equations by the ion-electron half-reaction method;
6. Define electrolysis, and label the parts of an electrolytic cell;
7. Define the terms coulomb, faraday, and ampere;
8. Recite Faraday's law of electrolysis;
9. Compute the amount of oxidation or reduction at an electrode from current and time information;
10. Describe a galvanic cell.

oxidation and reduction 13

Acid-base reactions do not change the oxidation numbers of the atoms. The charge on an ion changed only if it lost or acquired another charged atom. No matter how we shuffled atoms and ions around we could keep track of the charges by remembering the charge that each ion carried.

In this chapter we will examine some reactions in which the oxidation numbers of some of the atoms change during the reaction. An *increase* in the oxidation number of an atom indicates that it has been *oxidized*. The atom has fewer electrons assigned to it than before the reaction. An atom is said to undergo *reduction* when its oxidation number has been decreased or reduced. It has more negative electrons assigned to it than before the reaction.

Since there is no way to store chemically significant quantities of electrons, the number of electrons lost during oxidation must be exactly equal to the number gained during reduction. The two processes must occur together so these reactions are often called "redox" reactions.

Balancing oxidation-reduction reactions 13.1

Because charges are being transferred in oxidation-reduction reactions, there is an added complication to the problem of writing a balanced chemical equation. It is not only necessary that the numbers and kinds of atoms balance, but also that the numbers of moles of electrons gained by the oxidizing agent be the same as the number lost by the reducing agent. In other words, the charges on each side of the equation must balance.

Non-ionic reactions

Some oxidation-reduction equations are simple enough to balance by inspection. Generally non-ionic reactions in the gas phase or direct combinations and decompositions are represented by equations that can be balanced without worrying about the oxidation numbers.

EXERCISE 13.1

Balance these equations by inspection:

1. $CH_4 + Cl_2 = CH_2Cl_2 + HCl$
2. $C_3H_8 + O_2 = CO_2 + H_2O$
3. $Cu + Br_2 = CuBr_2$
4. $(NH_4)_2Cr_2O_7 = Cr_2O_3 + N_2 + H_2O$
5. $NH_3 + O_2 = NO + H_2O$

Answers: The coefficients are

1. 1,2 = 1,2
2. 1,5 = 3,4
3. 1, 1 = 1
4. 1 = 1, 1, 4
5. 4, 5 = 4, 6

Reactions of ions: net ionic equations

Ionic reactions in solution are difficult to balance by inspection, because there seem to be so many reactants and products. The first step in simplifying such an equation is to write a net ionic equation. Eliminate all the species that appear unchanged on both sides of the equation.

The method to be followed in writing net ionic equations is fairly simple. Molecular and insoluble species are written as neutral substances. Soluble ionic substances are written as separate ions, disregarding the coefficients. *The most common cause of error is to neglect to place the proper charge on the ion.*

The equation is then inspected to see which ions appear on both sides of the equation. These ions are left out of the net ionic equation.

EXAMPLE 13.1

An unbalanced molecular equation gives only the identity of the reactants and products.

$$KI + NaNO_2 + H_2SO_4 = I_2 + NO + H_2O + NaKSO_4$$

In ionic form,

$$K^+ + I^- + Na^+ + NO_2^- + H^+ + SO_4^{2-} = I_2 + NO + H_2O + K^+ + Na^+ + SO_4^{2-}$$

Sodium ion, potassium ion, and sulfate ion are unchanged in this reaction. We simplify the equation by writing

$$I^- + NO_2^- + H^+ = I_2 + NO + H_2O$$

This is the skeleton net ionic equation to be balanced.

Sometimes an ion has a dual role as a reactant and as a spectator ion. In those cases only the spectator ion on the product side of the reaction is eliminated.

EXAMPLE 13.2

Write a skeleton net ionic equation for this reaction:

$$KI + H_2SO_4 = K_2SO_4 + I_2 + H_2S + H_2O$$

Iodine, hydrogen sulfide, and water are molecular species. The rest are ionic in aqueous solution.

$$K^+ + I^- + H^+ + SO_4^{2-} = K^+ + SO_4^{2-} + I_2 + H_2S + H_2O$$

Sulfate ion is both a source of H_2S and a spectator ion, so only the product sulfate is eliminated.

$$I^- + H^+ + SO_4^{2-} = I_2 + H_2S + H_2O$$

Now that the unnecessary distractions have been eliminated, the problem of balancing the reactants, the products, and the charges of the net ionic equation remains. There are two common methods which, if carefully followed, ensure that mass and charge are conserved.

The oxidation state method

The oxidation state method determines the mole ratio of the oxidizing agent and the reducing agent only. It is based on the principle that the number of electrons lost (actually or virtually) by the reducing agent must equal the number gained by the oxidizing agent. To calculate this number it is necessary to compute the oxidation number change of each atom undergoing a change. When the change in oxidation numbers is readily apparent, as in the reaction of copper with silver ion, the steps leading to a balanced equation are few and easy.

$$Ag^+ + Cu = Ag + Cu^{2+}$$

Silver ion (+1) going to silver (0) gains one mole of electrons (1 e^-) per mole of silver reduced. Copper (0) going to copper ion (+2) loses two moles of electrons (2 e^-) per mole of copper oxidized. We evidently need two moles of silver ion to accept those electrons:

$$2 \text{ moles}(Ag^+) \times \frac{1\ e^-}{\text{mole}(Ag^+)} = 2\ e^- \text{ gained}$$

$$1 \text{ mole}(Cu) \times \frac{2\ e^-}{\text{mole}(Cu)} = 2\ e^- \text{ lost}$$

Therefore,

$$2\ Ag^+ + Cu = 2\ Ag + Cu^{2+}$$

EXAMPLE 13.3

$$NO_2^- + I^- + H^+ = I_2 + NO + H_2O$$

One mole of nitrite (N = +3) going to one mole of nitric oxide (N = +2) gains one mole of electrons (1 e^-). Two moles of iodide (I = −1) going to one mole of iodine (I_2 = 0) loses two moles of electrons (2 e^-). We need two moles of nitrite to oxidize two moles of iodide:

$$2 \text{ moles}(NO_2^-) \times \frac{1\ e^-}{\text{mole}(NO_2^-)} = 2\ e^- \text{ gained}$$

$$2 \text{ moles}(I^-) \times \frac{1\ e^-}{\text{mole}(I^-)} = 2\ e^- \text{ lost}$$

Therefore,

$$2\ NO_2^- + 2\ I^- + H^+ = I_2 + 2\ NO + H_2O$$

The remainder of the atoms must be balanced by inspection. Balance the oxygens first by adding water, then balance the hydrogens:

$$2\ NO_2^- + 2\ I^- + 4\ H^+ = I_2 + 2\ NO + 2\ H_2O$$

The ion-electron half-reaction method

The ion-electron half-reaction method has several advantages over the oxidation number method, especially if the equation being balanced is at all complicated. This method does *not* require a computation of the oxidation number change. The oxidation numbers of organic and biological compounds are obscure, to say the least, so this method is used for those reactions.

The ion-electron half-reaction method emphasizes the relationship between the reactant and the products formed *from it* rather than the relation between specific oxidizers and reducing agents. In many cases it is possible to carry out each half-reaction in a cell connected to the other half only by electrical and ionic conduits (see sections 13.3 and 13.4).

The method is best illustrated by an example.

Reactants and Products:

$$KMnO_4 + H_2C_2O_4 + H_2SO_4 = CO_2 + MnSO_4 + K_2SO_4 + H_2O$$

Step 1. Write a skeleton net ionic equation, ignoring the coefficients.

$$K^+ + MnO_4^- + H_2C_2O_4 + H^+ + SO_4^{2-} = CO_2 + Mn^{2+} + H_2O + K^+ + SO_4^{2-}$$

Step 2. Choose one of the reactants and write an equation showing only its product.

$$MnO_4^- = Mn^{2+}$$

Step 3. Balance the atoms.

In an aqueous acid solution water and hydrogen ions are available as products or reactants, so they can be added to the equation wherever necessary.

$$MnO_4^- + 8\ H^+ = Mn^{2+} + 4\ H_2O$$

Balance the oxygens by adding enough water to the side deficient in oxygen. Then add enough hydrogen ions to the other side to balance the hydrogens.

Step 4. Balance the charges.

The total charge on the left side of the equation is $+8-1=+7$, and the charge on the right side is $+2$. The addition of 5 e^- (each mole of electrons is -1) to the left side will make each side equal.

$$5\ e^- + MnO_4^- + 8\ H^+ = Mn^{2+} + 4\ H_2O$$
$$(-5) + (-1) + (+8) = (+2)$$

This is the completed half-reaction, showing the relationship between reactant, product, and associated ions and molecules. The source of the electrons must be the other half-reaction.

Step 2′. Choose the other reactant and write its products.

$$H_2C_2O_4 = CO_2$$

Step 3′. Balance the atoms.

$$H_2C_2O_4 = 2\ CO_2 + 2\ H^+$$

Step 4′. Balance the charges by adding electrons.

$$H_2C_2O_4 = 2\ CO_2 + 2\ H^+ + 2\ e^-$$
$$(O) = (O) + (+2) + (-2)$$

Step 5. The first half-reaction consumes five moles of electrons per mole of permanganate reduced. The second half-reaction produces two moles of electrons per mole of oxalic acid oxidized. We must reconcile the number of electrons gained and lost by multiplying each half-reaction by a whole-number factor such that the moles of electrons are equal. In this case we multiply the *entire* permanganate half-reaction by 2 and the oxalic acid half-reaction by 5.

$$10\ e^- + 2\ MnO_4^- + 16\ H^+ \longrightarrow 2\ Mn^{2+} + 8\ H_2O$$
$$5\ H_2C_2O_4 \longrightarrow 10\ CO_2 + 10\ H^+ + 10\ e^-$$

Step 6. Add the two half-reactions.

If hydrogen ions or water appear on both sides of the equation subtract the smaller number from both sides to obtain the net change.

$$2\ MnO_4^- + 5\ H_2C_2O_4 + 6\ H^+ \longrightarrow 10\ CO_2 + 2\ Mn^{2+} + 8\ H_2O$$

This is the completed net ionic equation, containing only those species that react. To obtain the molecular equation replace an equal number of spectator ions on each side.

EXAMPLE 13.4 Balance the equation for the reaction of hydrogen peroxide and potassium iodide in acid solution forming iodine, water, and the potassium salt of the acid as products.

Reactants and products:

$$H_2O_2 + KI + H^+ \longrightarrow I_2 + H_2O + K^+$$

Net ionic equation:

$$H_2O_2 + I^- + H^+ \longrightarrow I_2 + H_2O$$

First half-reaction:

$$H_2O_2 \longrightarrow H_2O$$

Balance atoms:

$$H_2O_2 + 2H^+ \longrightarrow 2H_2O$$

Balance charges:

$$2e^- + H_2O_2 + 2H^+ \longrightarrow 2H_2O$$

Second half-reaction:

$$I^- \longrightarrow I_2$$

Balance atoms:

$$2I^- \longrightarrow I_2$$

Balance charges:

$$2I^- \longrightarrow I_2 + 2e^-$$

Reconcile the moles of electrons in the two half-reactions. In this case the moles of electrons in the two half-reactions are the same, so no multiplication is necessary.

Add the two half-reactions:

$$H_2O_2 + 2I^- + 2H^+ \longrightarrow I_2 + 2H_2O$$

Aqueous alkaline solutions do not have a source of hydrogen ions as such, so the atom balance must proceed using hydroxide ion and water. The most general method for balancing the atoms in an alkaline solution is to balance the half-reaction using hydrogen ions, and then add sufficient hydroxide ions to *both* sides of the equation to neutralize all of the hydrogen ions.

EXAMPLE 13.5

Write a half-reaction for nitrate ion going to ammonia in alkaline solution.

1. $NO_3^- \longrightarrow NH_3$ (alkaline)

2. **(a)** $NO_3^- + 9\ H^+ \longrightarrow NH_3 + 3\ H_2O$

(b) $NO_3^- + 9\ H^+ + 9\ OH^- \longrightarrow NH_3 + 3\ H_2O + 9\ OH^-$

(c) $NO_3^- + 6^*\ H_2O = NH_3 + 9\ OH^-$ ($^*9H_2O - 3\ H_2O$)

3. $8\ e^- + NO_3^- + 6\ H_2O = NH_3 + 9\ OH^-$

Some facts about electricity you should know 13.2

An electric current is a movement of charge carriers. The charge carriers in metallic conductors are electrons, and their motions give rise to various mechanical, electromagnetic, and heat effects, which are the proper subject of physics. The charge carriers in solution and in molten salts are ions. The motions of ions and the chemical reactions that occur at the interface of the ionic and electronic conductors are the proper subject of chemistry.

A charged electronic conductor is called an *electrode.* If it is charged negatively, it is a *cathode;* if charged positively, it is an *anode.*

An ionic conductor is an *electrolyte.* We already know that anions are negative and cations are positive. It follows then that anions are attracted to the anode and cations to the cathode in an electrolyte.

Electrical current is measured by the amount or quantity of charge that passes a given point in a unit period of time. The practical quantity of charge is a *coulomb*. An electrical current of one coulomb per second is an *ampere*. The current in amperes is a property that relates the quantity of charge (coulombs) to time (seconds).

$$C \text{ coulombs} = \underset{\text{(amperes)}}{A \, \frac{\text{coulombs}}{\text{second}}} \times t \text{ seconds} \tag{13.1}$$

The charge on one electron—1.60×10^{-19} coulomb—is too small to be a practical unit of charge. On the other hand, a mole of electrons is too large to be a household quantity of charge. A mole of electrons is important chemically, however, and it is called a *faraday*. If there are 6.02×10^{23} electrons per faraday and 1.60×10^{-19} coulombs per electron, then there must be

$$\frac{6.02 \times 10^{23}\text{electrons}}{\text{faraday}} \times \frac{1.60 \times 10^{-19}\text{coulomb}}{\text{electron}} = \frac{96{,}500 \text{ coulombs}}{\text{faraday}}$$

One mole of electrons (a faraday) is 96,500 coulombs.[1]

EXAMPLE 13.6 If a current of 0.5 ampere flows for six hours, how many coulombs of electricity have passed? How many faradays? How many electrons?

Solution: **(a)** $6 \text{ hours} \times \frac{3600 \text{ seconds}}{\text{hour}} \times \frac{0.5 \text{ coulomb}}{\text{second}} = 10.800 \text{ coulombs}$

(b) $10{,}800 \text{ coulombs} \times \frac{1 \text{ faraday}}{96{,}500 \text{ coulombs}} = 0.112 \text{ faraday}$

(c) $0.112 \text{ faraday} \times \frac{6.02 \times 10^{23} \text{ electrons}}{\text{faraday}} = 6.74 \times 10^{22} \text{ electrons}$

13.3 *Faraday's law of electrolysis*

An electric current passing through an electrolyte will cause chemical reactions to occur at the interface of the electrode and the electrolyte. Oxidation occurs at the anode and reduction at the cathode. The reactions that occur depend on the voltage and the composition of the electrodes, the electrolyte, and the solvent. In general, a reaction caused by an external source of electrical energy is termed *electrolysis*. A diagram of an electrolytic cell is shown in Figure 13.1.

[1] It would have been nicer for chemists if a coulomb was an easy number like 10^{-5} mole or some such, but the coulomb is defined in terms of the ampere, which is defined in terms of its mechanical effect. To five significant figures, there are 96,487 coulombs per faraday.

Anode reactions

The anode is positively charged so it can act as a *source* of positively charged ions or as a *sink* for electrons.

(a) The anode may dissolve:

$$Ag = Ag^+ + 1e^-$$
$$Cu = Cu^{2+} + 2e^-$$
$$Al + 3\ H_2O = Al(OH)_3 + 3\ H^+ + 3\ e^-$$

(b) The anode may oxidize a negative ion or a neutral molecule:

$$2\ Br^- = Br_2 + 2\ e^-$$
$$H_2 = 2\ H^+ + 2\ e^-$$
$$2\ H_2O = O_2 + 4\ H^+ + 4\ e^-$$

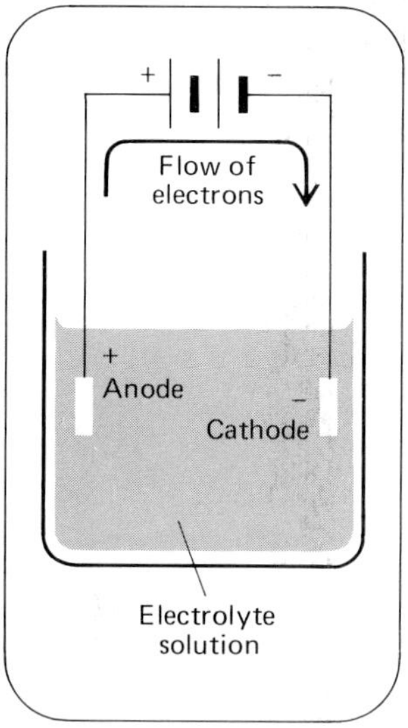

Figure 13.1
Electrolytic cell.

Cathode reactions

The cathode is negatively charged. Since electron conductors never form negative ions, the only reaction possible is reduction of the positive electrolyte or the solvent.

$$Ag^+ + 1\ e^- = Ag$$
$$Cr^{3+} + 3\ e^- = Cr$$
$$2\ H_2O + 2\ e^- = H_2 + 2\ OH^-$$

Michael Faraday first established that the amount of chemical reaction occurring during electrolysis depends exactly on the quantity of charge passing through the circuit. Faraday's law states that the passage of one mole of electrons (one faraday) through an electrolyte will produce or consume *one equivalent* of chemical reactants at each electrode.

An equivalent, like a camel, is easier to recognize than it is to describe. One equivalent of a reactant in an oxidation-reduction reaction is one mole divided by the change in oxidation number occurring in that half-reaction.

EXAMPLE 13.7

1. $Ag^+ + 1\ e^- = Ag$
   The passage of *one* faraday of charge (96,500 coulombs) through an electrolyte containing silver ions will deposit *one* mole of silver on the cathode. The change in oxidation number is *one*. There is *one* equivalent per mole of silver.
2. $Cu^{2+} + 2\ e^- = Cu$
   The passage of *two* faradays of charge through an electrolytic cell containing Cu^{2+} ions will deposit *one* mole of copper on the cathode. There are *two* equivalents per mole of copper(II) ion in this reaction.

If the half-reaction occurring at the electrode is known, Faraday's law can be used to predict the amount of reaction from information about the current and time.

EXAMPLE 13.8 How many moles of hydrogen gas are produced by a current of 5.0 amperes operating for 30 minutes in an electrolytic cell?

$$2\ H^+ + 2\ e^- = H_2$$

Want: moles of H_2

Have: 30 minutes of time

Know: $5.0\ \frac{\text{coulombs}}{\text{sec}}$; $60\ \frac{\text{sec}}{\text{min}}$; $96{,}500\ \frac{\text{coulombs}}{\text{faraday}}$; and from the equation, $\frac{1\ \text{mole}\ H_2}{2\ \text{faradays}}$

Solution: $30\ \text{min} \times \frac{60\ \text{sec}}{\text{min}} \times \frac{5.0\ \text{coulombs}}{\text{sec}} \times \frac{1\ \text{faraday}}{96{,}500\ \text{Cb}} \times \frac{1\ \text{mole}(H_2)}{2\ \text{faradays}}$

$= 0.047$ moles of hydrogen

13.4 *Galvanic cells*

A galvanic (or voltaic) cell is a device in which an oxidizing agent and a reducing agent exchange electrons through an external circuit. The circuit is completed by an electrolyte within the cell, but the oxidizer and the reducing agent are not in contact with each other.

The process is essentially the reverse of electrolysis. Instead of putting electrical energy into the system, a galvanic cell converts chemical potential energy into electrical energy.

A common example of a galvanic cell is a zinc/copper sulfate cell (Figure 13.2). A zinc electrode submerged in an electrolyte will have a certain tendency to dissolve, leaving an excess of electrons on the electrode.

$$Zn = Zn^{2+}(aq) + 2\ e^-$$

If the electrode is not connected to a circuit, this process reaches equilibrium after a very small amount of zinc has dissolved, because the negative charge on the electrode will prevent additional zinc ions from leaving. If the zinc electrode is electrically connected to a copper electrode in contact with a solution containing copper ions the electrons will flow from the zinc to the copper. There they attract copper ions from the solution and reduce them.

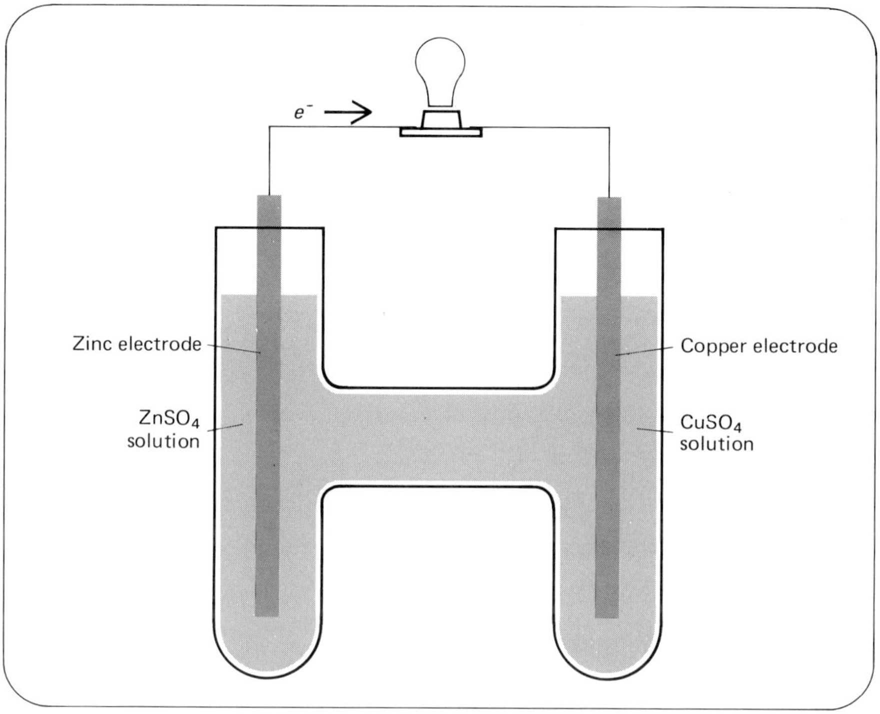

Figure 13.2 Zinc-copper galvanic cell.

$$Cu^{2+}(aq) + 2e^- = Cu$$

Negative ions will flow from the copper side through the electrolyte bridge to the zinc side, where they electrically balance the newly dissolved zinc ions.

Each half-reaction has a driving force measured as a voltage. The voltage obtained from the cell is the sum of the voltage of the two half-reactions. This cell develops 1.1 volt at 25°C when the electrodes are in 1*M* $ZnSO_4$ and 1*M* $CuSO_4$, respectively.

Summary 13.5

1. The oxidation numbers of some atoms change during oxidation-reduction reactions. An atom is oxidized if its oxidation number is increased. An atom is reduced if its oxidation number decreases in a reaction.

2. Balancing oxidation-reduction reactions is somewhat more complex than acid-base reactions because mass and charge must be balanced separately. One method conserves charge by balancing the changes in oxidation numbers. Another method balances the moles of electrons exchanged by the oxidizing and reducing agents.
3. The practical unit of electrical charge is the coulomb. One mole of electrons, a faraday, is 96,500 coulombs. A current of one coulomb per second is an ampere.
4. An electrolysis reaction is a chemical reaction at the interface of an electrode and an electrolyte, caused by an external source of electrical energy.
5. The passage of one faraday of electrical charge through an electrolyte will cause one equivalent of reaction at each electrode.
6. A galvanic cell converts chemical energy into electrical energy. The exchange of electrons between the oxidizing agent and the reducing agent take place through an external conductor.

EXERCISES

I. Define the following terms.

1. oxidation
2. reduction
3. coulomb
4. ampere
5. faraday
6. anode
7. cathode
8. galvanic cell
9. Faraday's law
10. an equivalent (oxidation-reduction)
11. electrolysis

II. Write skeleton net ionic equations from these reactants and products.

A.
1. $KMnO_4 + HCl = KCl + MnCl_2 + \overline{Cl_2} + H_2O$
2. $FeCl_2 + H_2O_2 + HCl = FeCl_3 + H_2O$
3. $\underline{CdS} + I_2 + HCl = CdCl_2 + HI + \underline{S}$
4. $\underline{Zn} + HNO_3(aq) = Zn(NO_3)_2 + NH_4NO_3 + H_2O$
5. $KBr + H_2SO_4 = K_2SO_4 + Br_2 + \overline{SO_2} + H_2O$

B.
1. $Cl_2 + KCN + KOH = KCNO + KCl + H_2O$
2. $I_2 + Na_2S_2O_3 = NaI + Na_2S_4O_6$
3. $KMnO_4 + FeSO_4 + H_2SO_4 = MnSO_4 + Fe_2(SO_4)_3 + K_2SO_4 + H_2O$
4. $K_2Cr_2O_7 + SnCl_2 + HCl = CrCl_3 + SnCl_4 + KCl + H_2O$
5. $\underline{PbO_2} + \underline{Pb} + H_2SO_4 = PbSO_4 + H_2O$

III. Write balanced half-reactions for the following:

A.

Oxidizing Agents	*Products*	*Conditions*
1. MnO_2 (s)	Mn^{2+}	acid
2. MnO_4^-	MnO_2	neutral
3. NO_3^- (dil. aq)	NO	acid
Reducing Agents		
6. HNO_2	NO_3^-	acid
7. $H_2C_2O_4$	CO_2	acid
8. Zn	Zn^{2+}	acid
9. HSO_3^-	HSO_4^-	acid
10. HCHO	HCO_2^-	basic

B.

Oxidizing Agents	*Products*	*Conditions*
1. Cl_2	Cl^-	acid
2. NO_3^- (conc)	NO_2	acid
3. NO_3^- (dil)	NH_3	basic
4. ClO_4^-	Cl^-	acid
5. CrO_4^{2-}	Cr^{3+}	acid
Reducing Agents		
6. N_2H_4	N_2	acid
7. BH_4^-	H_3BO_3	acid
8. H_2S	S	acid
9. Sn	Sn^{2+}	acid
10. H_2	H^+	acid

IV. Write balanced net ionic equations for these redox reactions.

A.
1. $HNO_3 + H_2S = NO + S + H_2O$
2. $K_2Cr_2O_7 + HCl = CrCl_3 + Cl_2 + H_2O$
3. $KTeO_3 + HI = Te + I_2 + H_2O + KI$
4. $Cu + HNO_3 = Cu(NO_3)_2 + H_2O + NO$
5. $C_6H_5CH_3 + KMnO_4 + H_2SO_4 = C_6H_5CO_2H + MnO_2 + K_2SO_4 + H_2O$

B.
1. $KI + H_2SO_4 + K_2CrO_4 = KIO_3 + Cr_2(SO_4)_3 + K_2SO_4 + H_2O$
2. $C_2H_4O + CuSO_4 + NaOH = C_2H_4O_2 + \underline{Cu_2O} + Na_2SO_4 + H_2O$
3. $Al + NaNO_3 + NaOH + H_2O = NaAlO_2 + NH_3$
4. $HClO + H_2S = HCl + S + H_2O$
5. $H_2S + H_2O_2 = H_2SO_4 + H_2O$

V. Electrolysis

1. How many grams of copper would be deposited from a copper(II) sulfate solution by a current of 5.0 amperes flowing for 5.0 minutes?
2. Aluminum is produced by the electrolysis of aluminum oxide dissolved in an electrolyte of molten cryolite (Na_3AlF_6). The reaction at the cathode is

$$Al^{3+} + 3e^- = Al$$

How long would it take to produce 10 kg of aluminum with a current of 500 amperes?

3. The reaction in a lead storage battery during discharge is

$$PbO_2 + Pb + 2H_2SO_4 = 2PbSO_4 + 2H_2O$$

How much lead is converted to lead(II) sulfate when an 80 ampere-hour battery is completely discharged?

4. Calcium metal and chlorine gas are produced by the electrolysis of molten calcium chloride. What volume of chlorine gas at standard conditions are produced by a current of 150 amperes for 24 hours?

5. What volume of hydrogen and what volume of oxygen at 25°C and 740 torr are produced by the electrolysis of water by a current of 30 amps for 30 minutes?

6. A current of 0.100 ampere for two hours deposited 0.727 g of platinum on the cathode. What was the oxidation number of the platinum ion in solution?

SOLUTIONS

II.A.

1. $MnO_4^- + H^+ + Cl^- = Mn^{2+} + Cl_2 + H_2O$
2. $Fe^{2+} + H_2O_2 + H^+ = Fe^{3+} + H_2O$
3. $CdS + I_2 + H^+ = Cd^{2+} + I^- + S$
4. $Zn + H^+ + NO_3^- = Zn^{2+} + NH_4^+ + H_2O$
5. $Br^- + H^+ + SO_4^{2-} = Br_2 + SO_2 + H_2O$

III.A.

1. $MnO_2 + 4H^+ + 2e^- = Mn^{2+} + 2H_2O$
2. $MnO_4^- + 4H^+ + 3e^- = MnO_2 + 2H_2O$
3. $NO_3^- + 4H^+ + 3e^- = NO + 2H_2O$
4. $Cr_2O_7^{2-} + 14H^+ + 6e^- = 2Cr^{3+} + 7H_2O$
5. $H_2O_2 + 2H^+ + 2e^- = 2H_2O$
6. $HNO_2 + H_2O = NO_3^- + 3H^+ + 2e^-$
7. $H_2C_2O_4 = 2CO_2 + 2H^+ + 2e^-$
8. $Zn = Zn^{2+} + 2e^-$
9. $HSO_3^- + H_2O = HSO_4^- + 2H^+ + 2e^-$
10. $HCHO + 3OH^- = HCO_2^- + 2H_2O + 2e^-$

IV.A.

1. $2NO_3^- + 3H_2S + 2H^+ = 2NO + 3S + 4H_2O$
2. $Cr_2O_7^{2-} + 6Cl^- + 14H^+ = 2Cr^{3+} + 3Cl_2 + 7H_2O$
3. $2TeO_3^- + 10\,I^- + 12H^+ = 2Te + 5I_2 + 6H_2O$
4. $3Cu + 8H^+ + 2NO_3^- = 3Cu^{2+} + 2NO + 4H_2O$
5. **a.** $C_6H_5CH_3 + 2H_2O = C_6H_5CO_2H + 6H^+ + 6e^-$

   b. $MnO_4^- + 4H^+ + 3e^- = MnO_2 + 2H_2O$ (times 2)

   $$C_6H_5CH_3 + 2MnO_4^- + 2H^+ = C_6H_5CO_2H + 2MnO_2 + 2H_2O$$

V. Electrolysis

1. $5 \text{ min} \cdot \frac{60 \text{ sec}}{\text{min}} \times \frac{5.0 \text{ Cb}}{\text{sec}} \times \frac{1 \text{ equiv}}{96500 \text{ Cb}} \times \frac{1 \text{ Cu}}{2 \text{ equiv}} \times \frac{63.5 \text{ g}}{\text{Cu}} = 0.49 \text{ g}$

3. $1 \text{ hr} \cdot \frac{3600 \text{ sec}}{\text{hr}} \times \frac{80 \text{ Cb}}{\text{sec}} \times \frac{1 \text{ equiv}}{96500 \text{ Cb}} \times \frac{1 \text{ Pb}}{2 \text{ equiv}} \times 208 \frac{\text{g}}{\text{Pb}} = 310 \text{ g}$

5. $30 \text{ min} \cdot \frac{60 \text{ sec}}{\text{min}} \times \frac{30 \text{ Cb}}{\text{sec}} \times \frac{1 \text{ equiv}}{96500 \text{ Cb}} \times \frac{1 H_2}{2 \text{ equiv}} \times \frac{62.4(298)}{740} \frac{\text{liters}}{\text{mol}}$
$= 7.0 \text{ liters}$

The volume of oxygen is one-half of 7.0 liters = 3.5 liters.

Change in the physical or chemical state of a system is caused by differences in intensity factors. The *amount* of change depends on the size of the system and the relationship between size and intensity. Put them all together and we have *energy*.

After reading this chapter you should be able to

1. Calculate work done and potential energy change under a constant force (in joules);
2. Calculate expansion work done at constant pressure;
3. Calculate heat quantities from temperature changes;
4. Describe a calorimeter, and measure a calorimeter constant;
5. Convert joules to calories and vice versa;
6. Calculate heats of phase changes;
7. Define enthalpy;
8. Define enthalpy of formation of compounds;
9. Calculate electrical work done;
10. Calculate the energy of a photon.

14 energy

A beam of sunlight falling upon a green leaf excites a chlorophyll molecule, enabling it to strip the hydrogens from a water molecule and begin the process of photosynthesis.

A drop of rain falls upon a mountain and momentarily pauses in its rush to the sea to turn a turbine.

A small boy consumes a peanut butter and jelly sandwich and later makes the dirt fly as he slides into second base.

A gasoline molecule is attacked by hordes of oxygen molecules and is shattered into hot fragments of carbon oxides and water; the hot products press upon a moving surface and convert their motion into the motion of a vehicle.

A lead atom leaves an electrode and becomes a lead ion, leaving behind its electrons. The electrons turn a motor on their way to the job of reducing lead dioxide.

The granite bowels of a mountain, tortured and distorted by opposed motion in their foundations, suddenly fracture; shock waves topple buildings and bridges; and people are afraid.

These changes or happenings are occurring in time and space. We not only want to know the *causes* of these changes but also *how much change* will occur. What is the capacity of a system to plow the fields, move the ships, planes, and trains, turn the generators, bake the bread, and do all the other countless actions and jobs that would not take place by themselves?

We also want to know what connection or relation exists between a beam of sunlight, a mountain reservoir, an electric storage battery, a ton of coal, a barrel of oil, and a peanut butter sandwich.

Why should we attempt what looks like a mind-boggling job? Because four billion people cannot live on this earth on "what Nature provides." We as citizens with more than our share of the goodies must understand where the goodies come from, and we must realize that Nature is about to close our account because of excessive withdrawals. If we understand the relationships, perhaps we can open a new account at Nature's bank.

14.1 *Work*

By using our senses we have observed that the *causes* of change are intensity factors—forces, pressures, temperatures, concentrations, and the like. To find out how much change can occur, we must somehow relate the intensity factors to the size of the system.

One of the first observations about laws was that a small weight falling a large distance could be made to lift a large weight a small distance (law of levers, Chapter 1). The interesting thing about that law is the fact that the product of the small weight and the distance it falls *equals* the product of the large weight and the distance it rises, if the action is done very slowly.

$$W_1 \times d_1 = W_2 \times d_2 \tag{14.1}$$

If the action is rapid, so that the weights gain a lot of motion and speed, the equation does not hold too well. The intensity factor that lifts the large weight must be the small weight. The *capacity* of the small weight to lift the large weight depends on the distance the small weight moves. The product of the two factors evidently is a measure of the capacity of something to cause change.

In general, the product of a force times the distance it moves in the direction of the force is called *work*.

$$\text{work} = \text{force} \times \text{distance} \tag{14.2}$$

When force is measured in newtons and distance in meters, the units of work are newton-meters, or *joules* (pronounced jools).

Work is the first recognized form of energy. It is calculated as the product of an intensity factor (force) and a capacity factor (distance). Work done on an object may increase its potential energy (e.g., in lifting a rock), or its kinetic energy (e.g., in throwing a baseball). Work done on a system may also increase the random average kinetic energy of the atoms of a system—its temperature.

EXAMPLE 14.1

1. If work is done on an object to change its position in a force field, the *potential energy* of the object is increased. The force field of the earth's gravity is 9.81 newtons per kilogram at the earth's surface.

(a) How much work is done to lift 1000 kilograms 30.0 meters?

Solution: $1000 \text{ kg} \times \frac{9.81 \text{ N}}{\text{kg}} \times 30.0 \text{ m} = 2.94 \times 10^5$ joules

(b) How much potential energy does the 1000 kilogram weight now have with respect to its original position?

Solution: 2.94×10^5 joules

2. (a) How much kinetic energy is given to a baseball if a constant force of 70 newtons is applied through a distance of one meter?

Solution: $70 \text{ N} \times 1.0 \text{ m} = 70$ joules

(b) The kinetic energy of a body is $\frac{1}{2}$ mass × velocity^2. What is the velocity of the baseball if the mass is 150 grams?

Solution: $\frac{1}{2} \times 0.150 \text{ kg} \times v^2 = 70$ joules

$$v^2 = \frac{2 \times 70}{0.150}$$

$$v = \sqrt{\frac{2 \times 70}{0.150}} = \frac{30.6 \text{ meters}}{\text{second}}$$

3. How much kinetic energy does a kilogram of water gain when it falls through a distance of 100 meters?

Solution: $1 \text{ kg} \times \frac{9.81 \text{ N}}{\text{kg}} \times 100 \text{ m} = 981$ joules

Expansion work

A force on a piston makes it move through a distance, and the work done can be calculated as mechanical work. The force exerted by a gas is usually given as a pressure, that is, force/area. The distance the piston moves multiplied by the area of the face of the piston is the volume increase of the gas.

$$\text{pressure} = \frac{\text{force}}{\text{area}}$$

$$\text{volume} = \text{area} \times \text{distance}$$

$$\text{work} = \frac{\text{force}}{\text{area}} \times \text{area} \times \text{distance} = P\Delta V$$

At constant pressure the expansion work done by a gas is

$$\text{work} = P(V_{\text{final}} - V_{\text{initial}}) = P\Delta V \tag{14.3}$$

When the pressure is in newtons per square meter and the volume is in cubic meters, the work done is in joules. More often the pressure is in atmospheres (1 atmosphere = 1.013×10^5 newtons/meter^2), and the volume change is in liters. One liter-atmosphere is 101.3 joules.

14.2 *Heat*

Benjamin Thompson (Count Rumford), an expatriate American,[1] was one of the first to demonstrate that heat and work were related. Heat was thought to be a massless fluid called "caloric," which flowed from hot things to cool ones. Thompson demonstrated that he could get an unlimited amount of "caloric" from a piece of steel by using a blunt boring tool. He concluded that heat was the random motion of the atoms of a substance, and random motion could be produced by friction.

It is relatively easy to measure work. One can measure forces with springs and weights and the like, and distance moved is easy to measure. Heat is something else. Not until the middle of the nineteenth century, after (sometimes heroic)[2] investigations by J. P. Joule, was the quantitative relationship of work and heat established.

Heat capacity

Work and heat are extensive properties that depend on the size (capacity) of the system. The most common evidence of the flow of heat is a change in temperature of some object. If 100 grams of water at 50°C is mixed with 100 grams of water at 20°C, it will surprise no one that the equilibrium temperature will be 35°C. If instead 100 grams of aluminum at 50°C is added to 100 grams of water at 20°C, the equilibrium temperature is only 25.3°C. Evidently 100 grams of aluminum contributes less heat to the cold water than 100 grams of hot water does.

Each substance differs in its *capacity* to store heat. It is an intensive property that relates the amount of heat absorbed to the temperature change. Because heat capacity is an intensive property, it does not tell us the *amount* of heat exchanged; we need to multiply the heat capacity and the temperature change by an extensive property like grams or moles of substance present.

Specific heat capacity (usually shortened to *specific heat*) is the property that relates the amount of heat absorbed to the mass of the system in *grams* and to the temperature change. The unit quantity of heat was originally defined as the calorie, which is the amount of heat necessary to raise the temperature of one gram of water one degree

[1] He wasn't expatriate by choice. He chose the losing side in the Revolutionary War.

[2] The story is told that he spent his honeymoon in Switzerland measuring the temperature at the top and bottom of waterfalls.

Celsius, from 14.5°C to 15.5°C. By definition, then, the specific heat of water is 1 cal/g-deg. Joule showed that work could be completely converted into heat and that the conversion factor is

$$1 \text{ calorie} = 4.184 \text{ joules}$$

The molar heat capacity of water is the specific heat times the molar weight.

$$\text{molar heat capacity of water} = \frac{1 \text{ cal}}{\text{g-deg}} \times \frac{18 \text{ g}}{\text{mole}} = \frac{18 \text{ calories}}{\text{mole-deg}}$$

Heat quantity measurements can be made if the quantity of water (the capacity factor) and the temperature change (the intensity factor) are known.

$$Q \text{ calories} = w \text{ grams} \times \frac{1 \text{ cal}}{\text{g-deg}} \times \Delta T \text{ degrees} \qquad (14.4)$$

As is customary, $\Delta T = T_{\text{final}} - T_{\text{initial}}$.

EXAMPLE 14.2

1. How many calories of heat are necessary to raise 500 grams of water from 20 to 80°C?

Solution: $Q \text{ calories} = 500 \text{ g} \times \frac{1 \text{ cal}}{\text{g-deg}} \times (80 - 20) \text{ deg} = 3 \times 10^4 \text{ cal}$

2. What will be the increase in temperature of 100 grams of water if an electrical heater supplies 4000 joules of energy to the water?

Solution: $\Delta T = \frac{4000 \text{ joules}}{100 \text{ grams}} \times \frac{1 \text{ calorie}}{4.184 \text{ joules}} \times \frac{1 \text{ g-degree}}{1 \text{ calorie}} = 9.56 \text{ degrees}$

The law of Dulong and Petit

Pierre Dulong and Alexis Petit proposed on theoretical grounds that the heat capacities of crystalline metals should be in the neighborhood of 6.0 calories per mole per degree. This was a valuable prediction in the days when the atomic weights of metals were not known with certainty. If the specific heat could be measured, the value 6.0 calories/mole-deg could be used to find an approximate atomic weight.

EXAMPLE 14.3

If the specific heat of iron is known to be 0.108 cal/g-deg, what is the approximate atomic weight of iron?

Solution: $6.0 \frac{\text{calories}}{\text{mole-deg}} \times \frac{1 \text{ gram-deg}}{0.108 \text{ cal}} = \frac{55.5 \text{ grams}}{\text{mole}}$

The heat capacities of some elements are given in Table 14.1.

Table 14.1

Heat capacities of common elements at 25°C

	Element	Specific Heat $\left(\frac{\text{cal}}{\text{g-deg}}\right)$	Molar Heat Capacity
SOLIDS	Aluminum	0.215	5.80
	Carbon (graphite)	0.172	2.06
	Copper	0.092	5.85
	Gold	0.030	5.91
	Iron	0.108	6.04
	Lead	0.031	6.33
	Phosphorus	0.19	5.90
	Sulfur	0.177	5.67
	Tin	0.052	6.16
	Zinc	0.092	6.00
LIQUIDS	Bromine	0.113	9.02
	Mercury	0.033	6.64
GASES	Helium	1.242	5.00
	Hydrogen	3.42	6.89
	Nitrogen	0.249	6.94
	Oxygen	0.219	7.05

14.3 *Calorimetry*

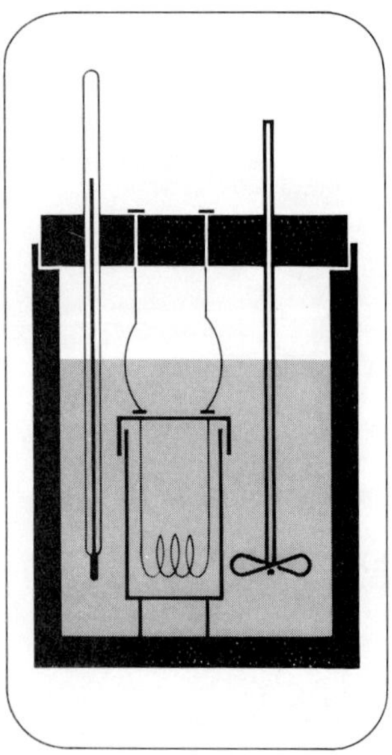

Figure 14.1 Bomb calorimeter

The measurement of heat changes is called calorimetry. The instrument used is a calorimeter. The most common type of calorimeter is an insulated can containing a known amount of water. The water surrounds an internal container in which a variety of heat-producing events may be carried out. The heat is distributed throughout the calorimeter by a stirrer, and the temperature before and after the event is measured with a precise thermometer (Figure 14.1).

The insulation serves as the boundaries of the system. If we do not add or remove energy from outside the boundaries, the sum of all energy changes within the system will be zero. A change in potential energy due to a chemical reaction must appear as heat, and the amount of heat can be measured if we know the temperature change and the heat capacity of the calorimeter and its contents.

Before measurements on unknown heats of reaction can be made, the heat capacity of the calorimeter must be found. A known quantity of heat is added by means of an electric heater. This quantity of heat must account for the temperature change of the measured amount of water in the calorimeter and the calorimeter itself. The heat absorbed by the calorimeter is the difference between the heat input and the heat absorbed by the water. This quantity divided by the temperature change is the calorimeter heat capacity, C.

EXAMPLE 14.4

To a calorimeter containing 1000 grams of water were added 2490 calories of heat by means of an electric heater. The initial temperature was 25.00°C and the final temperature was 27.20°C. Find the heat capacity of the calorimeter.

We know: The heat absorbed by the calorimeter equals the heat input minus the heat absorbed by the water.

$\Delta T = 2.20$ degrees

Solution: $Q_{\text{calorimeter}} = Q_{\text{input}} - Q_{\text{water}}$

$$\frac{C\text{ cal}}{\text{deg}} \times 2.20\text{ deg} = 2490\text{ cal} - 1000\text{ g} \times \frac{1\text{ cal}}{\text{g-deg}} \times 2.20\text{ deg}$$

$$\frac{C\text{ cal}}{\text{deg}} \times 2.20\text{ deg} = 2490\text{ cal} - 2200\text{ cal} = 290\text{ cal}$$

$$\frac{C\text{ cal}}{\text{deg}} = \frac{290\text{ cal}}{2.20\text{ deg}} = \frac{130\text{ cal}}{\text{deg}}$$

EXAMPLE 14.5

A chemical reaction is carried out in the inner container of the calorimeter. The calorimeter contains 1000 grams of water and the calorimeter itself has a heat capacity of 130 calories per degree. The temperature of the calorimeter and its contents rise 3.30 degrees. How much heat is released by the reaction?

Solution: Heat released = heat absorbed

$$\Delta H = 1000\text{ g} \times \frac{1\text{ cal}}{\text{g-deg}} \times 3.30\text{ deg} + \frac{130\text{ cal}}{\text{deg}} \times 3.30\text{ deg}$$

$$= 3730\text{ calories}$$

Heat quantities and phase changes 14.4

The temperature of a substance does not always change when heat is added to the substance. Melting ice is absorbing heat from its surroundings, but the temperature remains at 0°C until the ice is gone. Water boiling on a stove does not get any hotter if the heat is increased; it just boils faster. These facts may be contrary to your intuition, but they can readily be checked with a thermometer.

Heat of melting

When a crystalline substance melts, energy is required to produce the holes and general disorder characteristic of a liquid. As long as any solid crystals remain, the energy will be absorbed by the crystals.

No energy is left to increase the kinetic energy of the molecules, so the temperature remains constant. The heat required to disrupt the crystal structure is called the *latent heat of melting.* The reverse process holds too; this energy must be removed from the liquid in order to freeze it. The latent heat of melting of water is 80 calories per gram, or 1440 calories per mole.

Heat of vaporization

Liquids have little or no order to disrupt, but the molecules are close together so that they are acted upon by moderately strong cohesive forces. The heat necessary to convert a liquid to a gas is the energy required to overcome these forces until the molecules are far enough apart to have little affinity for each other. Note that the kinetic energy is not increased, because the temperature stays constant. The potential energy is increased. If the liquid is being vaporized under atmospheric pressure, additional energy is required to push back the atmosphere and make room for the vapor. The latent heat of vaporization is the sum of these two energies. Water absorbs 9700 calories per mole when it is evaporated at 100°C and one atmosphere pressure. Of that quantity 745 calories are necessary for expansion work, and 8955 calories are necessary to separate one mole of water molecules from each other.

14.5 *Enthalpy*

Most processes that occur in nature and in the laboratory do so under a constant external pressure. If a volume change occurs during the event, the heat effect due to the volume change and the external pressure is included in the transaction, just like a sales tax. If the system does work on the surroundings, as water does when it is converted to steam, an extra amount of work must be added to the system. Conversely, if the process involves a decrease in volume under constant external pressure, an additional amount of energy will be released by the system.

Chemists and heat engineers are not only interested in the changes in internal potential and kinetic energy, but also in the overall heat effect. They have therefore invented a new quantity called *enthalpy* (H), defined as the *heat gained by the system* when a process is carried out at constant pressure. The change in enthalpy is the sum of the change in internal energy and the expansion work done on the system.

$$\Delta H = \Delta E + P\Delta V \tag{14.5}$$

where ΔE is the change in internal potential and kinetic energy and

$P\Delta V$ is the expansion work. If no volume change occurs or if the external pressure is zero, ΔH is the same as ΔE.

Enthalpy is measured from the standpoint of the system, not the observer. If you as an observer see an evolution of heat, the reaction is said to be *exothermic*. The system is losing heat, so ΔH is negative. Conversely, if heat must be added to the system (endothermic reaction), the system is gaining enthalpy and ΔH is positive.

Observer sees	*System sees*
exothermic	$-\Delta H$
endothermic	$+\Delta H$

Heats of chemical reaction 14.6

Most energy available to us in an exploitable form comes from the combustion of fuels in air. Combustion is a chemical reaction, and the energy obtained is in the form of heat.

Because combustion was known from the time of the cavemen, and the benefits of heat were appreciated by all of us, it was naturally assumed that chemical reactions always produced heat, and that is why they took place.

It is certainly true that when two or more reactants combine to form products accompanied by a large release of heat energy the reaction spontaneously goes to completion, and the reverse reaction is highly unlikely. However, some reactions occur spontaneously with a positive increase in enthalpy, so we know now that an exothermic reaction is not the only criterion of whether a process will occur and how far it will go. We must also consider the entropy change of the reaction.

The enthalpy of chemical reactions is an extensive property and must be associated with a definite quantity of reactants. By convention, enthalpy is reported as the heat gained by the system when molar quantities of reactants are completely converted to the molar quantities of products on the right side of the equation. Both reactants and products are measured at 25°C and one atmosphere pressure. The conditions during the reaction do not matter, because ΔH depends only on the initial and final states, like any energy change.

EXAMPLE 14.6

$$C\ (g) + O_2\ (g) = CO_2\ (g) \qquad \Delta H = -94.5\ \text{kcal}$$

This equation says that one mole of carbon as graphite combines with one mole of oxygen gas to form one mole of carbon dioxide gas, and the heat released is 94.5 kilocalories. Both reactants and products are measured at 25°C and one atmosphere pressure. The fact that graphite and oxygen won't burn at 25°C makes no difference; if they are heated to combustion temperature that heat input will be recovered when the hot CO_2 is cooled to 25°C.

14.7 *Other forms of energy*

When an electron falls from an outer orbital in an atom to an unoccupied inner orbital, the difference in energy is sent out as a pulse of radiation—a photon. Photons travel at 3×10^8 meters per second, and in their pathways behave as waves. The science of optics makes it possible to measure the wavelength of light very precisely. The wavelength and the number of waves per second, the frequency, are related by the speed of light, 3×10^8 meters per second.

$$\text{wavelength (meters)} \times \text{frequency} \left(\frac{1}{\text{sec}}\right) = 3 \times 10^8 \frac{\text{m}}{\text{sec}}$$

In symbols the equation is

$$\lambda\nu = c \tag{14.6}$$

When a photon is absorbed by matter, it behaves as a particle with a definite energy. The energy is directly proportional to the frequency, and the proportionality constant, Planck's constant h, has the value 6.626×10^{-34} joule-seconds.

$$E \text{ (of a photon)} = h\nu \tag{14.7}$$

EXAMPLE 14.7 The wavelength of yellow light is about 540 nanometers (5.4×10^{-7} meters). Calculate the frequency and the energy of one photon of yellow light.

Solution: frequency (sec^{-1}) $= \dfrac{3 \times 10^8 \text{ m-sec}^{-1}}{5.4 \times 10^{-7}\text{m}} = 5.6 \times 10^{14} \text{ sec}^{-1}$

$$\begin{aligned}\text{energy} &= 6.6 \times 10^{-34} \text{ joule-sec} \times 5.6 \times 10^{14}\text{sec}^{-1}\\ &= 3.7 \times 10^{-19} \text{ joules}\end{aligned}$$

Electrical work

Electrical charges experience forces due to the presence of other electrical charges in a way similar to, but in some ways different from gravity forces acting on masses. Electrical charges at rest with respect to each other attract or repel each other with a force proportional to the product of the quantities of charge and inversely proportional to the square of the distance between them. If the quantities of charge are measured in coulombs and the distance is in meters, the force in a vacuum is

$$\text{electrical force (newtons)} = 8.99 \times 10^9 \frac{q_1 q_2}{r^2}$$

A concept that is very necessary for the description of electrical events is the idea that the presence of a charge causes a condition in space that interacts with another charge. This condition is called an *electric field*. The amount of work done in bringing a test charge of one coulomb a long distance to a particular point in that electric field is the *potential* at that point. The units of potential are joules/coulomb or *volts*. The work done in moving one coulomb of charge from one potential to another is simply the difference in potential. Since a volt is defined as one joule per coulomb, the electrical work done, in general, in moving q coulombs of charge through a potential of V volts is

$$\underset{\text{(joules)}}{\text{electrical energy}} = \underset{\text{volts}}{V} \times \underset{\text{coulombs}}{q} \tag{14.8}$$

An *electron-volt* is the energy change of one electron passing through a potential of one volt. The charge on an electron is 1.602×10^{-19} coulombs, so one electron-volt is 1.602×10^{-19} joules.

Power

The time rate of doing work is power. An energy rate of one joule per second is a *watt*. To find the amount of energy consumed at a given rate it is necessary to multiply the watts times the time.

$$\text{energy (joules)} = \text{watts} \times \text{time (seconds)}$$

One watt-second or joule is a rather small quantity of energy, so the power company sells energy by the kilowatt-hour. One kilowatt-hour is

$$\frac{10^3 \text{ joules}}{\text{second}} \times 3600 \text{ seconds} = 3.6 \times 10^6 \text{ joules}$$

Summary 14.8

1. Energy is the capacity of matter to cause change in the state of other matter. Energy is calculated as the product of a capacity factor and an intensity factor.
2. Kinetic energy is the energy a body possesses due to its motion with respect to a reference point. Potential energy is energy possessed due to a force and a position with respect to a reference point.

3. Mechanical work is calculated as the product of a force and the distance moved in the direction of the force.
4. Expansion work (at constant pressure) is the product of the pressure and the change in volume.
5. Heat is the random kinetic energy of the atoms of a substance. The quantity of heat necessary to change the temperature of a specified amount of matter one degree is the heat capacity of the substance. Specific heat (capacity) is the amount of heat needed to change the temperature of one gram of substance one degree Celsius. Molar heat capacity is the amount of heat needed to change the temperature of one mole of substance one degree Celsius.
6. The law of Dulong and Petit states that the molar heat capacities of solid elements are approximately 6.0 calories/mole-degree.
7. Heat changes are measured in a calorimeter. The quantities of heat exchanged or generated are calculated from the amounts of substances present, their heat capacities, and the temperature changes.
8. Heat is absorbed at constant temperature when substances melt or boil. The quantity of heat absorbed per gram or per mole of substance melted or vaporized is called the latent heat of fusion or the latent heat of vaporization.
9. The enthalpy of a process is the amount of heat absorbed by the system in going from an initial state to a final state at constant pressure. Enthalpy is positive when heat is absorbed by the system (endothermic) and negative when heat is evolved (exothermic).
10. The energy of a photon is proportional to its frequency. The proportionality constant is Planck's constant.
11. The unit of electrical energy is the joule. Electrical work is calculated as the product of the potential (volts) and the quantity of charge (coulombs).
12. Power is the rate of energy exchange. The units of power are watts, or joules/second.

EXERCISES

Define the following terms.

1. work
2. joule
3. expansion work
4. heat
5. heat capacity
6. calorie
7. potential energy
8. kinetic energy
9. latent heat of melting
10. latent heat of vaporization
11. enthalpy
12. heat of reaction
13. volt
14. electron-volt
15. energy of a photon

PROBLEMS

1. A 500 gram mass falls a distance of 1.0 meter. The gravitational constant is 9.81 newtons/kilogram.
   (a) What is the gravitational potential energy of the mass before it falls?
   (b) What is the kinetic energy of the mass after it has fallen one meter?
   (c) If the mass is abruptly brought to a halt by hitting the floor what happens to the kinetic energy of the mass?
2. Bridal Veil Falls in Yosemite National Park is 740 meters high from pool to lip. How much kinetic energy does a kilogram of water gain in falling that distance? What happens to the kinetic energy after the water lands in the pool at the bottom?
3. How much expansion work must be done to produce one mole of carbon dioxide gas from calcium carbonate at 25°C and 1 atmosphere pressure? Neglect the volume of the calcium carbonate.
4. How much expansion work must be done to condense 25 liters of ammonia gas to liquid ammonia if the vapor pressure of ammonia is 1.3 atmospheres? Neglect the volume of the liquid ammonia.
5. When 100 grams of water at 50°C is added to a container at 25°C the equilibrium temperature is 45°C. What is the heat capacity of the container?
6. 4000 calories of heat are added to a calorimeter containing 1000 grams of water. The temperature rises 3.00 degrees. What is the heat capacity of the calorimeter?
7. If 50 grams of copper at 90.0°C are dropped into 100 grams of water at 20.0°C in a container of heat capacity 4.6 calories per degree, what will be the equilibrium temperature? The heat capacity of copper is 0.092 calories/gram-degree.
8. A 30-gram sample of an unknown metal at 100°C is dropped into 100 grams of water at 20.0°C. The final temperature is 21.0°C. What is the specific heat of the metal? Assume the heat capacity of the container is negligible.
9. A calorimeter contains 500 grams of water and 100 grams of an unknown metal. The calorimeter heat capacity is 50 calories per degree. Exactly 1121.6 calories of heat are added to the calorimeter by means of a heater. The temperature rises 2.00 degrees.
   (a) What is the specific heat of the metal?
   (b) If the molar heat capacity of the metal is approximately 6.0 calories/mole-degree, what is the atomic weight of the metal?
10. 50.0 grams of a hydrochloric acid solution are added to 50 grams of a sodium hydroxide solution. Both solutions are at 25.0°C initially. The final temperature after reaction is 27.5°C. The container absorbs 5.0 calories/degree. How much heat was produced by the reaction? Assume the heat capacities of the solutions are 1.0 calories/gram-degree.
11. How much heat is necessary to convert 50 grams of ice at 0°C to 50 grams of water at 25°C?
12. How much heat is necessary to convert 50 grams of water at 25°C to 50 grams of steam at 100°C?
13. The heat of formation of $CO_2(g)$ is -94.5 kcal per mole. How much heat is

liberated when 50 grams of carbon (graphite) is burned in oxygen under standard conditions (25°C, 1 atmosphere)?

14. The heat of neutralization of H^+(aq) and OH^-(aq) is −13.9 kcal per mole. How much heat is liberated when 50 milliliters of 1.00*M* HCl solution reacts with 50.00 milliliters of 1.00*M* NaOH solution?

15. The positive charge on a carbon nucleus is six times 1.6×10^{-19} coulombs or 9.6×10^{-19} C. The electric potential 1 Angstrom unit (10^{-10}m) from the nucleus is 86.4 volts. At two Angstroms distance the potential is 43.2 volts. What potential energy change occurs when an electron "falls" from 2 Å to 1 Å? Give the answer in joules and in electron-volts.

16. If the energy required to ionize sodium vapor to sodium ions and free electrons is 5.12 electron-volts per atom, calculate the energy required to ionize one mole of sodium atoms in kilocalories.

17. If a 40-watt heater is turned on for 250 seconds how much heat is released? Give your answer in joules and calories.

SOLUTIONS

1. **(a)** $0.500 \text{ kg} \times 9.81 \dfrac{\text{N}}{\text{kg}} \times 1.0 \text{ m} = 4.90 \text{ joules}$

(b) 4.90 joules

(c) It turns into random kinetic energy (heat) of the mass and the floor.

3. $$\text{work} = 1 \text{ mol} \times \frac{(.0821)(298)}{1.0}\frac{\text{liters}}{\text{mol}} \times 1.0 \text{ atm} = 24.5 \text{ liter-atm}$$

$$24.5 \text{ liter-atm} \times 101.3 \frac{\text{joules}}{\text{l-atm}} = 2480 \text{ joules}$$

5. $$100 \text{ g} \times \frac{1 \text{ cal}}{\text{g-deg}} \times (45 - 50)\text{deg} + C\frac{\text{cal}}{\text{deg}}(45 - 25)\text{deg} = 0$$

$$C\frac{\text{cal}}{\text{deg}} = \frac{500 \text{ cal}}{20 \text{ deg}} = 25\frac{\text{cal}}{\text{deg}}$$

7. $$50 \text{ g} \times 0.092 \frac{\text{cal}}{\text{g-deg}} \times (T_f - 90) + 100 \text{ g} \times \frac{1 \text{ cal}}{\text{g-deg}} \times (T_f - 20)$$

$$+ 4.6\frac{\text{cal}}{\text{deg}} \times (T_f - 20) = 0$$

$$4.6T_f - 414 + 100T_f - 2000 + 4.6T_f - 92 = 0$$

$$109.2T_f - 2506 = 0$$

$$T_f = \frac{2506}{109.2} = 23.0°\text{C}$$

9. **(a)** $$500 \text{ g} \times \frac{1 \text{ cal}}{\text{g-deg}} \times 2.00 \text{ deg} + 100 \text{ g} \times \frac{q \text{ cal}}{\text{g-deg}} \times 2.00 \text{ deg}$$

$$+ 50\frac{\text{cal}}{\text{deg}} \times 2.00 \text{ deg} = 1121.6 \text{ cal}$$

$$1000 \text{ cal} + 200q \text{ cal} + 100 \text{ cal} = 1121.6 \text{ cal}$$

$$200q \text{ cal} = 21.6 \text{ cal}$$

$$q = \frac{21.6}{200} = 0.108 \frac{\text{cal}}{\text{g-deg}}$$

(b) $6.00\ \frac{\text{cal}}{\text{mol-deg}} \times \frac{1\ \text{g-deg}}{.108\ \text{cal}} = 55.5\ \frac{\text{g}}{\text{mole}}$

11. $50\ \text{g} \times 80\ \frac{\text{cal}}{\text{g}} + 50\ \text{g} \times \frac{1\ \text{cal}}{\text{g-deg}} \times 25\ \text{deg} = 5250\ \text{cal}$

13. $50\ \text{g} \times \frac{1\ \text{mol(C)}}{12\ \text{g}} \times -94.5\ \frac{\text{kcal}}{\text{mol}} = -394\ \text{kcal}$

15. $(86.4\ \text{volts} - 43.2\ \text{volts}) \times 1.6 \times 10^{-19}\ \text{coulombs} = 6.91 \times 10^{-18}\ \text{joules}$

in electron-volts: 43.2 volts × 1 electron = 43.2 e.v.

17. $40\ \frac{\text{joules}}{\text{sec}} \times 250\ \text{seconds} = 10^4\ \text{joules}$

$10^4\ \text{joules} \times \frac{1\ \text{cal}}{4.184\ \text{joules}} = 2390\ \text{cal}$

APPENDIX I physical constants and conversion factors

Physical constants

Avogadro's Number $N = 6.023 \times 10^{23}\ \frac{\text{molecules}}{\text{mol}}$

Boltzmann constant $k = 1.380 \times 10^{-23}\ \frac{\text{joule}}{\text{deg K}}$

Faraday's constant $F = 96{,}487\ \frac{\text{coulombs}}{\text{faraday}}$

Gas constant $R = 0.0821\ \frac{\text{liter-atm}}{\text{mol-deg}}$

$R = 62.4\ \frac{\text{liter-torr}}{\text{mol-deg}}$

$R = 8.314\ \frac{\text{joules}}{\text{mol-deg}}$

$R = 1.987\ \frac{\text{cal}}{\text{mol-deg}}$

Gravity constant (surface of earth) $g = 9.81\ \frac{\text{newtons}}{\text{kg}}$

Planck's constant $h = 6.626 \times 10^{-34}$ joule-seconds

Pi $\pi = 3.1416$

Conversion factors

Length

1 inch	= 2.54 cm
1 yard	= 91.44 cm
1 meter	= 39.37 in.
1 mile (stat)	= 5280 ft = 1609.3 m
1 mile (naut)	= 6080 ft = 1853.2 m

Volume

1 cubic inch = 16.39 cm^3
1 gallon (US, liq) = 231 cu. in.
4 quarts = 1 gal
8 pints = 1 gal
1 quart = 32 fl oz
1 quart = 0.946 liter

Mass

1 pound (avdp) = 453.6 g
1 kilogram = 2.205 lb (avdp)
1 pound (avdp) = 16 oz
1 ounce (avdp) = 28.35 g

Energy

1 joule = 1 newton-meter
1 joule = 10^7 ergs
1 joule = 1 volt-coulomb
1 joule = 0.239 cal
1 joule = 9.87×10^{-3} liter-atmosphere
1 electron-volt = 1.602×10^{-19} joules

APPENDIX II algebra

In an earlier course you may have learned that logic can be carried out without numbers, using the concepts of sets, subsets, and the union and intersection of sets. The members of a set were defined by some rule, so that by applying the rule you could tell whether a certain object was a member of the set. The purpose was to carry out certain operations on the sets so that the truth of statements could be found without fear of contradictory results.

Algebra is a game of logic played with numbers. It is a useful tool in physical science because it is often possible to put real objects in a one-to-one correspondence with the numbers. We can then presume that the logic learned with numbers also applies to the real objects.

Right here we must be cautious. The idea of equality is fundamental to logic, and it means that if two things are equal in some respect then any statement that is true of the one is also true of the other *in that same respect.*

It is absurd to say that "50 cents = 1 gallon of gasoline," because you know that if you drop 50 cents into your gas tank it will go "clunk" and the car won't run. Whenever we express equality with physical units, we *must* have the same units on both sides of the equation. In mathematical language, two sets are equal if and only if they have the same members.

Rules of algebra of real numbers

We start our development of the rules of algebra with the set of ordered natural numbers (1,2,3,4, . . . etc.). We *define* the operation of addition by saying that if we start with the number 5, say, and count successively 3 more numbers we get 8. That is,

$$5 + 3 = 8$$

or, in general,

$$a + b = c$$

where *a, b,* and *c* are all members of the same set of natural numbers. We also find that if we start with 3 and count 5 more numbers we still get 8, so the order in which we add numbers makes no difference. Addition is commutative.

Likewise, we *define* the operation of multiplication by starting with nothing and adding 5 to it 3 times, so that 3×5 is the sum of three 5's. Multiplication, therefore, is simply the addition of a number a specified number of times.

$$a + a + a + a = 4 \times a$$

By experience we find that $a \times 4$ also gives the same number, so multiplication is commutative.

If we can add a number (a) to itself b times and get ($a \times b$), why can't we multiply (a) times itself b times? We can, and the operation is called "raising to a power."

$$a \times a \times a \times a = a^4$$

The number of times a is multiplied by itself is called an *exponent,* and it is placed as a superscript on a.

With only the set of ordered natural numbers and the defined operations of addition and multiplication, the following relationships can be shown to be true:

1. $a + b = b + a$ **2.** $a \times b = b \times a$	Addition and multiplication are commutative.
3. $(a + b) + c = a + (b + c)$ **4.** $(ab)c = a(bc)$	Addition and multiplication are associative.
5. $a \times (b + c) = ab + ac$	Multiplication is distributive with addition.
6. If $a = b$, and $b = c$, then $a = c$	Transitive property of equality
7. If $a + c = b + c$ then $a = b$	Cancellation property of addition
8. If $a \times c = b \times c$ then $a = b$	Cancellation property of multiplication
9. $a + 0 = a$	Additive identity
10. $a \times 1 = a$ **11.** $a^1 = 1 \times a = a$	Multiplicative identity
12. $a^0 = 1$	$1 \times$ (a zero times) is 1
13. $a^b \times a^c = a^{(b+c)}$	

The natural numbers are not enough to answer the question "What number when added to 5 gives 3?"

$$5 + ? = 3$$

So negative numbers were invented.

$$5 + (-2) = 3$$

In effect, negative numbers are counting backward, and there is nothing to prevent "counting" less than zero.
So,

$$5 + (-8) = -3$$

Negative numbers can be multiplied, too:

$$5 \times (-3) = -15$$

To keep the commutative property we must also say that 5 taken (-3) times is also -15, even though we might puzzle about what it means to take something -3 times.

If $5 \times (-3) = -15$, then $(-5) \times (-3)$ cannot also be -15. To be consistent it must be $+15$. In general, then,

$$a \times (-1) = -a$$
$$-a \times (-1) = +a$$

We do not yet have a number that answers the question "What number when multiplied by a gives 1?" We answer by setting up the *inverse* or *reciprocal* of a, defined as $\frac{1}{a}$. The reciprocal of a, then, is the number which, when multiplied by a, gives 1.

$$a \times \frac{1}{a} = 1$$

Generalizing, we can say that

$$a \times \frac{1}{b} = \frac{a}{b}$$

Roots

Subtraction is the inverse of addition, and division is the inverse of multiplication. We do not yet have an inverse to the operation of raising to a power, so let's make one up.

What number when multiplied by itself gives 2?

$$x^2 = 2$$
$$x = ?$$

No rational number (ratio of integers) answers that question. We say that x is the *square root* of 2 ($\sqrt{2}$), and we can perform an operation that will approximate the number to as many decimal places as we choose ($\sqrt{2} = 1.4142\ldots$).

Likewise $\sqrt[3]{2}$ is the answer to the equation

$$x^3 = 2$$

Logarithms

Finally we may ask the question "To what power must I raise (a) to get (b)?"

$$a^x = b$$

Observe that this is the inverse of raising a to a known power and finding b. The exponent x is called "the logarithm of b to the base a." It means we must multiply a times itself x times to get b.

We know $10^2 = 100$, so the logarithm of 100 to the base 10 is 2. This is abbreviated

$$\begin{aligned} 10^2 &= 100; & \log_{10}(100) &= 2 \\ 10^3 &= 1000; & \log_{10}(1000) &= 3 \\ 10^4 &= 10{,}000; & \log_{10}(10{,}000) &= 4 \end{aligned}$$

This is easy as long as the number is a whole number power of ten, but what do we do if the number is something else?

$$\text{If } 10^a \times 10^a = 10^{(a+a)} \text{ then } 10^{1/2} \times 10^{1/2} = 10^1$$

But that is the definition of the square root of 10! Evidently roots are fractional powers of numbers. By taking enough roots of 10 and adding them together we can express any number between 1 and 10 as a fractional power of ten.

For example,

$$\begin{aligned} 10^{1/2} &= 3.1623 \\ (10^{1/2})^{1/2} &= \sqrt{3.1623} = 10^{1/4} = 1.7783 \\ (3.1623)(1.7783) &= 10^{1/2} \times 10^{1/4} = 10^{0.75} = 5.6235 \end{aligned}$$

The fractional power of ten that equals a number between one and ten is called the *mantissa* of the number.

Every positive number can be expressed as the product of an integral power of ten and a fractional power of ten; thus,

$$5{,}623.5 = 5.6235 \times 10^3$$

The integral power of ten tells us where to put the decimal point. The whole number exponent is called the *characteristic*. If we express both numbers as powers of ten, we have

$$5{,}623.5 = 5.6235 \times 1000 = 10^{0.75} \times 10^{3} = 10^{3.75}$$

The number 3.75 is the logarithm of 5,623.5 to the base 10. TO FIND THE LOGARITHM OF A NUMBER first express the number as a decimal between 1 and 10 times an integral power of 10.
Thus,

$$2{,}380 = 2.380 \times 10^{3}$$

and

$$0.0123 = 1.23 \times 10^{-2}$$

Then look up the mantissa of the number in a table of logarithms.

$$\log 2.380 = 0.3766$$
$$\log 1.23 = 0.0899$$

The logarithm of the integral power of ten is simply the exponent. By the law of exponents, when numbers are multiplied we *add* the logarithms.

$$\log 2380 = \log 2.380 + \log 10^{3} = 0.3766 + 3 = 3.3766$$
$$\log 0.0123 = \log 1.23 + \log 10^{-2} = 0.0899 - 2 = -1.9101$$

Antilogarithms ($\log^{-1}$)

TO FIND A NUMBER WHOSE LOGARITHM IS KNOWN perform the same steps in reverse order.

Find the number whose logarithm is 2.700.

(a) Look up the mantissa in a table of logarithms. The number whose logarithm is 0.700 is 5.01.

(b) The number whose logarithm is 2 is 10^2 (You don't have to look that up!)

(c) So the $\log^{-1}(2.700)$ is 5.01×10^2 or 501

If the logarithm whose number you want is negative, you must first convert the logarithm to a *positive mantissa* plus a negative integer.

$$-3.60 = 0.40 - 4$$
$$-5.75 = 0.25 - 6$$

Look up the positive mantissa to find the number, then multiply it by ten raised to the negative power.

$$\log^{-1}(.40 - 4) = 2.5 \times 10^{-4}$$

Equations

Algebraic equations are statements about the relationships of numbers in terms of the operations of addition and multiplication and the inverse operations of subtraction and division. Whatever is true of the left side of the equation is also true of the right side.

Equations are "solved" by placing all known quantities on one side of the equality and all unknown quantities on the other side.

1. Let a and b be known quantities and let x be an unknown.

Suppose $$a + x = b$$

Since a is added to x, perform the inverse operation. Subtract a from *both* sides of the equation to get x by itself:

$$\begin{aligned} a + x - a &= b - a \\ x &= b - a \end{aligned}$$

2. Suppose $$ax = b$$

Since x is multiplied by a, perform the inverse operation. Divide both sides of the equation by a.

$$\frac{ax}{a} = \frac{b}{a}$$

$a/a = 1$; so

$$x = \frac{b}{a}$$

3. Suppose $$ax + b = c$$

b is added to ax. Subtract b from both sides.

$$ax = c - b$$

x is multiplied by a. Divide both sides by a.

$$x = \frac{c - b}{a}$$

4. Suppose $$a(b - x) = c$$

$(b - x)$ is multiplied by a. Divide both sides by a.

$$(b - x) = \frac{c}{a}$$

x is subtracted from b. Add x to both sides.

$$b = \frac{c}{a} + x$$

Now c/a is added to x. Subtract c/a from both sides.

$$b - \frac{c}{a} = x$$

5. Suppose $$ax^2 = b$$

x^2 is multiplied by a. Divide both sides by a.

$$x^2 = \frac{b}{a}$$

If x is multiplied by itself, then b/a must be the square of some number. Take the square root of both sides.

$$x = \sqrt{\frac{b}{a}}$$

6. If the equation has the form $ax^2 + bx + c = 0$,

$$x = \frac{-b \pm \sqrt{b^2 - 4ac}}{2a}$$

logarithm tables APPENDIX III

FOUR-PLACE LOGARITHMS
ANTILOGARITHMS

FOUR-PLACE LOGARITHMS

N	0	1	2	3	4	5	6	7	8	9
10	0000	0043	0086	0128	0170	0212	0253	0294	0334	0374
11	0414	0453	0492	0531	0569	0607	0645	0682	0719	0755
12	0792	0828	0864	0899	0934	0969	1004	1038	1072	1106
13	1139	1173	1206	1239	1271	1303	1335	1367	1399	1430
14	1461	1492	1523	1553	1584	1614	1644	1673	1703	1732
15	1761	1790	1818	1847	1875	1903	1931	1959	1987	2014
16	2041	2068	2095	2122	2148	2175	2201	2227	2253	2279
17	2304	2330	2355	2380	2405	2430	2455	2480	2504	2529
18	2553	2577	2601	2625	2648	2672	2695	2718	2742	2765
19	2788	2810	2833	2856	2878	2900	2923	2945	2967	2989
20	3010	3032	3054	3075	3096	3118	3139	3160	3181	3201
21	3222	3243	3263	3284	3304	3324	3345	3365	3385	3404
22	3424	3444	3464	3483	3502	3522	3541	3560	3579	3598
23	3617	3636	3655	3674	3692	3711	3729	3747	3766	3784
24	3802	3820	3838	3856	3874	3892	3909	3927	3945	3962
25	3979	3997	4014	4031	4048	4065	4082	4099	4116	4133
26	4150	4166	4183	4200	4216	4232	4249	4265	4281	4298
27	4314	4330	4346	4362	4378	4393	4409	4425	4440	4456
28	4472	4487	4502	4518	4533	4548	4564	4579	4594	4609
29	4624	4639	4654	4669	4683	4698	4713	4728	4742	4757
30	4771	4786	4800	4814	4829	4843	4857	4871	4886	4900
31	4914	4928	4942	4955	4969	4983	4997	5011	5024	5038
32	5051	5065	5079	5092	5105	5119	5132	5145	5159	5172
33	5185	5198	5211	5224	5237	5250	5263	5276	5289	5302
34	5315	5328	5340	5353	5366	5378	5391	5403	5416	5428
35	5441	5453	5465	5478	5490	5502	5514	5527	5539	5551
36	5563	5575	5587	5599	5611	5623	5635	5647	5658	5670
37	5682	5694	5705	5717	5729	5740	5752	5763	5775	5786
38	5798	5809	5821	5832	5843	5855	5866	5877	5888	5899
39	5911	5922	5933	5944	5955	5966	5977	5988	5999	6010
40	6021	6031	6042	6053	6064	6075	6085	6096	6107	6117
41	6128	6138	6149	6160	6170	6180	6191	6201	6212	6222
42	6232	6243	6253	6263	6274	6284	6294	6304	6314	6325
43	6335	6345	6355	6365	6375	6385	6395	6405	6415	6425
44	6435	6444	6454	6464	6474	6484	6493	6503	6513	6522
45	6532	6542	6551	6561	6571	6580	6590	6599	6609	6618
46	6628	6637	6646	6656	6665	6675	6684	6693	6702	6712
47	6721	6730	6739	6749	6758	6767	6776	6785	6794	6803
48	6812	6821	6830	6839	6848	6857	6866	6875	6884	6893
49	6902	6911	6920	6928	6937	6946	6955	6964	6972	6981
50	6990	6998	7007	7016	7024	7033	7042	7050	7059	7067
51	7076	7084	7093	7101	7110	7118	7126	7135	7143	7152
52	7160	7168	7177	7185	7193	7202	7210	7218	7226	7235
53	7243	7251	7259	7267	7275	7284	7292	7300	7308	7316
54	7324	7332	7340	7348	7356	7364	7372	7380	7388	7396
N	0	1	2	3	4	5	6	7	8	9

FOUR-PLACE LOGARITHMS

N	0	1	2	3	4	5	6	7	8	9
55	7404	7412	7419	7427	7435	7443	7451	7459	7466	7474
56	7482	7490	7497	7505	7513	7520	7528	7536	7543	7551
57	7559	7566	7574	7582	7589	7597	7604	7612	7619	7627
58	7634	7642	7649	7657	7664	7672	7679	7686	7694	7701
59	7709	7716	7723	7731	7738	7745	7752	7760	7767	7774
60	7782	7789	7796	7803	7810	7818	7825	7832	7839	7846
61	7853	7860	7868	7875	7882	7889	7896	7903	7910	7917
62	7924	7931	7938	7945	7952	7959	7966	7973	7980	7987
63	7993	8000	8007	8014	8021	8028	8035	8041	8048	8055
64	8062	8069	8075	8082	8089	8096	8102	8109	8116	8122
65	8129	8136	8142	8149	8156	8162	8169	8176	8182	8189
66	8195	8202	8209	8215	8222	8228	8235	8241	8248	8254
67	8261	8267	8274	8280	8287	8293	8299	8306	8312	8319
68	8325	8331	8338	8344	8351	8357	8363	8370	8376	8382
69	8388	8395	8401	8407	8414	8420	8426	8432	8439	8445
70	8451	8457	8463	8470	8476	8482	8488	8494	8500	8506
71	8513	8519	8525	8531	8537	8543	8549	8555	8561	8567
72	8573	8579	8585	8591	8597	8603	8609	8615	8621	8627
73	8633	8639	8645	8651	8657	8663	8669	8675	8681	8686
74	8692	8698	8704	8710	8716	8722	8727	8733	8739	8745
75	8751	8756	8762	8768	8774	8779	8785	8791	8797	8802
76	8808	8814	8820	8825	8831	8837	8842	8848	8854	8859
77	8865	8871	8876	8882	8887	8893	8899	8904	8910	8915
78	8921	8927	8932	8938	8943	8949	8954	8960	8965	8971
79	8976	8982	8987	8993	8998	9004	9009	9015	9020	9025
80	9031	9036	9042	9047	9053	9058	9063	9069	9074	9079
81	9085	9090	9096	9101	9106	9112	9117	9122	9128	9133
82	9138	9143	9149	9154	9159	9165	9170	9175	9180	9186
83	9191	9196	9201	9206	9212	9217	9222	9227	9232	9238
84	9243	9248	9253	9258	9263	9269	9274	9279	9284	9289
85	9294	9299	9304	9309	9315	9320	9325	9330	9335	9340
86	9345	9350	9355	9360	9365	9370	9375	9380	9385	9390
87	9395	9400	9405	9410	9410	9420	9425	9430	9435	9440
88	9445	9450	9455	9460	9465	9469	9474	9479	9484	9489
89	9494	9499	9504	9509	9413	9518	9523	9528	9533	9538
90	9542	9547	9552	9557	9562	9566	9571	9576	9581	9586
91	9590	9595	9600	9605	9609	9614	9619	9624	9628	9633
92	9638	9643	9647	9652	9657	9661	9666	9671	9675	9680
93	9685	9689	9694	9699	9703	9708	9713	9717	9722	9727
94	9731	9736	9741	9745	9750	9754	9759	9763	9768	9773
95	9777	9782	9786	9791	9795	9800	9805	9809	9814	9818
96	9823	9827	9832	9836	9841	9845	9850	9854	9859	9863
97	9868	9872	9877	9881	9886	9890	9894	9899	9903	9908
98	9912	9917	9921	9926	9930	9934	9939	9943	9948	9952
99	9956	9961	9965	9969	9974	9978	9983	9987	9991	9996
N	0	1	2	3	4	5	6	7	8	9

ANTILOGARITHMS

	0	1	2	3	4	5	6	7	8	9
.00	1000	1002	1005	1007	1009	1012	1014	1016	1019	1021
.01	1023	1026	1028	1030	1033	1035	1038	1040	1042	1045
.02	1047	1050	1052	1054	1057	1059	1062	1064	1067	1069
.03	1072	1074	1076	1079	1081	1084	1086	1089	1091	1094
.04	1096	1099	1102	1104	1107	1109	1112	1114	1117	1119
.05	1122	1125	1127	1130	1132	1135	1138	1140	1143	1146
.06	1148	1151	1153	1156	1159	1161	1164	1167	1169	1172
.07	1175	1178	1180	1183	1186	1189	1191	1194	1197	1199
.08	1202	1205	1208	1211	1213	1216	1219	1222	1225	1227
.09	1230	1233	1236	1239	1242	1245	1247	1250	1253	1256
.10	1259	1262	1265	1268	1271	1274	1276	1279	1282	1285
.11	1288	1291	1294	1297	1300	1303	1306	1309	1312	1315
.12	1318	1321	1324	1327	1330	1334	1337	1340	1343	1346
.13	1349	1352	1355	1358	1361	1365	1368	1371	1374	1377
.14	1380	1384	1387	1390	1393	1396	1400	1403	1406	1409
.15	1413	1416	1419	1422	1426	1429	1432	1435	1439	1442
.16	1445	1449	1452	1455	1459	1462	1466	1469	1472	1476
.17	1479	1483	1486	1489	1493	1496	1500	1503	1507	1510
.18	1514	1517	1521	1524	1528	1531	1535	1538	1542	1545
.19	1549	1552	1556	1560	1563	1567	1570	1574	1578	1581
.20	1585	1589	1592	1596	1600	1603	1607	1611	1614	1618
.21	1622	1626	1629	1633	1637	1641	1644	1648	1652	1656
.22	1660	1663	1667	1671	1675	1679	1683	1687	1690	1694
.23	1698	1702	1706	1710	1714	1718	1722	1726	1730	1734
.24	1738	1742	1746	1750	1754	1758	1762	1766	1770	1774
.25	1778	1782	1786	1791	1795	1799	1803	1807	1811	1816
.26	1820	1824	1828	1832	1837	1841	1845	1849	1854	1858
.27	1862	1866	1871	1875	1879	1884	1888	1892	1897	1901
.28	1905	1910	1914	1919	1923	1928	1932	1936	1941	1945
.29	1950	1954	1959	1963	1968	1972	1977	1982	1986	1991
.30	1995	2000	2004	2009	2014	2018	2023	2028	2032	2037
.31	2042	2046	2051	2056	2061	2065	2070	2075	2080	2084
.32	2089	2094	2099	2104	2109	2113	2118	2123	2128	2133
.33	2138	2143	2148	2153	2158	2163	2168	2173	2178	2183
.34	2188	2193	2198	2203	2208	2213	2218	2223	2228	2234
.35	2239	2244	2249	2254	2259	2265	2270	2275	2280	2286
.36	2291	2296	2301	2307	2312	2317	2323	2328	2333	2339
.37	2344	2350	2355	2360	2366	2371	2377	2382	2388	2393
.38	2399	2404	2410	2415	2421	2427	2432	2438	2443	2449
.39	2455	2460	2466	2472	2477	2483	2489	2495	2500	2506
.40	2512	2518	2523	2529	2535	2541	2547	2553	2559	2564
.41	2570	2576	2582	2588	2594	2600	2606	2612	2618	2624
.42	2630	2636	2642	2649	2655	2661	2667	2673	2679	2685
.43	2692	2698	2704	2710	2716	2723	2729	2735	2742	2748
.44	2754	2761	2767	2773	2780	2786	2793	2799	2805	2812
.45	2818	2825	2831	2838	2844	2851	2858	2864	2871	2877
.46	2884	2891	2897	2904	2911	2917	2924	2931	2938	2944
.47	2951	2958	2965	2972	2979	2985	2992	2999	3006	3013
.48	3020	3027	3034	3041	3048	3055	3062	3069	3076	3083
.49	3090	3097	3105	3112	3119	3126	3133	3141	3148	3155
	0	1	2	3	4	5	6	7	8	9

ANTILOGARITHMS

	0	1	2	3	4	5	6	7	8	9
.50	3162	3170	3177	3184	3192	3199	3206	3214	3221	3228
.51	3236	3243	3251	3258	3266	3273	3281	3289	3296	3304
.52	3311	3319	3327	3334	3342	3350	3357	3365	3373	3381
.53	3388	3396	3404	3412	3420	3428	3436	3443	3451	3459
.54	3467	3475	3483	3491	3499	3508	3516	3524	3532	3540
.55	3548	3556	3565	3573	3581	3589	3597	3606	3614	3622
.56	3631	3639	3648	3656	3664	3673	3681	3690	3698	3707
.57	3715	3724	3733	3741	3750	3758	3767	3776	3784	3793
.58	3802	3811	3819	3828	3837	3846	3855	3864	3873	3882
.59	3890	3899	3908	3917	3926	3936	3945	3954	3963	3972
.60	3981	3990	3999	4009	4018	4027	4036	4046	4055	4064
.61	4074	4083	4093	4102	4111	4121	4130	4140	4150	4159
.62	4169	4178	4188	4198	4207	4217	4227	4236	4246	4256
.63	4266	4276	4285	4295	4305	4315	4325	4335	4345	4355
.64	4365	4375	4385	4395	4406	4416	4426	4436	4446	4457
.65	4467	4477	4487	4498	4508	4519	4529	4539	4550	4560
.66	4571	4581	4592	4603	4613	4624	4634	4645	4656	4667
.67	4677	4688	4699	4710	4721	4732	4742	4753	4764	4775
.68	4786	4797	4808	4819	4831	4842	4853	4864	4875	4887
.69	4898	4909	4920	4932	4943	4955	4966	4977	4989	5000
.70	5012	5023	5035	5047	5058	5070	5082	5093	5105	5117
.71	5129	5140	5152	5164	5176	5188	5200	5212	5224	5236
.72	5248	5260	5272	5284	5297	5309	5321	5333	5346	5358
.73	5370	5383	5395	5408	5420	5433	5445	5458	5470	5483
.74	5495	5508	5521	5534	5546	5559	5572	5585	5598	5610
.75	5623	5636	5649	5662	5675	5689	5702	5715	5728	5741
.76	5754	5768	5781	5794	5808	5821	5834	5848	5861	5875
.77	5888	5902	5916	5929	5943	5957	5970	5984	5998	6012
.78	6026	6039	6053	6067	6081	6095	6109	6124	6138	6152
.79	6166	6180	6194	6209	6223	6237	6252	6266	6281	6295
.80	6310	6324	6339	6353	6368	6383	6397	6412	6427	6442
.81	6457	6471	6486	6501	6516	6531	6546	6561	6577	6592
.82	6607	6622	6637	6653	6668	6683	6699	6714	6730	6745
.83	6761	6776	6792	6808	6823	6839	6855	6871	6887	6902
.84	6918	6934	6950	6966	6982	6998	7015	7031	7047	7063
.85	7079	7096	7112	7129	7145	7161	7178	7194	7211	7228
.86	7244	7261	7278	7295	7311	7328	7345	7362	7379	7396
.87	7413	7430	7447	7464	7482	7499	7516	7534	7551	7568
.88	7586	7603	7621	7638	7656	7674	7691	7709	7727	7745
.89	7762	7780	7798	7816	7834	7852	7870	7889	7907	7925
.90	7943	7962	7980	7998	8017	8035	8054	8072	8091	8110
.91	8128	8147	8166	8185	8204	8222	8241	8260	8279	8299
.92	8318	8337	8356	8375	8395	8414	8433	8453	8472	8492
.93	8511	8531	8551	8570	8590	8610	8630	8650	8670	8690
.94	8710	8730	8750	8770	8790	8810	8831	8851	8872	8892
.95	8913	8933	8954	8974	8995	9016	9036	9057	9078	9099
.96	9120	9141	9162	9183	9204	9226	9247	9268	9290	9311
.97	9333	9354	9376	9397	9419	9441	9462	9484	9506	9528
.98	9550	9572	9594	9616	9638	9661	9683	9705	9727	9750
.99	9772	9795	9817	9840	9863	9886	9908	9931	9954	9977
	0	1	2	3	4	5	6	7	8	9

CREDITS FOR CHAPTER OPENINGS:

Chapter 1	The Bettmann Archive, New York
Chapter 2	Droemersche Verlagsanstalt, Munich
Chapter 3	The Bettmann Archive, New York
Chapter 4	New York Public Library Picture Collection
Chapter 5	Science Museum, London
Chapter 6	University of California Lawrence Berkeley Laboratory
Chapter 7	Bibliothèque de Genève
Chapter 8	*Symbols, Signs and Signets* by Ernst Lehner, Dover Publications, New York
Chapter 9	*Symbols, Signs and Signets* by Ernst Lehner, Dover Publications, New York
Chapter 10	*Symbols, Signs and Signets* by Ernst Lehner, Dover Publications, New York
Chapter 11	Frontispiece, *Tractat von Bier-Brau-Recht* by Schöpffer, Germany, 1732
Chapter 12	*Symbols, Signs and Signets* by Ernst Lehner, Dover Publications, New York
Chapter 13	The Bettmann Archive, New York

CREDITS FOR PHOTOS:

	Laboratory equipment courtesy of Owens-Illinois	21
	Photo courtesy of Ohaus Scale Company	23
Figure 3.1	American Museum of Natural History	47
Figure 3.2	Reynolds Metals Company	48
Figure 3.3	American Museum of Natural History	49
Figure 3.7A	Photo courtesy of Sargent-Welch Scientific Company	52

index*

* Page numbers in *italics* refer to figures and tables.

A 5
B 6
C 7
D 8
E 9
F 0
G 1
H 2
I 3
J 4

oxidation #'s code # 24

IA	IIA	IIIB	IVB	VB	VIB	VIIB	
1 H 1.0079							
3 Li 6.941	4 Be 9.01218						
11 Na 22.98977	12 Mg 24.305						
19 K 39.098	20 Ca 40.08	21 Sc 44.9559	22 Ti 47.90	23 V 50.9414	24 Cr 51.996	25 Mn 54.9380	26 Fe 55.847
37 Rb 85.4678	38 Sr 87.62	39 Y 88.9059	40 Zr 91.22	41 Nb 92.9064	42 Mo 95.94	43 Tc 98.9062	44 Ru 101.07
55 Cs 132.9054	56 Ba 137.34	57 La* 138.9055	72 Hf 178.49	73 Ta 180.9479	74 W 183.85	75 Re 186.207	76 Os 190.2
87 Fr (223)	88 Ra 226.0254	89 Ac** (227)	104 (260)	105 (260)			

*Lanthanum Series

58 Ce 140.12	59 Pr 140.9077	60 Nd 144.24	61 Pm (147)

**Actinium Series

90 Th 232.0381	91 Pa 231.0359	92 U 238.029	93 Np 237.0482